Architectures for Digital
Signal Processing

Architectures for Digital Signal Processing

Peter Pirsch
University of Hannover, Germany

John Wiley & Sons
Chichester · New York · Weinheim · Brisbane · Singapore · Toronto

First published under the title Architekturen der digitalen Signalverarbeitung
© B.G. Teubner Verlag, Stuttgart 1996

Translation arranged with the approval of the original publisher
Copyright © 1998 by John Wiley & Sons Ltd,
                      Baffins Lane, Chichester,
                      West Sussex PO19 1UD, England

            National   Chichester (01243) 779777
            International (+44) 1243 779777

e-mail (for orders and customer service enquiries): cs-books@wiley.co.uk

Visit our Home Page on http://www.wiley.co.uk or http://www.wiley.com

Other Wiley Editorial Offices

John Wiley & Sons, Inc., 605 Third Avenue,
New York, NY 10158-0012, USA

Weinheim·Brisbane·Singapore·Toronto

Library of Congress Cataloging-in-Publication Data

Pirsch, P. (Peter)
    [Architekturen der digitalen Signalverarbeitung. English]
    Architectures for digital signal processing / Peter Pirsch.
        p. cm.
    Includes bibliographical references and index.
    ISBN 0-471-97145-6 (ppc : alk. paper)
    1. Signal processing—Digital techniques. 2. Computer
architecture. I. Title.
    TK5102.9.P57 1998
    621.382'2'02855422—dc21   98-10613
                                                CIP

British Library Cataloguing in Publication Data

A catalogue record for this book is available from the British Library

ISBN 0 471 97145 6

Produced from camera-ready copy supplied by the author
Printed and bound in Great Britain by Biddles Ltd, Guildford and King's Lynn
This book is printed on acid-free paper responsibly manufactured from sustainable forestry in which at least two trees are planted for each one used for paper production

Contents

Preface

Digital signal processing is a rapidly growing field. It comprises the presentation, evaluation, transformation and manipulation of signals. These signal processing measures can, for instance, serve the efficient storage and transmission of signals. Other applications are the controlling and quality assurance of manufacturing processes, employment in autonomous mobile systems, the analysis of biomedical processes and the evaluation of seismic data.

In particular, temporal processing criteria lead to high digital signal processing requirements in terms of processing power and access rates. These high requirements are frequently the cause of implementation problems. Often, programmable standard processors cannot provide the necessary signal processing power. Application specific circuits are an obvious choice for solving this problem. Nonetheless, such dedicated hardware has the disadvantage of lacking flexibility, which hinders a later modification of the algorithms, and furthermore, the employment of the same hardware for different applications is not possible. Special programmable signal processors (DSPs) have thus been developed since 1979. Through the separation of programs from data, the adaption of the computational unit to typical signal processing operations in addition to an improvement in data access through the aid of local memory and several bus systems, signal processing power has been substantially increased over microprocessors.

The standard DSPs currently available fulfil the requirements for the evaluation of measurement signals in addition to those of speech and audio signal processing. Nonetheless, applications with real-time requirements, such as the evaluation of image sequences or of radar signals, require signal processing performance that cannot be provided by standard DSPs. Such applications are thus implemented with programmable multiprocessor systems or with circuits that have been adapted to the task. This can be accomplished on the basis of commercially available standard components or through application specific integrated circuits (ASICs).

A large number of textbooks exist on signal processing algorithms and on the implementation and design of microelectronic circuits. This book was conceived to close the gap between these two named fields. The focus of this book is the transition from the specification of algorithms to architectures for VLSI implementations. Neither the derivation of signal processing algorithms nor the techniques for developing integrated circuits will be treated in detail. It will be shown that, through exact analysis of the algorithm to be implemented, circuit structures and architectures adapted to the task can be derived that allow the efficient implementation of signal processing systems with high computational power. Generally known, fundamental signal processing algorithms are assumed as the basis for the derivation of alternative architectures.

After an introduction and a review of digital circuit techniques for VLSI implementations in CMOS technology, this book treats four main topics in eight chapters. The main topics are the implementation of basic operations and elementary functions, measures for increasing the signal processing performance, the mapping of algorithms to application specific array processors and architectures of application specific, programmable signal processors. Circuit techniques and architectures for implementing the basic operations of addition, subtraction, multiplication and division are shown in Chapter 3. Furthermore, CORDIC structures for generating elementary functions are presented. Subsequently, general parallel processing and pipelining concepts follow in Chapter 4 for increasing the throughput, i.e. the signal processing power. In addition, criteria for assessing alternative architectures are discussed. As a basis for the derivation of application specific architectures for the implementation of regular algorithms, methods for mapping algorithms onto array processors are described in Chapter 5. Structures for implementing linear filters and the discrete Fourier transform are presented in Chapters 6 and 7 as special, important applications. Chapter 8 treats application specific, programmable signal processors. The characteristics and features of currently available DSPs are presented, and the architectural measures applied for increasing the throughput of signal processing algorithms are explained. Simple models as a basis for optimizing the architectures are included. Multiprocessor systems are treated in a separate chapter (Chapter 9). General methods for data and task division in addition to diverse control strategies are covered. Finally, Chapter 10 gives an overview of implementation strategies and summarizes the essential fundamentals of deriving alternative architectures.

This book is a translation of a book that was published in German by the B. G. Teubner Verlag in 1996. In comparison to the German version, a number of modifications and revisions were carried out. The basis of the book was a graduate course for electrical and computer engineering students held by the author at the University of Hannover. Basic knowledge of signal processing algorithms and the implementation of digital circuits were assumed as a prerequisite. For those engineers who have since entered the job market, this book can serve both to deepen their knowledge and as a reference book. It was conceived primarily as a textbook and then as a reference book. Thus, it is not always the original publications that are listed, but frequently simpler, accessible standards and textbooks instead. I apologize to those whose valuable work has not been referenced.

For reinforcement of the material, exercises are listed at the end of the individual chapters. A solution set can be ordered from the author. I am responsible for all omissions and for any errors found in the book. Readers are encouraged to contact me directly (email: pirsch@mst.uni-hannover.de) regarding error correction, content and extensions for future editions.

I thank all who were substantially involved in the creation of this book and who assisted me with their valuable suggestions, critical reviews and technical assistance. My special thanks go to David Dickerson for carrying out the translation and to my secretary, Ms Regina Nowakowski, for the typing and illustration work. Many of my PhD students assisted me. Here I would like to name Martin Ohmacht, Klaus Grüger, Winfried Gehrke, Markus Schwiegershausen, Marco Winzker, Klaus Gaedke and Achim Freimann. I also thank the publisher, John Wiley and Sons, for adopting this book.

Peter Pirsch

1　Introduction

Signals are a means of communication between people, between people and machines, and between machines. Signals can be speech, music or video signals but also measurement, sonar and radar signals. Signal processing encompasses signal acquisition, preparation, analysis and synthesis. Originally, signal processing was solely analog. Advances in digital computing and microelectronics have made digital signal processing economically viable. The term digital signal processing refers to the manipulation of discrete-time, discrete-amplitude signals. A particular advantage of digital technology is the nearly complete elimination of the influence of production tolerances, ageing and random processes such as noise. Digital signal processing has been so well defined that implementations are reproducible. This allows algorithms to be simulated on digital computers before being implemented in the target hardware. This, in turn, assists the optimization of signal processing systems through extensive analysis using true-to-life representative signals.

1.1　Signal Processing Algorithms

The theoretical foundation for the conversion of analog signals into time-discrete signals and vice versa is the sampling theorem [1], [2], [3], [4]. The actual sampling itself is carried out by analog to digital (A/D) converters. Likewise, analog signals can be extracted from digital data via digital to analog (D/A) converters. Signals whose bandwidth exceeds half the maximum sampling rate of the given A/D converter are prefiltered with analog lowpass filters to avoid the problem of aliasing. Similarly, the reconstructed analog output of D/A converters is also put through an analog lowpass filter to improve the interpolation process. All of the signal processing techniques, architectures and circuitry discussed in this book are limited to a digital representation of the signals. Conversion to analog signals will not be covered.

The first step in signal processing is usually a form of preprocessing, for example signal improvement via noise reduction achieved through linear or adaptive non-linear filtering. Another type of preprocessing is the emphasis of signal characteristics. A typical representative here is edge sharpening in video signals through use of Laplace filters, gradient filters or the like.

An important area of signal processing is signal interpretation. In this case, particular characteristics are to be extracted from the signal. Take, for instance, a speech recognition system that is to ascertain from a voice signal the basic phonetic elements upon which the language is based. Another example is the extraction of basic geometric elements such as straight lines from picture data. In signal processing, basic elements such as these are called symbols. An additional, a relatively new form of signal processing is the manipulation of symbols. Individual symbols are grouped to form abstract data objects consisting of linked lists of symbols. The true results of the signal processing can then be computed through object-oriented processing with a programming language such as LISP.

From the aforementioned one can see that digital signal processing is carried out on a number of levels of abstraction. The classical methods of manipulating signal sequences are called low-level methods. The input sequence undergoes a processing step, for instance filtering, the result of which is a sequence with altered characteristics. A common characteristic of such low-level algorithms is that they follow a predetermined sequence of operations, independent of the amplitudes contained in the input sequence. The extraction of symbols from a preprocessed input sequence is called medium-level processing. In general, symbols are identified by measuring their distance relative to reference patterns. In the simplest case this corresponds to a sequence of transforms followed by a threshold comparison. The sequence of operations needed to identify symbols depends on the data; thus the number of operations cannot be completely predetermined. High-level processing describes both the identification of data objects based on symbols and the evaluation of final results based on the relationships between the data objects. High-level processing is nearly always formulated in a high-level programming language and implemented on programmable processors.

Low-level algorithms consist of simple relationships based on just a few types of operations. Medium and high-level processes, on the other hand, are composed of complex sequences of operations. The data volume to be processed must be taken into account when determining the necessary signal processing power. The corresponding data volume (samples, symbols, data objects) sinks inversely to the level of processing. Low-level pro-

cessing consumes the majority of the processing power in most signal processing tasks. This book pays special attention to circuit techniques and
architectures for the implementation of fundamental signal processing algorithms, most of which can be classified as low-level algorithms. A number
of typical signal processing algorithms are presented in the following.

These algorithms demonstrate requisite operations and special operation sequences. Only the resulting algorithms and important computational
steps are presented. Derivation and justification of these algorithms can be
found in the standard digital signal processing literature [1], [2], [3], [4], [5].

Linear, time-invariant filtering is a classic example of signal processing. Such filters can be classified as having either an infinite impulse response (IIR filters) or a finite impulse response (FIR filters).

An FIR filter algorithm can be described by the equation

$$y(i) = \sum_{k=0}^{N-1} h(k) \; x(i-k) \tag{1.1.1}$$

where x is the input sequence, y the output sequence and h the impulse response of length N. The output sequence of an FIR filter depends solely on
the input sequence. The output sequence of an IIR filter, on the other hand,
is influenced by previous values in the output sequence. An IIR filter is given
by the equation

$$y(i) = \sum_{k=1}^{M-1} a(k) \; y(i-k) + \sum_{k=0}^{N-1} b(k) \; x(i-k) \tag{1.1.2}$$

An important analytical method in signal processing is the spectral
analysis through discrete transforms [1], [6]. A discrete transform of N input
values x into N output values is given by

$$y(k) = \sum_{n=0}^{N-1} c(k,n) \; x(n) \quad 0 \leq k \leq N-1 \tag{1.1.3}$$

This can be written as a matrix–vector product

$$\mathbf{y} = \mathbf{C} \, \mathbf{x} \tag{1.1.4}$$

The corresponding matrix elements of $\mathbf{C}$ for the discrete Fourier transform
(DFT) are as follows:

$$c(k,n) = W_N^{nk} \quad \text{where} \quad W_N = e^{-j2\pi/N}, \; j = \sqrt{-1} \tag{1.1.5}$$

In addition to the DFT, a number of discrete transforms exist. The discrete cosine transform (DCT), for instance, is of special importance in source coding [3], [7]. DCT is a special, real-valued, spectral transform.

In pattern recognition and coding, the problem of finding representatives for a segment within a signal sequence often crops up [7], [8]. From a set of K representatives $r_k \in R$, the representative r_k is sought whose distance to the input vector x is minimal.

$$r_i = \min_{r_k \in R}^{-1} \| x - r_k \| \tag{1.1.6}$$

This notation defines $\min \| \cdot \|$ to be the minimal distance and $\min^{-1} \| \cdot \|$ to be the representative corresponding to the minimal distance. A measure of distance is used in the meaning of the L_1 norm

$$\frac{1}{N} \sum_{i=0}^{N-1} | x(i) - r_k(i) | \tag{1.1.7}$$

and in the meaning of the L_2 norm

$$\frac{1}{N} \sum_{i=0}^{N-1} \left[x(i) - r_k(i) \right]^2 \tag{1.1.8}$$

In signal modelling procedures it is essential to estimate parameters. A number of estimation techniques require the optimal parameters to be found by solving a system of equations or by computing the inverse of a matrix [9]. Such techniques are also used in system parameter optimization [10]. A possible method for solving a system of equations is the Gauss–Jordan method, in which a matrix A is triangularized [11]. It uses the following operations over the range of valid indices:

$$a'_{ik} = a_{ik} - \frac{a_{ij} a_{nk}}{a_{nj}} \qquad \forall k, \forall i \text{ and } i \neq n$$

$$a'_{nk} = \frac{a_{nk}}{a_{nj}} \qquad \forall k \tag{1.1.9}$$

where a_{ik} is the previous element, a'_{ik} the new element and a_{nj} the pivot (the maximal available element). This reduction of the matrix can also be carried out by a sequence of numerically stable rotations, the so-called Given's method [11]:

$$a'_{ik} = a_{ik} \cos \Theta - a_{jk} \sin \Theta$$

$$a'_{jk} = a_{ik} \sin \Theta + a_{jk} \cos \Theta \qquad \Theta = \tan^{-1} \frac{a_{ij}}{a_{jj}} \qquad (1.1.10)$$

Other matrix manipulations such as inversion, eigenvalue transforms, and so on consist of operational procedures similar to the above methods.

From the methods presented, one can derive the following core operations. Filtering, transforms and matrix multiplications carry out dot product computations of the form

$$\sum Op_1 \cdot Op_2 \qquad (1.1.11)$$

in which a sequence of products of the operands Op_1 and Op_2 are summed. The term MAC-Operation (MAC = Multiply and ACcumulate) was coined to describe such operations. Matching procedures use core operations such as

$$\sum |Op_1 - Op_2| \qquad (1.1.12)$$

or

$$\sum (Op_1 - Op_2)^2 \qquad (1.1.13)$$

for distance measurement. Solutions to systems of equations and the inverse of matrices can be found with core operations of the form

$$Op_1 - Op_2 \cdot Op_3 / Op_4 \qquad (1.1.14)$$

The alternative implementation based on rotations leads to

$$\pm \, Op_1 \cos \Theta \pm Op_2 \sin \Theta \qquad (1.1.15)$$

Economical signal processing hardware requires low overhead implementations of the core operations addition, subtraction, division and multiplication. Furthermore, frequently used functions such as the trigonometric functions, for instance, must be efficiently implemented. High throughput can be achieved when the core operations and other frequently used functions are implemented in fast circuitry.

1.2   Implementation Aspects

When implementing signal processing algorithms, one must pay particular interest to the hardware resources required. Here the term hardware resources describes the number of logic gates, memory cells and wires connecting the various modules. A relationship is sought between the characteristics of the algorithm and the implied hardware expense. A particularly simple estimate can be made through use of a considerably simplified processor model. It is assumed that all operations have equal logic and time requirements. In this case the computational rate, which is defined as the number of necessary operations per unit of time, is a direct measure for the hardware expense in the operative portion of the processor. The computational rate R_C is proportional to the sampling rate R_S. This relationship is given by

$$R_C = R_S \, n_{OP} \qquad\qquad (1.2.1)$$

where n_{OP} is the average number of operations per sample.

This assumed processor model is approximately equal to that of programmable processors with an ALU (arithmetic logic unit) in their operative path. An ALU can carry out practically all operations and requires only one clock cycle for most operations. Examples of such operations are addition, subtraction, comparison, increment, decrement, negation and logical bit operations. Operations such as multiplication, division and most shift operations, however, require several clock cycles.

Using filtering as an example, the computational rate needed to process various signals will be compared. Based on (1.1.1), an FIR filter with an impulse response of length N requires N multiplications and N additions. Accordingly, the requirements for a 1-D (one dimensional) FIR filter are

$$n_{OP} = 2N \qquad\qquad (1.2.2)$$

operations and for a 2-D (two dimensional) FIR filter with equal impulse response length in both dimensions

$$n_{OP} = 2N^2 \qquad\qquad (1.2.3)$$

Table 1.2.1 gives the performance requirements in MOPS (mega operations per second) for speech, music and video signals based on typical sampling rates and impulse response lengths. A comparison of the computational rate of several computers and processors is listed in Table 1.2.2. One can see that speech and music signals can be processed on standard signal processors. Current standard signal processors and even most supercomputers are not capable of real-time video signal processing. This implies that hardware specially adapted to the problem must be used. This can be achieved by either combining several off-the-shelf ICs or the fabrication of an application specific IC (ASIC). The term ASIC is broadly used to describe ICs developed for a specific application and customer. ASICs are usually fabricated using semi-custom processes such as gate-array or standard-cell technology. Application specific standard products (ASSP) are available for many signal processing applications, for instance filtering.

Table 1.2.1: Required computational rate for FIR filtering

Signal type	Frequency	Impulse response length	Performance
Speech	8 kHz	$N = 128$	2 MOPS
Music	48 kHz	$N = 256$	24 MOPS
Video telephone	6.75 MHz	$N^2 = 81$	1090 MOPS
TV	27 MHz	$N^2 = 81$	4370 MOPS
HDTV	144 MHz	$N^2 = 81$	23300 MOPS

Table 1.2.2: Performance of computers and processors

Processor	Type	Performance
CISC	MC 68040	1.5 MOPS
RISC	i 860	80 MFLOPS
Signal processor	MC 56100	80 MOPS
Supercomputer	CRAY 2	1200 MFLOPS

It follows from the results of Tables 1.2.1 and 1.2.2 that special hardware structures (architectures) must be developed for systems with particularly high requirements, in terms of computational rate, for instance. The development process and its possible alternatives are briefly discussed below.

The complexity of signal processing systems requires description and implementation on a number of hierarchical levels. A possible implementation hierarchy in accord with Figure 1.2.1 contains the levels

Processing method
Algorithm
Architecture
Circuitry

Here the term "Processing method" stands for the general description of the signal processing task at hand, whereas "Algorithm" reflects the actual computational steps and the functional relationships. "Architecture" describes the implementation as an interconnection of blocks, each of which deals with a set of related sub-tasks relevant to signal processing, computation or storage. The "Circuitry" level specifies the basic modules as a circuit of basic elements such as logic gates and transistors.

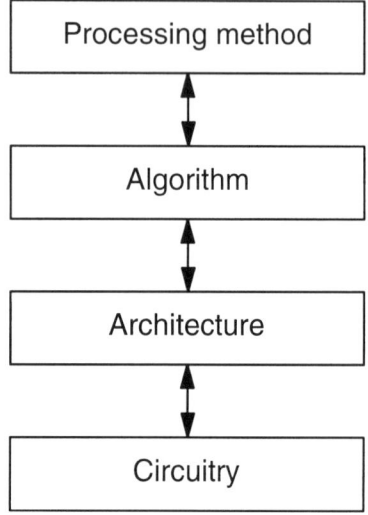

Figure 1.2.1: Implementation hierarchy

One problem with the development of complex systems is the number of alternatives at each level. Frequently, through algebraic manipulation, various algorithms can be found to represent a processing method. Assuming a linear function f, the following identities hold true:

$$f(x_1 + x_2) = f(x_2 + x_1) \qquad (1.2.4)$$

$$f(ax) = a\,f(x) \qquad (1.2.5)$$

Thus neither the processing of the operands nor the operations are constrained to a particular order. This has consequences in both the hardware structure and the numeric behaviour, assuming limited computational accuracy of both the operands and the results.

If data dependency is respected, the projection of the algorithms to the architectural level can be carried out as a direct conversion of the operations. Alternative implementations result through the use of flexible modules whose functions can be modified via control lines.

Alternatives at the circuit level are the result of the various technologies, for example bipolar, NMOS and CMOS. On the other hand, diverse circuit techniques can be applied to a particular technology. CMOS, for instance, allows the basic modules to be implemented in static gate logic, dynamic gate logic, switch logic, programmable function blocks, and so on.

The diversity of design indicated above leads to the problem of finding the best solution within given specifications. Approximations of performance and cost serve as the optimization criteria. For the sake of simplicity, the throughput is taken to be a measure of the performance and the chip area to be a measure of the implementation's cost.

1.3 Overview

This book deals with circuit techniques and architectures for the implementation of fundamental signal processing algorithms. Both dedicated and flexible, programmable architectures are derived. The objective is the transition from algorithm to architecture and the subsequent bridging of the gap between algorithmic specifications and VLSI implementations. Thus, ASIC development will not be treated in detail. Nor does the determination of algorithms for signal processing tasks belong among the themes of this book. Generally known algorithms serve as the preferred foundation for the derivation of architectural alternatives.

Basic knowledge of digital circuit implementation for a given semiconductor technology is a prerequisite for comprehension of the circuit techniques presented. With this in mind, a brief refresher is given in Chapter 2. Due to its particular importance, Chapter 2 is dedicated to CMOS technology. In order to approximate a circuit's computational rate, a simplified model is introduced for evaluating circuit delay in digital logic.

Chapter 3 demonstrates circuit techniques for the implementation of basic operations such as addition, subtraction, multiplication and division. Furthermore, implementations of trigonometric and hyperbolic functions based on the CORDIC structure (COordinate Rotation DIgital Computer) are presented. Starting at the definition of operation at the bit level, initial circuit techniques are derived. A number of alternatives to these basic structures are elaborated. One of the foundations of these alternatives is the linearity of numeric representations. This allows both hierarchical and recursive structures to be derived. Another method is the parallel computation of all alternative results, from which the correct result can then be quickly chosen.

In Chapter 4, general measures to improve the performance of signal processing systems such as parallel processing and pipelining are presented. For the assessment of alternative, high performance structures, a measure of efficiency is introduced that measures the throughput in relation to the hardware costs.

Many signal processing techniques are based on regular and recursive algorithms. This is true in particular for low-level algorithms. An obvious implementation technique for dedicated hardware is the use of regular architectures based on identical processor elements, so-called array processor architectures. The extraction of application specific, array processor architectures is described in Chapter 5. The properties of systolic arrays are expanded upon, and a technique for the systematic projection of regular algorithms onto array processors is introduced. The variety of possible architecture based on this method of projection is demonstrated.

Chapters 6 and 7 show the corresponding, application specific architectures for two particularly important methods: linear, time-invariant filtering and the discrete Fourier transform (DFT). In addition to one- and two-dimensional filter structures that can be derived directly from the algorithm, special structures for decimation and interpolation filters are elaborated. Structures with relaxed time-critical paths are derived for recursive filters. For the DFT, the results of direct conversion to array processors are compared with the results gleaned from the fast Fourier transform (FFT) based on butterfly processors.

In addition to dedicated implementations, implementations using flexible, programmable processors are often desired. Due to the special requirements affiliated with signal processing, special digital signal processors (DSP) have been developed. Chapter 8 deals with the architectures of programmable, standard signal processors. The particular features of DSP chips, CISC and RISC processors are compared. Using FIR filtering and the DFT as implementation examples, the speed gain of DSP chips is brought to light. Methods for optimizing architectures are expanded upon using simple models.

Signal processing techniques that demand particularly high performance cannot be implemented on a single, programmable DSP, and thus require a multiprocessor system. Chapter 9 elucidates programmable multiprocessor systems. General strategies for data and task distribution are discussed, as are diverse control strategies. This discussion includes both homogeneous and heterogeneous multiprocessor systems and their particular advantages.

Finally Chapter 10 looks at various signal processing implementations. The implementation alternatives that result from global requirements such as performance, flexibility, fabrication and development costs are weighed against one another. The range of applications for programmable and function-oriented implementations are included. In addition to implementation alternatives the design process of application specific signal processing systems is presented in brief.

2 Basic CMOS Circuits

There are countless ways of implementing a given digital function as an integrated circuit. The alternatives pertain to both the semiconductor technology and the circuit technique. Most contemporary ICs use silicon as a substrate, regardless of whether bipolar or field effect transistors (FET) build the foundation of the circuit. Primarily, however, field effect transistors with insulated gate (MOS transistors) are used in integrated digital circuits. The abbreviation MOS stems from the original fabrication process, in which the gate was of metal, the insulating layer oxide (SiO_2) and the substrate silicon. In present-day MOS transistors the gate is made of polycrystalline silicon (polysilicon, for short). For the implementation of highly complex, application specific integrated circuits, CMOS (complementary MOS) technology is of particular importance. For this reason, the following fundamental digital circuit techniques are shown in CMOS technology. More specific details concerning semiconductor technologies, their affiliated circuit techniques and aspects of design and fabrication can be taken from more specialized textbooks [12], [13], [14], [15], [16], [17].

2.1 Transistor Modelling

Circuits in CMOS technology consist of two transistor types, namely n-channel transistors and p-channel transistors. The principal operation of transistors will be explained using an n-channel transistor. Figure 2.1.1 shows a cross-section of an n-channel MOS transistor. The foundation for the transistor is silicon base material that has been doped so as to create a high concentration of holes (p-material). The active area of the transistor consists of a thin layer of SiO_2 that isolates the gate from the substrate and two diffusion zones with a high concentration of electrons (n-zones). MOS transistors have four terminals: gate (G), substrate (B for bulk or body), source (S) and drain (D), the last two of which are located in the diffusion zones. The gate serves to control the current flowing between the source and drain. The actual circuitry is produced by networking the transistors via one or several metal layers used for wiring.

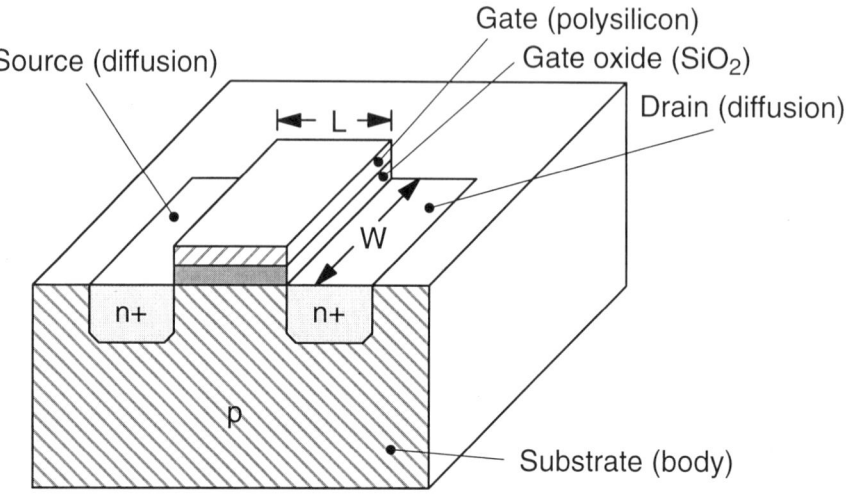

Figure 2.1.1: Structure of an n-channel MOS transistor

In order to expound the principles of transistor operation, it will be initially assumed that the terminal body and source are connected ($V_{SB} = 0V$). By means of a negative gate–source voltage V_{GS}, majority carriers (holes) are attracted to the silicon surface under the gate, resulting in an increase of majority carriers (accumulation). The p-n junctions form diodes with reverse bias; only a negligible leakage current flows between the drain and source. A positive gate–source voltage, on the other hand, induces an electric field that leads to a reduction in the number of holes along the silicon surface under the gate (depletion). A space-charge region is created that inhibits current flow between the source and drain. Once the critical positive voltage (threshold voltage V_T) has been passed, a conducting "channel" is formed through the injection of mobile electrons from the diffusion regions. This condition is called inversion since the substrate minority carriers (electrons) reach a concentration that is in the order of the majority carriers in the substrate.

If a conductive channel has formed under the gate on account of $V_{GS} > V_T$ any non-zero voltage between drain and source ($V_{DS} \neq 0$) will result in a current flow between these two terminals. For small voltages, the drain–source current I_{DS} is proportional to V_{DS}. With increasingly larger voltages, the factor of proportionality between V_{DS} and I_{DS} reduces. This is due to the voltage drop along the channel, which, in turn, affects the effective value of the channel forming voltage. If V_{DS} reaches $V_{GS} - V_T$, then the gate–body voltage next to the drain is insufficient to create a channel. This is called the saturation point. Any further increase of the voltage V_{DS} simply extends this "pinch-off" region. Through injection of mobile electrons from the rest of the channel, however, current continues to flow, increasing only slightly with V_{DS} (saturation region).

The behaviour of n-channel transistors can be easily modelled if divided into three regions of operation: cutoff region, linear region and saturation. The equations for the drain current based on this simplified model are as follows.

Cutoff:

$$I_{DS} = 0 \qquad V_{GS} - V_T \leq 0 \qquad (2.1.1)$$

Linear region (nonsaturation):

$$I_{DS} = \beta_n \left[(V_{GS} - V_T)V_{DS} - \frac{V_{DS}^2}{2} \right] \qquad 0 < V_{DS} \leq V_{GS} - V_T \qquad (2.1.2)$$

Saturation:

$$I_{DS} = \frac{\beta_n}{2} (V_{GS} - V_T)^2 \qquad 0 < V_{GS} - V_T \leq V_{DS} \qquad (2.1.3)$$

The device transconductance is specified by

$$\beta_n = \mu_n \cdot \frac{\varepsilon_{ox}}{d_{ox}} \cdot \frac{W}{L} \qquad (2.1.4)$$

where μ_n is the mobility of electrons, ε_{ox} the dielectric constant of silicon dioxide, d_{ox} the oxide thickness, L the channel length and W the channel width.

The derivation of these equations can be found in textbooks on the subject of MOS technology [15], [17]. Figure 2.1.2 gives a graphic representation of the drain current. The IEEE symbol for an n-channel MOS-transistor is shown in Figure 2.1.3a. The arrows give the direction of forward bias for the pn-junctions between the substrate and the diffusion zones. A dashed line between source and drain marks enhancement type transistors, i.e. those which are turned off for $V_{GS} = 0$. This book uses the simplified symbol shown in Figure 2.1.3b. For most practical purposes, it is moot to distinguish between terminal source and drain. When referring to n-channel transistors, it is general practice to call the terminal with the higher voltage the drain. In such a case the source is the "source" of mobile charge carriers.

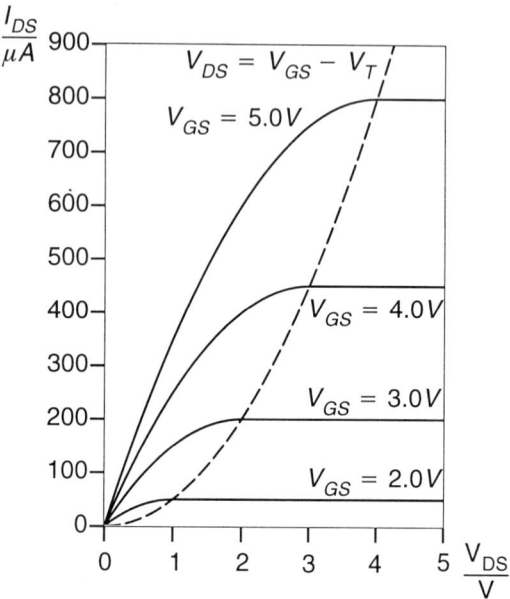

Figure 2.1.2:   Current–voltage characteristics based on a model given by (2.1.1) through (2.1.3) with $\beta_n = 100\,\mu\mathrm{A/V^2}$ and $V_{Tn} = 1\mathrm{V}$

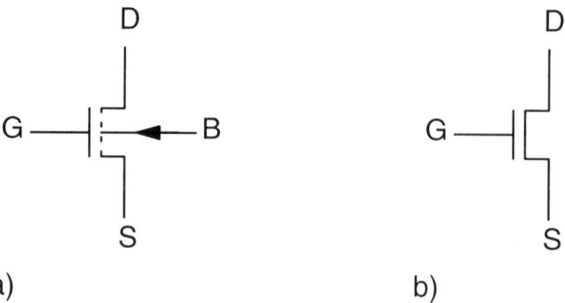

a)   b)

Figure 2.1.3:   Symbols for n-channel MOS transistors (enhancement mode): a) IEEE symbols, b) symbols used in this book

The CMOS fabrication parameters are generally chosen such that the threshold voltage V_T is approximately 20% of the supply voltage V_{DD}.

$$V_T \approx 0.2V_{DD} \qquad \text{for} \quad V_{SB} = 0 \text{ V} \qquad (2.1.5)$$

The threshold voltage varies whenever the source and body are not tied to the same potential. Take, for instance, a substrate bias as large as the supply voltage.

$$V_T \approx 0.3V_{DD} \qquad \text{for} \quad V_{SB} = V_{DD} \qquad (2.1.6)$$

The variation of the threshold voltage as a result of a voltage between source and substrate is called the substrate effect [15].

The model given by (2.1.1) through (2.1.4) can only approximate the exact behaviour of a real transistor. The transition from cutoff to linear operation, for example, is not abrupt, but rather is given by an exponential function. In addition, the current in saturation is not constant. It increases slightly due to channel length modulation. Yet exact modelling of individual transistors is not particularly meaningful for the evaluation of complex circuits. On the contrary, it is best to simplify the relations.

To simplify the relationships, the transistor will be modelled by a voltage controlled resistance. From equation (2.1.2), the following relationship can be derived for the channel resistance in nonsaturated operation:

$$R_n = \frac{1}{\beta_n[V_{GS} - V_T - V_{DS}/2]} \qquad 0 < V_{DS} \le V_{GS} - V_T \qquad (2.1.7)$$

As a result, the current channel resistance depends not only on the gate voltage V_{GS} but also on the operating voltage V_{DS}. In digital circuits, the behaviour for the discrete levels LOW and HIGH is of particular interest. Whenever not otherwise specified, it is assumed that LOW and HIGH correspond to ground V_{SS} and the power supply voltage V_{DD}, respectively.

$$\begin{aligned} LOW: & \qquad V_L = V_{SS} \\ HIGH: & \qquad V_H = V_{DD} \end{aligned} \qquad (2.1.8)$$

Whenever numerical values are used in association with these levels, the power supply voltage is assumed to be $V_{DD} = 5$ V and ground $V_{SS} = 0$ V. Let it be noted, however, that digital integrated circuits with channel length smaller than 0.8 µm are increasingly often driven with 3.3 V and that it can be assumed future processes work with even lower supply voltages.

When $V_{GS} = V_L$, the transistor is off and the following relationship holds:

$$R_{n,OFF} = \infty \qquad \text{for} \quad V_{GS} = V_L \qquad (2.1.9)$$

When $V_{GS} = V_H$, the transistor will be in either the linear region or saturation. From equations (2.1.2) and (2.1.3), it follows that the channel resistance is

$$R_{n,ON} = \frac{\alpha}{\beta_n(V_H - V_T)} \qquad \text{for} \quad V_{GS} = V_H \qquad (2.1.10)$$

where

$$1 \leq \alpha \leq \frac{2V_H}{V_H - V_T} \qquad (2.1.11)$$

The value of α depends on the drain–source voltage. The smallest value of α corresponds to $V_{DS} = 0$ V and linear operation. The transition from linear operation to saturation occurs at $\alpha = 2$. The largest value of α is reached in the saturation region when $V_{DS} = V_H$. It can be shown that the channel resistance in saturation is approximately 2.5 times larger than in linear operation. The relationship between the geometry of the gate, specifically the L/W (length to width) ratio, and the channel resistance is also useful. It is given by

$$R_{n,ON} = r_{S,n} \cdot \frac{L}{W} \qquad (2.1.12)$$

where

$$r_{S,n} = \frac{1}{\mu_n} \frac{d_{ox}}{\varepsilon_{ox}} \cdot \frac{\alpha}{V_H - V_T} \qquad (2.1.13)$$

the sheet resistance in its ON-state. Assuming a 1 μm CMOS process, a typical value is

$$r_{S,n} \approx 2.5 \text{ k}\Omega \qquad (2.1.14)$$

Logic circuits consist of a network of many individual MOS-transistors. When they are modelled as voltage-controlled switches, the switch state (ON or OFF) must be recognizable. This is normally done based on the transit voltage $v_{in} \rightarrow v_{out}$. For an ideal switch, $v_{in} = v_{out}$ when closed. Figure 2.1.4 shows the model used to measure the transit voltage of an n-channel transistor. The charge/discharge behaviour of the output capacitance in this model is asymmetric.

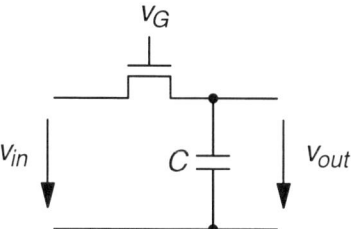

Figure 2.1.4: The transfer of logic states using MOS transistors

Before discharging, the following holds:

$$v_{out} = V_H , \qquad v_{in} = V_L$$

The gate voltage v_G changes from V_L to V_H during switching. The transistor is initially in saturation, and the capacitor discharges with a constant current. Once the capacitor's voltage has fallen under the threshold voltage, the transistor goes into linear operation. The capacitor adopts a final value of $v_{out} = V_L$.

Before recharging,

$$v_{out} = V_L , \qquad v_{in} = V_H$$

At the point of switching, v_G goes from V_L to V_H. The transistor is constantly in saturation since $v_D = v_G = V_H$. Nonetheless, due to $v_{GS} = v_G - v_{out}$, the saturation current decreases as the capacitor charges. The saturation current, and subsequently the charging current, reduces to zero when $V_{GS} = V_T$; i.e. charging is complete when the output voltage reaches $v_{out} = V_H - V_T$. If the substrate effect (2.1.6) is taken into account, the output voltage does not exceed 0.7 V_H. If several transistors are connected in series, the output level will remain the same since only the first transistor is in saturation and dictates the current.

It follows that, since the transistor is constantly in saturation, the charging process is more time-consuming than discharging. Thus it is initially modelled by a resistance in accordance with (2.1.10) and a large value for α. During charging, the value of V_{GS} decreases, and since the current is proportional to the square of the voltage $V_{GS} - V_T$, the corresponding resistance increases. Exact analysis shows that the charging process takes approximately twice the time to reach half the supply voltage than discharging.

Until now only n-channel transistors have been treated. P-channel transistors are structured as shown in Figure 2.1.1 except that the substrate is n-

material and the diffusion zones p-material. Due to the inverse doping profile, the channel of mobile holes is created in a p-channel transistor when a
negative gate–source voltage is applied. The current equations are modelled
analogous to equations (2.1.1) to (2.1.4), whereby all currents and voltages
are given the opposite sign and all < symbols replaced with > in range specifications. In addition, the symbols β and μ are given the index p. In order to
differentiate the threshold voltages of the two transistor types, these two are
given an additional index: V_{Tp} and V_{Tn}. Figure 2.1.5 gives the symbols for
p-channel transistors. The small circle at the gate signifies the inverse behaviour. On account of this inverse behaviour, when dealing with p-channel
transistors the diffusion zone with the more negative voltage is specified to
be the drain.

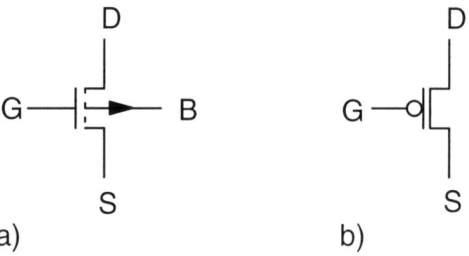

a)   b)

Figure 2.1.5:   Symbols for p-channel MOS transistors (enhancement mode):
a) IEEE symbols,   b) symbols used in this book

The inverse behaviour of p-channel transistors relative to n-channel
transistors is also seen when passing logic signals. If the n-channel transistor
in Figure 2.1.4 is substituted with a p-channel transistor, the discharging process is the unfavourable case and not the charging process. When a LOW signal is passed, a p-channel transistor is constantly in saturation and the output
voltage v_{out} only discharges to the threshold voltage $|V_{Tp}|$. In terms of speed,
however, circuits based on p-channel transistors tend to be somewhat slower
than n-channel circuits with the same transistor geometry. This is due to the
better mobility of electrons compared to holes. A rule of thumb for long
channel transistors is

$$\mu_n \approx 2.5 \ \mu_p \tag{2.1.15}$$

From (2.1.13), it follows that the sheet resistance in the 'ON'-state is
similarly proportional.

$$r_{S,p} \approx 2.5 \ r_{S,n} \tag{2.1.16}$$

If this difference in speed is undesirable, the *L/W* ratio of the transistor gate can be altered so as to compensate for the original difference in channel resistance. Yet since the minimal channel length is dictated by the fabrication process, the channel width must be widened by a factor of 2.5. Nonetheless, if only one transistor is used to pass a logic signal, one of the two states will be sub-optimal. A combination of n- and p-channel transistors as in Figure 2.1.6 has a transistor for each logic state. Such a configuration is called a transmission gate. The transfer characteristics of the various configurations are summarized in Table 2.1.1.

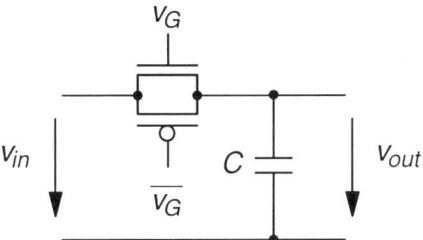

Figure 2.1.6: The transfer of logic signals using a transmission gate

During the fabrication of an integrated circuit in CMOS technology, both n- and p-channel transistors are created on the substrate. This is made possible through the use of special regions called wells (or tubs) that have doping opposite to that of the substrate. These regions act as the substrate for the transistors of "the other type". Both p-wells in n-substrate and n-wells in p-substrate can be fabricated in addition to twin-well processes.

Table 2.1.1: Transfer characteristics of MOS transistors

		Transfer response	
Switch	Transfer level	Output level	Transfer speed
n-channel transistor	LOW	V_L	good
	HIGH	$V_H - V_{Tn}$	poor
p-channel transistor	LOW	$V_L + |V_{Tp}|$	poor
	HIGH	V_H	good
Transmission gate	LOW	V_L	good
	HIGH	V_H	good

2.2 Basic Logic Circuits

The logical values (0 and 1) of control signals and transfer signals can be represented through either a switch state (open / closed) or a voltage level (LOW / HIGH). Accordingly, logic functions may be created through a network of voltage controlled switches. The following correspondence to logical values is assumed:

Switch logic: *open*   0

                    closed   1

Voltage levels: *LOW*   0

                    HIGH   1

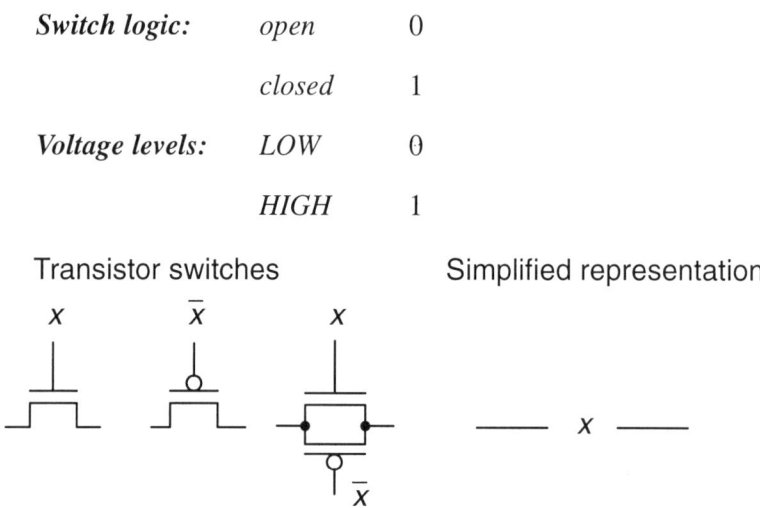

Figure 2.2.1: Simplified switch symbols

The use of MOS transistors as voltage controlled switches was discussed in the previous section. A simplified representation for the description of switch logic is shown in Figure 2.2.1. It signifies that the path is closed (logical 1) when the control value x is 1.

The switch logic equivalents of several basic logic operations are given in Figure 2.2.2. Two switches in parallel create a logical OR ($\vee$) since one switch suffices to close the path. Two switches in series are equivalent to a logical AND ($\wedge$); both switches must be closed to connect the path. A logical complement function is created through the use of a negative switch that closes the path when the control signal is zero.

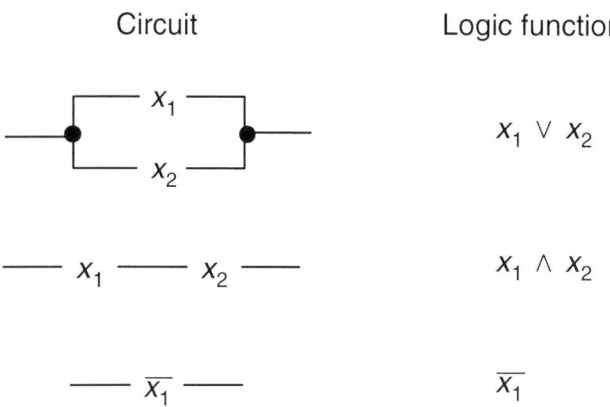

Figure 2.2.2: Simple switch networks

The logical function of a network of switches can be readily deduced when viewed in terms of its basic structures. Figure 2.2.3 shows an example. From the basic functions

$$parallel\ circuit\quad =\quad logical\ OR$$
$$series\ circuit\quad =\quad logical\ AND$$

it is easy to derive the functionality of this combinational logic.

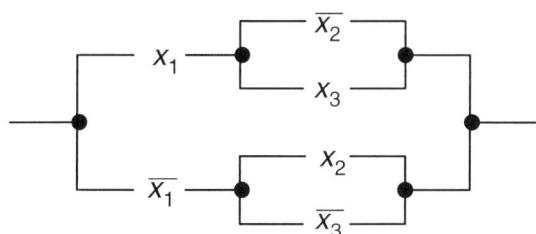

Figure 2.2.3: Switch logic for the function $x_1(\overline{x_2} \lor x_3) \lor \overline{x_1}(x_2 \lor \overline{x_3})$

The relationships between complementary logic functions are of special importance in the creation of such complex logic [18], [19]. Given the logic function

$$f(\,\boldsymbol{x},\, \land,\, \lor\,)\qquad \boldsymbol{x} = (x_1, x_2, \dots\ x_n) \qquad (2.2.1)$$

its so-called dual function

$$f_D(\,\boldsymbol{x},\, \lor,\, \land\,)\qquad \boldsymbol{x} = (x_1, x_2, \dots\ x_n) \qquad (2.2.2)$$

can be produced by exchanging the operators AND and OR. According to Shannon's theorem, a generalization of DeMorgan's theorem, the complement $\bar{f}$ of a function f can be formed by complementing all literals x_i in the dual function f_D of f.

$$\bar{f} = f(\bar{x}, \vee, \wedge) \qquad \bar{x} = (\overline{x_1}, \overline{x_2}, ...\overline{x_n}) \qquad (2.2.3)$$

This theorem can also be used for direct creation of logic functions, as illustrated in Figure 2.2.4. The exchange of ANDs and ORs converts all parallel branches to a serial configuration and vice versa. Taking the complement of the function literals is equivalent to replacing the switches with negative switches.

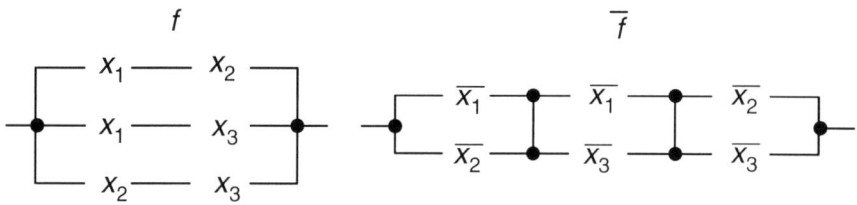

Figure 2.2.4: Example for the dual structure between f and $\bar{f}$

The given associations are sufficient to create all forms of combinatoric logic. A 4:1 multiplexer will be studied for further illustration.

Given the input variable $\boldsymbol{u} = (u_0, u_1, u_2, u_3)$ and the control variable $s = (s_0, s_1)$, for a multiplexer

$$
\begin{aligned}
f &= \overline{s_1}\,\overline{s_0}\,u_0 \vee \overline{s_1}\,s_0\,u_1 \vee s_1\,\overline{s_0}\,u_2 \vee s_1\,s_0\,u_3 \\
&= \overline{s_1}\,(\overline{s_0}\,u_0 \vee s_0\,u_1) \vee s_1\,(\overline{s_0}\,u_2 \vee s_0\,u_3)
\end{aligned}
\qquad (2.2.4)
$$

The corresponding switch logic and its physical implementation are shown in Figure 2.2.5. The fact that the output function f adopts the value of the input variable when the path is closed leads to the additional AND effect of the circuit. The physical implementation of logic functions requires at least one path to be complete to ensure a defined output signal. If the output node is disconnected, capacitive effects cause it to retain its previous value for a period of time.

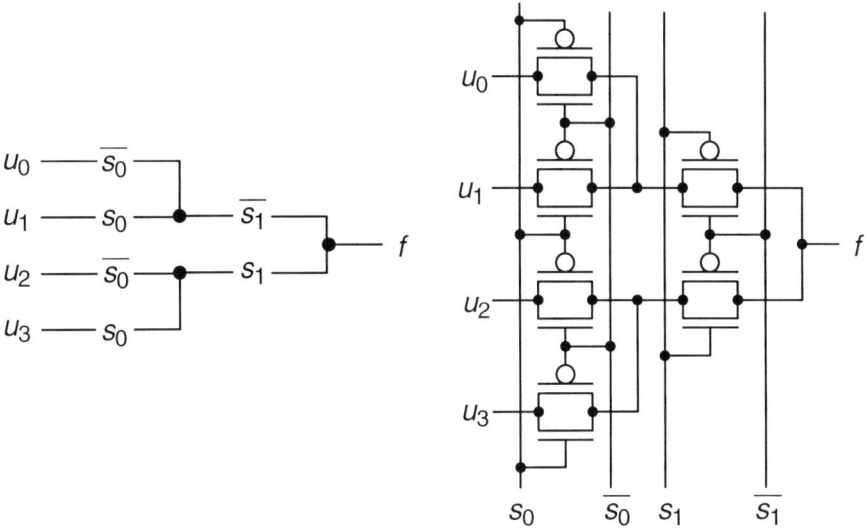

Figure 2.2.5: Switch logic implementation of a 4:1 multiplexer

In principle, any logic function can be created in switch logic based on the structure given in Figure 2.2.6. The input values can be logical variables $u = (u_0, ... u_{n-1})$ and constants $\{0, 1\}$. A set of control variables $s = (s_0, ... s_{k-1})$ is then needed to select the desired function. The implemented logic function f is then a function of the input variables and the control variables.

$$f(x) = f(u, s) \tag{2.2.5}$$

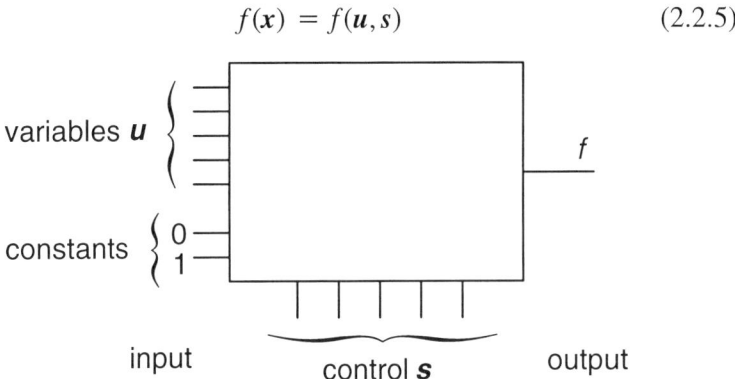

Figure 2.2.6: General structure for realization of logic functions in switch logic

The division of the variables x into input variables u and control variables s is arbitrary. Thus many implementations are possible through different divisions of the variables.

A special form of switch logic is obtained if only constants (the set levels 0 and 1) and no variables are used as input "variables". This is called static gate logic, or gate logic for short. Figure 2.2.7 shows its principal configuration. The logic levels 0 and 1 are implemented through connections to ground (V_{SS}) or the supply voltage (V_{DD}). The term p-network is used to describe circuits consisting solely of p-channel transistors. Likewise, the term n-network is used for n-channel transistor circuits.

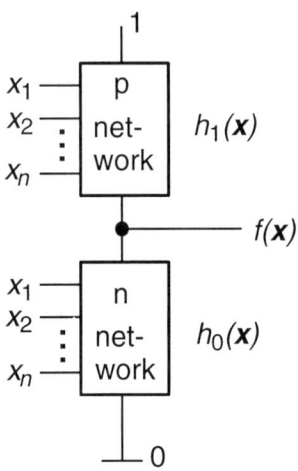

Figure 2.2.7: Principal structure of gate logic

The circuits are simplified to consist of solely one transistor type in accord with Table 2.1.1. If $h_0(x)$ is the switching function for the n-network and $h_1(x)$ the circuit for the p-network (the index of the function h specifies which logical value should be switched), then given the logic function

$$f(x) \qquad x = (x_1, \dots x_n) \qquad\qquad (2.2.6)$$

the following holds:

$$h_0(x) = \bar{f}(x)$$

$$h_1(x) = f(x)$$

$$(2.2.7)$$

The simplest logical function is an inverter, whereby

$$f(x) = \bar{x}$$

$$h_0(x) = x \qquad\qquad (2.2.8)$$

$$h_1(x) = \bar{x}$$

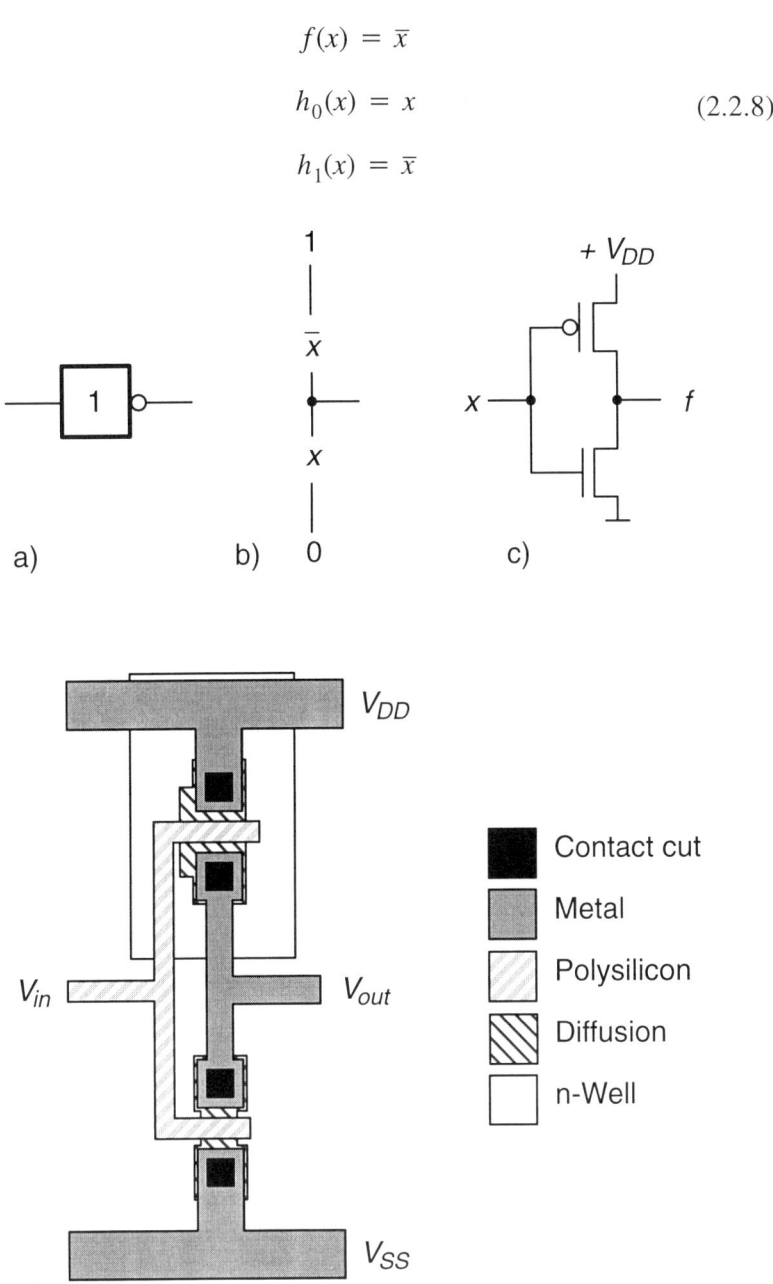

a) b) c)

d)

Figure 2.2.8: Inverter: a) symbol, b) switch representation, c) CMOS implementation, d) layout

Its switch representation and a physical implementation are shown in Figure 2.2.8. In the explanations above, only the behaviour for the logical values 0 and 1 was observed. In some cases, the behaviour for voltages between the limits V_{SS} and V_{DD} is of interest for a more exact description. Figure 2.2.9 illustrates the output characteristics $v_{out}(v_{in})$ of an inverter. For small voltages below V_{Tn}, only the p-channel transistor is closed, and the output voltage has the constant level V_{DD}. Once the input voltage surpasses the threshold voltage V_T, both transistors are active. If the voltage is less than $V_{DD}/2$, the n-channel transistor is in saturation; above that value, the p-channel transistor is in saturation. Near $V_{DD}/2$, both transistors are in saturation, resulting in a sharp transition. For input voltages above $V_{DD} + V_{Tp}$, the p-channel transistor is open, and the output voltage equals the LOW level V_{SS}.

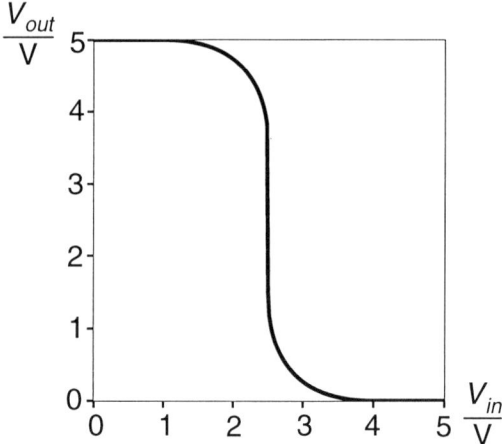

Figure 2.2.9: Output characteristics of an inverter (V_{DD} = 5 V)

These output characteristics are responsible for CMOS technology's low sensitivity to noise. An extraneous signal of considerable amplitude ($< 0.3V_{DD}$) can be superimposed on the input signal without affecting its output. If the transition is to be located at $0.5V_{DD}$, the p- and n-channel transistors must be symmetric in terms of their amplification factor. Thus, based on (2.1.15), the channel of the p-channel transistor must be widened in proportion with the carrier mobility.

In the literature on the design of logic circuits it is shown that all logic functions can be implemented by two-level NAND and NOR gates [18],

[19], [20]. This proves the special importance of these particular logic functions. A NAND gate with n inputs is given by

$$f_{NAND} = \overline{x_1 \, x_2 \, \dots \, x_n} \tag{2.2.9}$$

Given (2.2.7), it follows that

$$h_{0NAND} = x_1 \, x_2 \, \dots \, x_n$$
$$\tag{2.2.10}$$
$$h_{1NAND} = \overline{x_1} \vee \overline{x_2} \vee \dots \vee \overline{x_n}$$

The term for h_1 is derived by application of De Morgan's theorem [18], [19]. Likewise, the NOR function is given by

$$f_{NOR} = \overline{x_1 \vee x_2 \vee \dots x_n} \tag{2.2.11}$$

and it follows that

$$h_{0NOR} = x_1 \vee x_2 \vee \dots x_n$$
$$\tag{2.2.12}$$
$$h_{1NOR} = \overline{x_1} \, \overline{x_2} \, \dots \, \overline{x_n}$$

Figure 2.2.10 presents NAND and NOR gates for two input variables. The correlation series circuit with AND and the parallel circuit with OR are clearly visible.

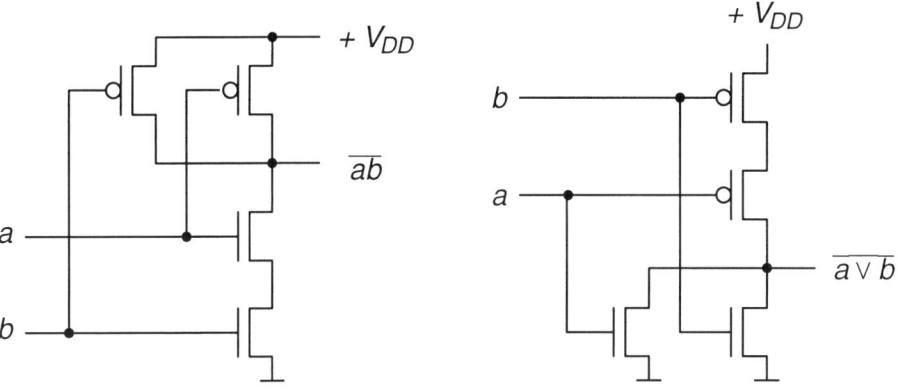

Figure 2.2.10: NAND and NOR gates with two inputs

Any logic functions can be implemented through a two-level NAND–NAND or NOR–NOR configuration. Single level implementations (generally called complex gates) are also possible, as illustrated in Figure 2.2.7. The following example of an exclusive OR (XOR) is further discussed.

The desired functionality is to be given by

$$f = a \oplus b$$

$$
\begin{aligned}
h_0 &= \bar{f} = a \odot b = a\,b \vee \bar{a}\,\bar{b} = (\bar{a} \vee b) \wedge (a \vee \bar{b}) \\
h_1 &= f = a \oplus b = \bar{a}\,b \vee a\,\bar{b} = (a \vee b) \wedge (\bar{a} \vee \bar{b})
\end{aligned}
\tag{2.2.13}
$$

whereby the $\oplus$ operator signifies exclusive OR (antivalence) and $\odot$ equivalence. Figure 2.2.11 shows an example XOR implementation.

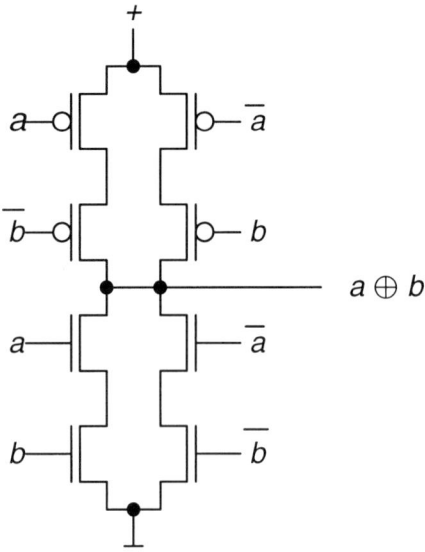

Figure 2.2.11: XOR complex gate implementation

In addition to implementations in switch logic and static gate logic, special structures exist with fewer transistors. One such structure is the pseudo-NMOS shown in Figure 2.2.12 in which the pull-up path consists of a single p-channel transistor. This form of logic requires special dimensioning of the transistors in the pull-up and pull-down paths to guarantee an output voltage of about $0.1V_{DD}$ when the n-network is closed. Pseudo-NMOS are often used in application specific PLA implementations due to their low transistor count and high speed.

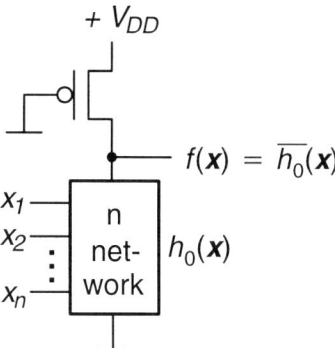

Figure 2.2.12: Pseudo-NMOS logic

Dynamic logic (Figure 2.2.13) is an implementation particular to clocked logic. During the pre-charge phase (clock $\Phi = $ LOW), the output capacitor is charged via the pull-up path (p-channel transistor) to HIGH level. During the following evaluation phase (clock $\Phi = $ HIGH), the pull-down block is activated. Provided that the input signals cause this logic block to connect to ground, the pre-charged capacitor is discharged; if not, the HIGH level output signal remains.

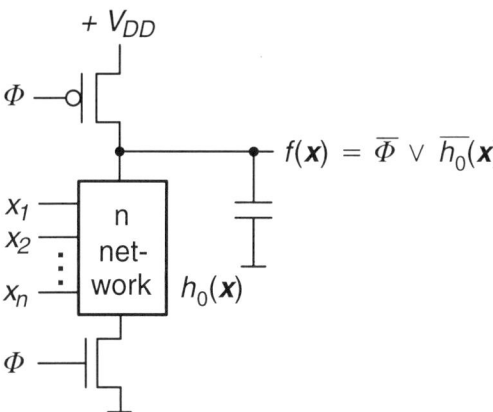

Figure 2.2.13: Dynamic CMOS logic

In addition to combinational logic, sequential circuits require a certain percentage of storage elements such as flip-flops. Delay flip-flops (D-FF) are a typical foundation for which Figure 2.2.14 illustrates a special CMOS implementation. This particular case consists of a master–slave based on D-latches.

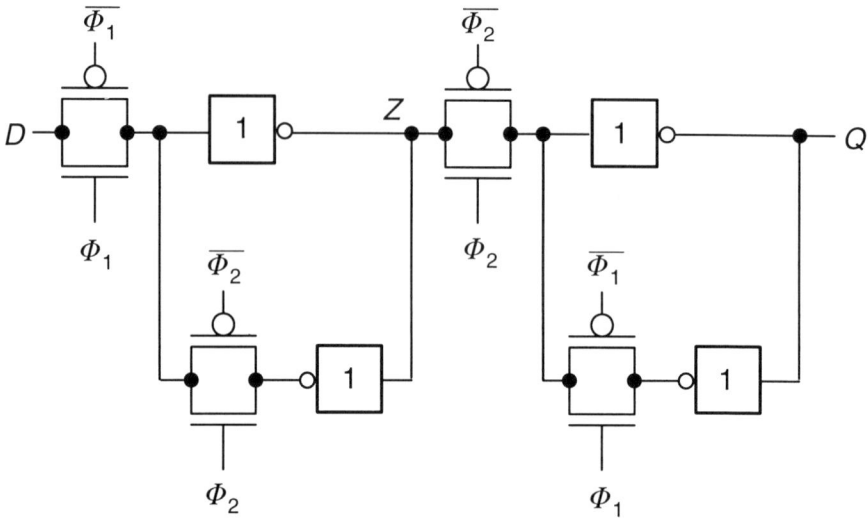

Figure 2.2.14: Delay flip-flop (D-FF)

It is clocked with two, non-overlapping signals, i.e. between each clock pulse there is a short period in which both clock signals are LOW. During the clock phase Φ_1 = HIGH, the input D is connected to the following inverter. The intermediate signal Z equals $\overline{D}$. During the short period between the two clock pulses Φ_1 and Φ_2, the input level is held by the input capacitance of the first inverter. During the clock phase Φ_2 = HIGH, the original of this level value of D (in the form of $\overline{Z}$) is passed to the output Q. Simultaneously, the signal level is held via feedback within the first block. The feedback loop is created through the transmission gate clocked to Φ_2. When the master is reading the input data (Φ_1 = HIGH), a similar feedback loop is created in the slave. When cyclically clocked as illustrated in Figure 2.2.15, the output signal Q adopts the input value D after a delay of one clock cycle. The clock phases Φ_1 and Φ_2 correspond to the signals LOAD and HOLD, respectively.

The quasi-stationary D-FF of Figure 2.2.14 can be further simplified for cyclic clocking at high clock rates. In this case, the storage effect of the gate capacitance within the inverters is sufficient to hold the logic levels without the aid of the feedback loops, which can then be removed. Such dynamic D-FFs require only eight transistors as opposed to the 16 of the quasi-stationary implementation.

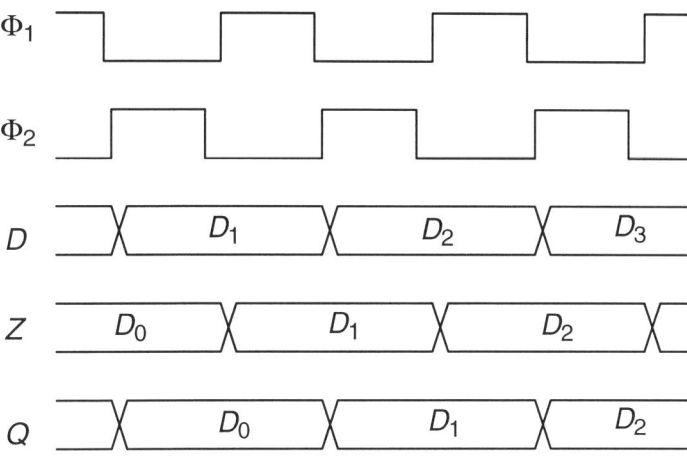

Figure 2.2.15: Timing diagram for a D-FF

2.3 Silicon Area

Once a particular volume has been produced, the production costs for integrated circuits depend mainly on the chip size $A_{Si,Chip}$ and the package. It can be shown that the area of silicon occupied by a logic module (cell) is more or less proportional to the transistor count, i.e.

$$A_{Si,Cell} = a_{Tr}\, n_{Tr} \qquad (2.3.1)$$

This has been researched for both full custom and semi-custom ICs. Experience in full custom design using $1\,\mu m$ CMOS technology has shown that the factor of proportionality lies in the range

$$a_{Tr} = 100...400\ \mu m^2 \qquad (2.3.2)$$

The relationship between cell size and transistor count is strongly dependent on the design type. For a given transistor count, the cell size can span a range 1:4, depending on the cell design. Nonetheless, the relationship (2.3.1) has proven to be a helpful first estimate.

Full custom circuits are made up of transistors with various geometries. Driving transistors, in particular, possess significantly wider channels. In this case, the chip size is more accurately described as being proportional to the active transistor regions:

$$A_{Si,Cell} = a_{act} \cdot \sum_{i=1}^{n_{Tr}} L_i \, W_i \qquad (2.3.3)$$

For full custom designs,

$$a_{act} \approx 25 \qquad (2.3.4)$$

is a good approximation. The summand $L_i W_i$ represents the active area of the transistor i. The summation should be carried out for all transistors in the circuit. Since the transistors generally all have the same length (the minimal length L_0), the following modified approximation can be derived from (2.3.1) and (2.3.3):

$$A_{Si,Cell} = a'_{Tr} \, n_{Tr} \, w_{av} \qquad (2.3.5)$$

where w_{av} is the mean relative channel width given by

$$w_{av} = \frac{1}{n_{Tr}} \sum_{i=1}^{n_{Tr}} \frac{W_i}{W_0} \qquad (2.3.6)$$

where W_0 gives the minimal channel width of the particular transistor type. This accommodates for the difference in channel width between n- and p-type transistors. The value

$$a'_{Tr} \approx 50 \; \mu\text{m}^2 \qquad (2.3.7)$$

was ascertained from a full custom design of a filter module based on a $1\,\mu m$ CMOS process.

The total area of silicon required by a chip is determined by the logic cells, wiring and contacts including their input and output drivers. Their relative size depends on factors such as regularity of the total circuit, design style, the number of metal layers and the number of pads. The relationships for various technologies and design styles will not be discussed. One should remember the essential result that the silicon area is proportional to the number of transistors as a reasonable first approximation. Nonetheless, the space required for pads and drivers must be taken into consideration. For other technologies, the given numeric values should be modified accordingly.

2.4 Delay Estimation

The evaluation of a digital circuit's performance requires an examination of
the signal delay of the individual elements. The abstract modelling of transistors as switches as in section 2.2 only allows for stationary analysis of their
behaviour. Better models must be used to examine their signal transition
characteristics, whereby the voltage dependent parameters of transistor and
capacitor models are a particular problem. Thus, truly analytic examinations
are only carried out for simple configurations. Normally, simplifying assumptions are made about the circuit configurations such that only one non-
linear transistor function remains. More detailed attention to the dependencies in more complex configurations requires simulation using programs
such as SPICE [21].

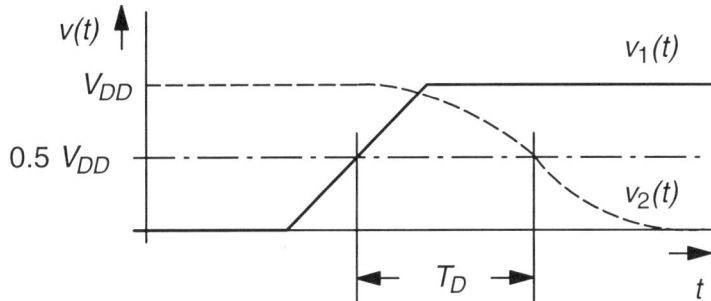

Figure 2.4.1: Definition of signal delay

Due to the non-linear, threshold-like response of digital circuits to the
input signal, exact knowledge of the signal transition is not necessary. In
most cases it suffices to know the delay time relative to a known point within
the transfer process. The 50% level is normally taken as a reference point
(Figure 2.4.1). In this manner, a simplified model of resistors and capacitors
can be used for signal delay analysis of a circuit instead of a simulation program such as SPICE. This will be illustrated for an inverter chain.

The transfer behaviour is determined using exact transistor models in
a SPICE simulation. An RC configuration is then sought that induces the
same delay for the 50% level. Figure 2.4.2 shows such a configuration for
an inverter. The resistors R_p and R_n model the channel resistors of the transistors. C_I is the inner capacitance of the inverter as seen from the output. It consists primarily of the depletion layer capacitances of both diffusion zones and
the wiring to the inverter's output. For an active transistor, the gate oxide ca-

pacitance between channel and gate contributes to the inner capacitance of the inverter. The active transistor channel is a distributed RC network that can be modelled by the channel resistance and half of the gate oxide capacitance at source and drain plus the diffusion capacitances. The diffusion capacitances are junction capacitances resulting from the reverse-biased source–bulk and drain–bulk pn-junctions. C_W is the capacitance of the connection wiring and C_G the load capacitance of the following inverter. C_G consists of the gate capacitances of the n- and p-channel transistors and a short segment of wiring.

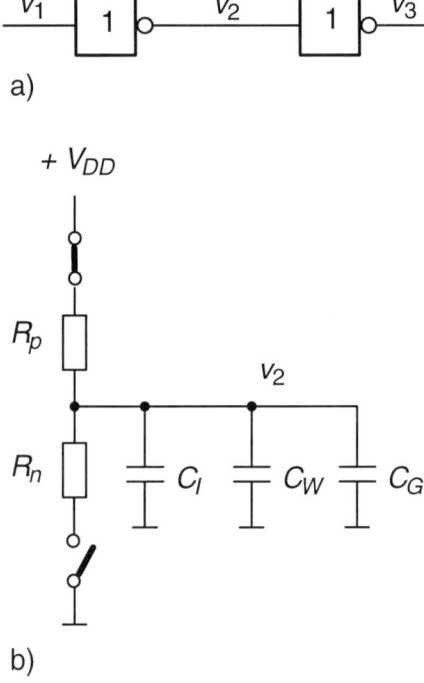

Figure 2.4.2:   Inverter model for delay time determination: a) inverter chain, b) equivalent circuit for $v_1 \rightarrow V_L$

With

$$C = C_I + C_W + C_G \tag{2.4.1}$$

the charging process is given by

$$v_2(t) = V_{DD}(1 - e^{-t/R_pC}) \tag{2.4.2}$$

and the discharging process by

$$v_2(t) = V_{DD}e^{-t/R_nC} \tag{2.4.3}$$

Thus the delay time T_D for the 50% level is

$$\begin{aligned} T_{D,LH} &= (ln2)R_pC \\ T_{D,HL} &= (ln2)R_nC \end{aligned} \tag{2.4.4}$$

Assuming symmetric inverters (equal charge and discharge time) and through normalization based on an inverter of minimal size, the delay time follows from (2.4.1) and (2.4.4) as

$$T_D = \tau_I + ZF\tau_L \tag{2.4.5}$$

where τ_I and τ_L are delays of the normalized inverter that result from capacitive charging and discharging, τ_I is the delay that depends on the inner capacitance $C_{I,0}$ of the normalized inverter, and τ_L represents the delay caused by the external reference capacitance $C_{L,0}$. Changes in the delay time due to deviations from the minimal size or enlargement of the load capacitance are accommodated by the resistance factor Z and the fan-out factor F. In the model, it is assumed that a reduction of the channel resistance through widening of the channel leads to an equivalent increase in the capacitance C_I, such that the product $R_0 C_I$ remains constant.

Normalization is carried out relative to the effective channel resistance R_0 of a symmetric inverter based on transistors of minimal size and the reference capacitance $C_{L,0}$ representing the load capacitance of an inverter with minimal geometry ($L_{0,n}$, $W_{0,n}$, $L_{0,p}$, $W_{0,p}$). The effective channel resistance is assessed via the response delay of a minimally sized inverter to a sloped edge. The influence of voltage dependent changes in the channel resistances (saturation, linear operation) and the brief short-circuit current through both transistors during the transitions LH and HL is also taken into account. Furthermore, the fan-out factor F and the resistance factor Z are given by

$$F = \frac{C_L}{C_{L,O}} \tag{2.4.6}$$

$$Z = \frac{R}{R_0} = \frac{L/W}{L_0/W_0} \tag{2.4.7}$$

The load capacitance C_L is the sum of the wiring and input capacitances of subsequent gates.

Characteristic values for a $1\,\mu m$ CMOS process are in the vicinity of

$$
\begin{array}{llll}
C_{I,0} & = 12.7\ \text{fF} & C_{L,0} & = 6.5\ \text{fF} \\
R_0 & = 10.7\ \text{k}\Omega & C_W & = 0.05\ \text{fF}/\mu\text{m} \\
W_{0,n} & = 1\ \mu\text{m} & W_{0,p} & = 2.6\ \mu\text{m} \\
L_{0,n} & = 1\ \mu\text{m} & L_{0,p} & = 1\ \mu\text{m} \\
\tau_I & = 94\ \text{ps} & \tau_L & = 48\ \text{ps}
\end{array}
\tag{2.4.8}
$$

It is often desirable to determine the corresponding relations for more complex gates based on the delay behaviour of an inverter. The individual transistors in an inverter can be substituted by a network of transistors as in Figure 2.2.7. If it is assumed that the network is made up of transistors with identical geometry, a path of several transistors in series will lead to an increase in the resistance proportional to the number of transistors. In turn, the delay characteristics of gates can be generally approximated by

$$
T_D = m(\tau_I + ZF\tau_L)
\tag{2.4.9}
$$

where m is the maximal number of transistors connected in series along the path to ground or the supply voltage. Since in most cases the n-network and the p-network do not lead to same number of transistors in series (see NAND and NOR gates), the delay for the LH and HL transitions will differ. The model given by (2.4.9) was chosen for its simplicity, but for complex gate configurations it can lead to large errors since the individual influences of the capacitances within the n- and p-networks are not sufficiently regarded. More exact modelling requires that the source and drain capacitances of all active and inactive switch paths be included [22], [23]. An inactive switch can be modelled by the diffusion capacitances at the terminals, whereas an active switch must be modelled by the channel resistance and half of the gate oxide capacitance at the terminals plus the diffusion capacitances.

Example 2.4.1 Consider a NAND gate with three inputs consisting of transistors of minimal geometry as in a symmetrical inverter. In the worst case, the LH-transition charges through one transistor ($m = 1$). Thus

$$
T_{D,LH} = \tau_I + F\tau_L
$$

The HL-transition discharges via a series circuit of three transistors ($m = 3$). It follows that

$$
T_{D,HL} = 3\tau_I + 3F\tau_L
$$

Through the reduction of the resistance factor Z by a factor of three (increase in the channel width W) in the serial path, the coefficient of τ_L can be compensated, but not the coefficient of τ_I. The reason is that a widening of the channel increases the inner capacitance by approximately the same factor as the reduction in resistance.

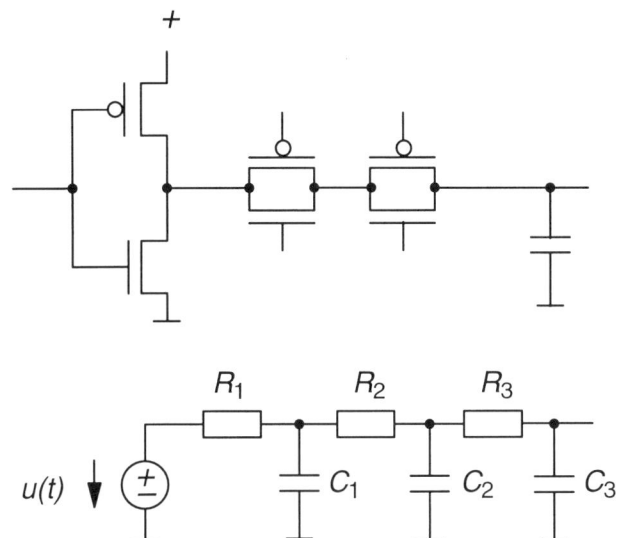

Figure 2.4.3: Modelling of transmission gates

A special model is required for the modelling of switch logic based on transmission gates. Figure 2.4.3 shows such a configuration and its equivalent RC circuit. It can be shown that a chain of several RC circuits, in terms of its delay response to a step function, can be replaced by an equivalent circuit consisting solely of R_E and C_E [24]. For a chain of three RC subcircuits

$$R_E C_E = R_1(C_1 + C_2 + C_3) + R_2(C_2 + C_3) + R_3 C_3 \quad (2.4.10)$$

This delay modelling of a cascade of RC elements is called an Elmore delay [25].

The general solution for n subcircuits is

$$R_E C_E = \sum_{k=1}^{n} R_k \sum_{i=k}^{n} C_i \quad (2.4.11)$$

The solution can be interpreted as a sum of terms, where each term consists of one resistor of the chain multiplied by the sum of all subsequent capacitors along any path to the output.

A transmission gate consisting of transistors as in an inverter of minimal geometry can be modelled approximately as a Π-circuit with two capacitors C_I and a resistor ηR_0 with $\eta = 0.6$ (Figure 2.4.4).

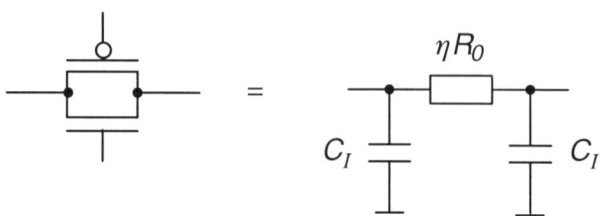

Figure 2.4.4: Delay response model of a transmission gate

The internal resistance of the driving inverter of the transmission gate chain must be modelled differently than with purely capacitive load. The reason lies in the chronological order of the charging and discharging. Although the driver output undergoes a considerable change in voltage, the voltages in the capacitors behind the first transmission gate change only slightly. During the main charge transfer phase between the capacitors, the driving inverter fell below the saturation point (mid-voltage) of the transistors, and the active transistor is now in linear operation. A low ohmic resistance ηR_0 should be used in the active circuit region. The value $\eta = 0.6$ has been found to be appropriate for the evaluation of transmission gates. The impact of the driver's non-linear inner resistance on the delay response can be modelled by giving the first capacitor C_1 an inner resistance of $R_1' = R_0$ and all subsequent capacitors C_2, etc. an inner resistance of $R_1'' = \eta R_0$. This can be carried out in equations (2.4.10) and (2.4.11) by using R_1' in all products containing C_1, and R_1'' with all the other capacitors. Considering that the delay times τ_I and τ_L result from RC circuits with R_0, $C_{I,0}$ and $C_{L,0}$, it is possible to express the delay of transmission gate configurations in terms of τ_I and τ_L.

Example 2.4.2 A symmetric inverter with minimal geometry drives a load capacitor C_L through a transmission gate based on transistors of the same size. This configuration can be modelled by an RC chain with two RC circuits. Resistances and capacitances have the following values:

$$R_1' = R_0 \qquad R_1'' = 0.6R_0 \qquad C_1 = 2C_I$$
$$R_2 = 0.6R_0 \qquad\qquad\qquad\quad C_2 = C_I + C_L$$

Using (2.4.11), an equivalent RC constant is computed:

$$
\begin{aligned}
R_E C_E &= R_1(C_1 + C_2) + R_2 C_2 = R_1' C_1 + R_1'' C_2 + R_2 C_2 \\
&= R_0 C_1 + 0.6 R_0 C_2 + 0.6 R_0 C_2 \\
&= 3.2 R_0 C_I + 1.2 R_0 C_L
\end{aligned}
$$

According to the definitions of τ_I and τ_L,

$$
(\ln 2) R_0 C_I = \tau_I, \qquad (\ln 2) R_0 C_L = F \tau_L
$$

Thus

$$
T_D = 3.2 \tau_I + 1.2 F \tau_L
$$

It follows that for a configuration of two transmission gates as in Figure 2.4.3

$$
T_D = 6.2 \tau_I + 1.8 F \tau_L
$$

In the two identities given above, it is assumed that no addition capacitive loads are present at the output of the inverter or between the transmission gates.

From equations (2.4.5) and (2.4.9) it follows that gate delay depends directly upon the load capacitance. The delay can also be affected by varying the transistors' channel width. For large load capacitances, the delay must be limited through appropriate reduction of the resistance factor. The corresponding increase in the gate capacitance affects the driving gates, however. Thus, isolated optimization of the delay of singular gates is not possible. In respect to their delay response, gates are mutually dependent. As a particular example, the delay optimization of a driver follows.

A single inverter with minimal geometry and symmetric switching characteristics has a delay of

$$
T_D = \tau_I + F \tau_L \tag{2.4.12}
$$

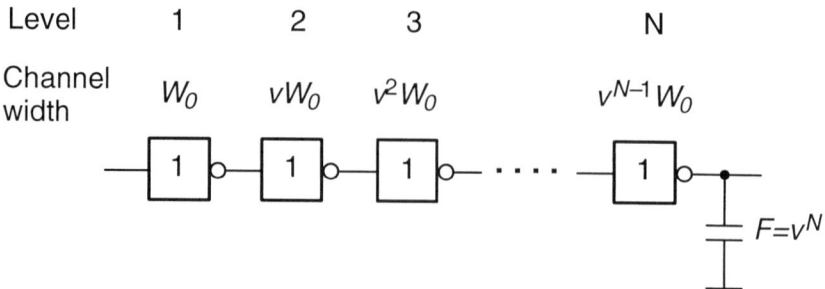

Figure 2.4.5: Driver circuit for a fan-out factor of F

For large fan-out factors, for instance 1000, this denotes an approximately 1000-fold increase in delay in comparison to the case of a minimal inverter as load. This delay can be reduced using a driver circuit consisting of a chain of several inverters, each of which has wider transistor channels than its predecessor (Figure 2.4.5) [15]. It can be shown that in an optimal driver the channel width increases from level to level by a constant factor v starting from the original width W_0. In the case of N inverter levels

$$v = F^{1/N} \tag{2.4.13}$$

For each inverter level, the fan-out factor is v times larger than the resistance factor. Thus the delay for N levels is

$$
\begin{aligned}
T_{D,N} &= N(\tau_I + v\tau_L) \\
&= \frac{\ln F}{\ln v}(\tau_I + v\tau_L)
\end{aligned}
\tag{2.4.14}
$$

The optimal v for minimizing the delay $T_{D,N}$ can be determined via the zeroes in its derivative with respect to v. The result is the requirement

$$v(\ln v - 1) = \frac{\tau_I}{\tau_L} = \gamma \tag{2.4.15}$$

The derivation of the solution $\gamma = 0$ for $v = e$ can be found in many sources [15]. For the more realistic value $\gamma = 2$, $v \approx 4.3$. In the multi-level solution, the delay grows with the logarithm of the fan-out; in the single level solution, it is directly proportional to the fan-out. Nonetheless, the improvement in the delay must be paid for with a considerable increase in silicon area. In accordance with (2.3.3), the area of each level grows by a factor of v. The total sum is a geometric series resulting in

$$\frac{A_{Si,N}}{A_{Si,1}} = \frac{v^N - 1}{v - 1} = \frac{F - 1}{v - 1} \qquad (2.4.16)$$

The delay model used here makes it possible to describe the delay characteristics of complex circuits based on knowledge of the delay characteristics of an inverter. Knowing the current proportions relative to the reference inverter and the configuration of the transistors, any circuit's delay behaviour can be modelled in accordance with

$$T_D = T_0 + FT_1 \qquad (2.4.17)$$

where T_0 and T_1 are functions of the reference values τ_I and τ_L of the inverter. As previously discussed, this general relationship can be demonstrated using Examples 2.4.1 and 2.4.2.

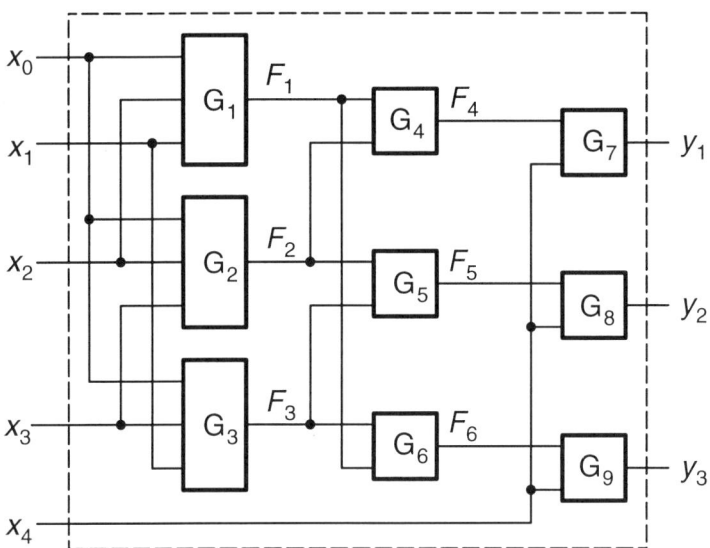

Figure 2.4.6: Example of a configuration of basic logic functions

Complex logic functions are implemented through a configuration of several basic circuits. For each basic circuit, a delay function as given by (2.4.17) must be constructed. When the internal fan-out values are taken into account, the total signal path delay can be evaluated through addition of its component delays. Within complex circuits, the delay of many component paths must be analysed in order to determine the overall time-critical path. Figure 2.4.6 shows an example configuration with nine gates; 21 paths must be compared to determine the delay. The maximal delay of all paths is assigned to the total configuration, whereby it is assumed that all input signals lie in the same time frame.

Often the task arises to compare alternative implementations of equivalent logic in regard to their delay characteristics. In this case it must be noted that, due to the various input capacitances, dissimilar delays can appear between the neighbouring blocks and the logic block under analysis. In order to account for this effect when the neighbouring blocks are undefined, it is assumed that simple inverters drive the inputs and are the output loads. Additional delays at the inputs due to fan-out factors larger than one are assigned to the block under analysis.

In the following chapters, delay times are computed according to a fictitious 1 μm CMOS process, whereby the simple relationship

$$\tau_I = 2\tau_L \qquad (2.4.18)$$

is assumed between the values τ_I and τ_L. This makes it possible to express the delay times in terms of just the one parameter τ_L. The reader is not expected to justify the example delay times given in the following chapters. They are simply comparative values based on the delay model introduced here. Assuming a 1 μm CMOS process, numeric values can be derived using

$$\tau_L = 50 \text{ ps}$$

This value can be adjusted for smaller scaled processes according to the formula

$$\tau_L{}' = \tau_L \cdot \left(\frac{L_0{}'}{L_0}\right)^2 \cdot \frac{V_{DD}}{V_{DD}{}'} \qquad (2.4.19)$$

which also accounts for the possibility of a different supply voltage.

2.5   Power Dissipation

Four sources can be named that contribute to power dissipation in digital CMOS circuits:

$$P = P_{Sw} + P_{SC} + P_{Leak} + P_{Stat} \qquad (2.5.1)$$

The switching power P_{Sw} describes the losses due to the charging and discharging of capacitances. The short circuit power loss P_{SC} is caused by the momentary direct path between the supply voltage and ground during switching. P_{Leak} is induced by leakage currents of reverse-biased diodes and transistors in their sub-threshold region. P_{Stat} are the losses due to constant currents, such as occur in pseudo-NMOS circuits. In general, P_{Sw} contributes most significantly. These losses will be analysed in the following.

For the analysis of power dissipation, the simplest CMOS gate, an inverter, will be assumed. Figure 2.5.1 shows a circuit diagram for the classification of the components of dynamic power dissipation. In accordance with (2.4.1), the capacitance C consists of the three components C_I, C_W and C_G. It is assumed that the controlling input voltage is a periodic pulse with period T. The corresponding steady-state voltage and current waveforms are depicted in Figure 2.5.2. When the input voltage v_{in} changes from HIGH to LOW, the capacitor is charged via the p-channel transistor (peak of current $i_p(t)$). Due to the finite slope of the input voltage flanks, a small current flows through both transistors directly from the supply voltage to ground. This short-circuit component is measurable in the n-channel transistor during the HL-transition of v_{in}. During changes in the input voltage v_{in} from LOW to HIGH, the capacitor is discharged via the n-channel transistor (peak of current $i_n(t)$). Again a short-circuit is induced, as can be detected in the p-channel transistor.

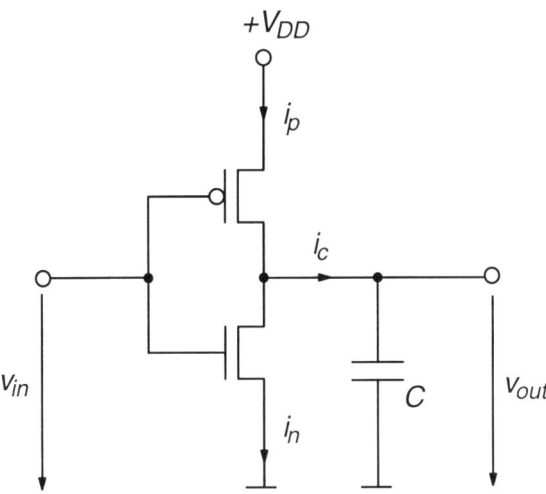

Figure 2.5.1: Inverter circuit for determination of power dissipation

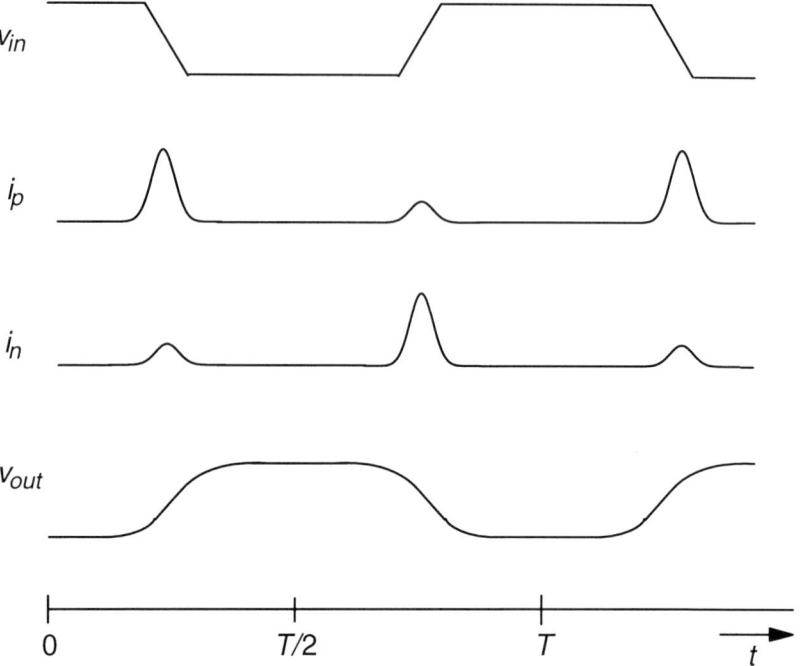

Figure 2.5.2: Voltage and current waveforms of an inverter

The dissipation power is defined as the mean value of the instantaneous power in both transistors. The instantaneous power is the product of the voltage and current in the p- and n-channel transistors.

$$P = \frac{1}{T} \int_0^T [p_p(t) + p_n(t)]\, dt$$

$$= \frac{1}{T} \int_0^T \left\{ [V_{DD} - v_{out}(t)]\, i_p(t) + v_{out}(t)\, i_n(t) \right\} dt$$

(2.5.2)

Through substitution according to the Kirchhoff sum of the currents in the central junction

$$i_p(t) = i_c(t) + i_n(t)$$

(2.5.3)

into equation (2.5.2), it follows that

$$P = \frac{1}{T}\int_0^T \left[V_{DD}\, i_n(t) + V_{DD} i_c(t) - v_{out}(t)\, i_c(t) \right] dt \qquad (2.5.4)$$

The mean of the capacitive current is zero because in a periodic switching the current $i_c(0)$ equals $i_c(nT)$. The third term represents the power dissipation in the capacitor, which is also zero. This can be formally proven by substituting the relationship between voltage and current in a capacitor

$$i_c(t) = C\, \frac{d\, v_{out}(t)}{dt} \qquad (2.5.5)$$

and integrating over a full period T. Thus the first term represents the dissipation power, which can be determined from the product of the supply voltage V_{DD} and the mean value of the current in the n-channel transistor $I_{M,n}$.

$$P = V_{DD}\, \frac{1}{T}\int_0^T i_n(t)\, dt = V_{DD}\, I_{M,n} \qquad (2.5.6)$$

Similarly, by solving (2.5.3) for $i_n(t)$ and through substitution into (2.5.2), it can be shown that the dissipation power can be derived from the product of the supply voltage V_{DD} and the mean value of the current in the p-channel transistor $I_{M,p}$. This means that the mean value of the currents in both transistors must be equal.

$$I_{M,p} = I_{M,n} \qquad (2.5.7)$$

Two components comprise the mean value of the currents. The first component comes from the currents for charging and discharging the capacitor. The other component is produced by the short-circuit current. Through appropriate transistor sizing, a short-circuit share can be achieved that is significantly smaller than the contribution created by the charging / discharging of the capacitor. Assuming the short-circuit components to be negligible, the mean values $I_{M,p}$ and $I_{M,n}$ can each be determined via a current impulse during the HL- or the LH-transition respectively. The currents in the transistors are then identical to the capacitive currents such that their mean values can be derived.

$$I_{M,p} = \frac{1}{T} \int_0^{T/2} i_p(t) \, dt = \frac{1}{T} \int_0^{T/2} i_c(t) \, dt \qquad (2.5.8)$$

Substitution of (2.5.5) into the previous relation it follows that

$$
\begin{aligned}
I_{M,p} &= \frac{1}{T} \int_0^{T/2} C \frac{dv_{out}}{dt} \, dt \\
&= \frac{C}{T} \left[v_{out}(T/2) - v_{out}(0) \right] \\
&= \frac{C}{T} \Delta V
\end{aligned}
\qquad (2.5.9)
$$

This means that the mean value of the current depends on the size of the capacitance and the voltage swing that results from a periodic input signal. Correct dimensioning of the transistor resistances for CMOS gates results in a voltage swing in the order of the supply voltage V_{DD}. Taking (2.5.6) into account, the dissipation power due to charging cycles of the capacitance is given by

$$P_{Sw} = C V_{DD}^2 f_{CLK} \qquad (2.5.10)$$

The relation assumes that the periodic input signal is a square wave, and thus the cycle frequency f_{CLK} is equivalent to the inverse of the period T. With each charge / discharge cycle, twice the temporarily stored capacitor energy is lost as dissipation power – 50% during charging, the other 50% during discharging.

Only the capacitively stored energy is necessary for the computation of the dissipation power. This means that as long as the maximal voltage swing occurs within the given clock period, the transistor resistances within the pull-up and pull-down paths have no influence on the power dissipation. Thus the dissipation power in (2.5.10) holds true for complex logic gates as well as for inverters.

Integrated circuits consist of several subcircuits. Not all capacitors within the subcircuit are charged or discharged in each clock cycle. To describe the influence of such less frequent voltage swings, statistically based, characteristic values for circuits are introduced [26], [27]. Examples are the switching factor σ and the transition activity factor α. The switching factor describes the relative frequency of the capacitor charge cycles to the pulse frequency. The transition activity factor, on the other hand, specifies the frequency of the voltage changes. Since a complete charging cycle encompasses two transitions (LH and HL), the transition activity factor is twice as large as the switching factor. Using the transition activity factor, the dissipation power is

$$P_{Sw} = \frac{\alpha}{2} \, C \, V_{DD}{}^2 f_{CLK} \qquad (2.5.11)$$

A complex circuit contains a large number of capacitors. Using n_C as the number of nodes with capacitive loads, the following holds for the complete circuit:

$$P_{Sw,total} = \frac{1}{2} \, V_{DD}{}^2 f_{CLK} \sum_{i=1}^{n_c} \alpha_i \, C_i \qquad (2.5.12)$$

where α_i and C_i denote the specific values of the individual nodes.

The transition activity factor α of the individual subcircuits depends on its function and the pattern of the input signals [26], [27]. For small circuits, the transition activity factor α can be analytically derived. Complex circuits require the use of special simulation tools. Within complex circuits, some capacitors do not undergo a full voltage swing. In these cases, the specific voltage swing must be regarded in the computation of the dissipation power.

$$P_{Sw,total} = \frac{1}{2} \, V_{DD} f_{CLK} \sum_{i=1}^{n_c} \alpha_i \, C_i \varDelta V_i \qquad (2.5.13)$$

These are only generalized derivations. Further information concerning the computation of the dissipation power of specific circuits and measures for the reduction of power dissipation can be found in the literature [26], [27], [28].

2.6 Exercises

1. An implementation is sought for the logic functions $x = ab \lor ac \lor bc$
    and $y = a\overline{b}\overline{c} \lor \overline{a}b\overline{c} \lor \overline{a}\overline{b}c \lor abc$.
    a. Find the appropriate switch logic using n-channel transistors assuming
        0 and 1 as input signals and $a, \overline{a}, b, \overline{b}, c, \overline{c}$ as control signals. The configuration with the fewest transistors is sought. Hint: factor out common variables.

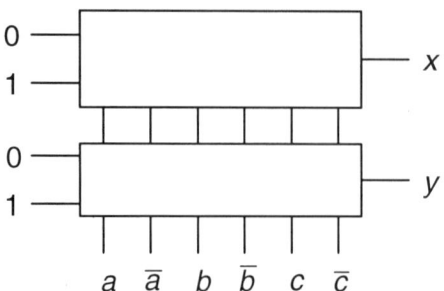

b.   Simplify the configuration using $c, \bar{c}, 0$ and 1 as input signals and $a, \bar{a}, b, \bar{b}$ as control signals.

c.   Derive complex gate implementations in CMOS and pseudo-NMOS.

d.   Compare the hardware costs, the advantages and the disadvantages of the various circuits.

2.   The following circuit is to be used for the implementation of an AND-gate $y = ab$.

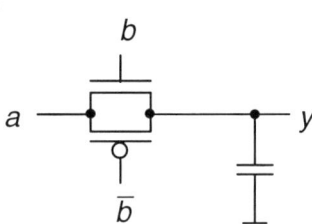

a.   Check the behaviour for all possible values of a and b. For which values does an error occur that depends on the previous output state?

b.   This erroneous behaviour can be avoided by the following circuit modification. Explain the function of the additional transistor.

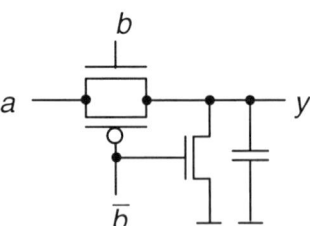

c.   Justify why a single transmission gate is insufficient for an AND-gate implementation, but that for a 2:1 multiplexer the two transmission gates are sufficient for implementation of two AND functions.

3.   The charging of the output capacitor is to be studied in the following configu-
     ration. At time $t = 0$, $v_{out} = 0$ and the voltages v_{in}, v_{G1}, v_{G2} jump to the supply
     voltage V_{DD}.

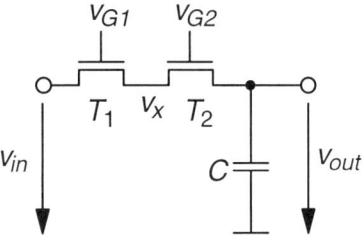

a.   Give the state (saturation or linear operation) of the transistors T_1 and
     T_2 immediately after $t = 0$. Justify your answers.

b.   Give the current equations in terms of the voltages for both transistors.
     In which range must v_x lie so that neither transistor is off? Compute the
     voltage v_x at the moment when the circuit is first switched on. For which
     voltage are both transistors off ($I_{DS} = 0$)? What is then the final value
     of v_{out}?

c.   Using the results of part b and the modified circuit below, compute the
     final output voltage of the charging process. Which general result can
     be deduced for the charging of capacitors?

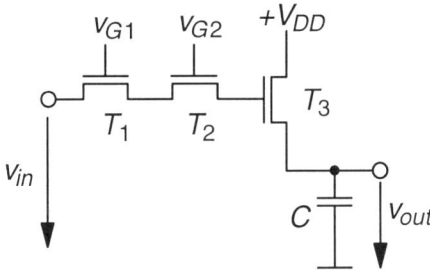

4.   The transistor current equations are simply modelled by two straight lines.
     For an n-channel transistor and $(V_{GS} - V_T) > 0$

$$I_{DS} = \begin{cases} \beta (V_{GS} - V_T)V_{DS} & V_{DS} \le (V_{GS} - V_T)/2 \\ (\beta/2) (V_{GS} - V_T)^2 & V_{DS} > (V_{GS} - V_T)/2 \end{cases}$$

The first relation represents a constant resistance; the second represents a
constant current (saturation current). V_{GS} is the function variable. The thresh-

old voltage is assumed to be $V_T = 0.2V_{DD}$. Compute the discharge process of a capacitor C through a symmetric inverter.

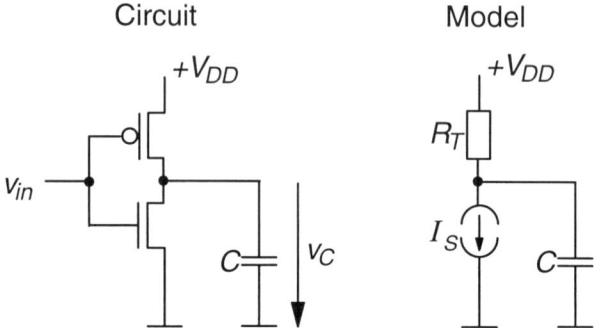

Circuit   Model

a. Which conditions must the input voltage v_{in} fulfil in order for the n-channel transistor to be in saturation? How large is the saturation current as a function of the input voltage?

b. Which conditions must be fulfilled by the input voltage v_{in} and the output voltage v_C in order for the p-channel transistor to be in linear operation? Give the equation for the transistor resistance R_T in terms of v_{in}.

c. The capacitor is charged to the supply voltage V_{DD}. A sudden change in the voltage v_{in} from GRD to V_{DD} triggers the discharge process. Give the time-dependent equation for the capacitor voltage. At which point in time has the voltage sunk to $V_{DD}/2$?

d. Determine an equivalent resistor for an RC circuit such that it takes equal time to discharge to $V_{DD}/2$. What is the relationship between R_E and the transistor resistance (linear operation) for $V_{GS} = V_{DD}$?

e. The input voltage for discharging now undergoes a linear transition of duration T_R.

$$v_{in}(t) = \begin{cases} 0 & t \leq 0 \\ V_{DD}\dfrac{t}{T_R} & 0 < t \leq T_R \\ V_{DD} & t > T_R \end{cases}$$

Compute the discharge process according to the given input voltage. The current through the p-channel transistor is negligible. In addition, the duration T_R is smaller than the time T necessary for the capacitor to discharge to $V_{DD}/2$. Three states for the current through the n-channel transistor are to be observed: no current, time-dependent saturation current, constant saturation current. Compute the delay until $V_{DD}/2$ and compare it with the solution from part c.

5. Circuits are to be designed to drive large capacitive loads FC_L. Consider a symmetric CMOS inverter with minimal geometry modelled by an input capacitor C_L, an output capacitor $C_I = 2C_L$ and a channel resistance R_0. Delay times in this section are to be given as a multiple of the reference value $\tau_L = (ln2)R_0C_L$. In order to reduce the delay, additional driver inverters should be inserted (see Figure 2.4.5). The additional inverters have channel width $w_i = k^i w_0$ where the index i denotes the inverter position. It is assumed that the capacitance grows in proportion to increasing k^i in channel width and that the channel resistance decreases in inverse proportion to k^i.

   a. Compute the delay under the assumption that an inverter with minimal geometry must drive a load capacitance FC_L.

   b. What is the total delay for a driver with N levels ($N-1$ additional drivers)? For a specified value of N, what is the optimal value for k? What is the optimal value for k for $N \in \{3,5,7\}$ if $F = 1000$? Compare the corresponding delay times $T_{D,N}$, the hardware costs $A_{Si,N}$ (see (2.4.16)) and the products $A_{Si,N} \cdot T_{D,N}$. Which generalization about the minimization of the product $A_{Si,N} \cdot T_{D,N}$ can be deduce from these specific results?

6. The power dissipation of digital circuits is influenced by the transition activity factor (equation (2.5.12)). For a simple basic gate this factor should be investigated in more detail. According to the following schematic, a gate with two inputs x_1, x_2 and output y is assumed.

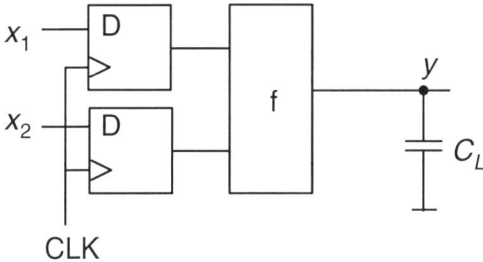

   The probability of occurrence of levels 0 and 1 at input and output is specified by $P(x_1), P(x_2), P(y)$. The joint probability of x_1 and x_2 is described by $P(x_1, x_2)$.

   a. Determine the probability $P(y)$ for a NAND2 gate when the levels 0 and 1 at the input are equal distributed and statistically independent. In the case of statistical independence $P(x_1, x_2) = P(x_1) \cdot P(x_2)$.

   b. For a NAND2 gate all quadruples $(x_1(t), x_1(t-1), x_2(t), x_2(t-1))$ at the input are to be determined, which result in a change of the output level.

c.   The transition activity function at the output of a NAND2 gate is to be specified. First, a statistical independence in the time direction is assumed, which means $P\big(x_i(t),\ x_i(t-1)\big) = P\big(x_i(t)\big) \cdot P\big(x_i(t-1)\big)$.

d.   Now a correlation in the time sequence is assumed. The probability γ specifies that the levels are unchanged, where $1-\gamma$ specifies the change of a level.

$$P\big(x_i(t), x_i(t-1)\big) = \begin{cases} \gamma P(x_i) & x_i(t) = x_i(t-1) \\ (1-\gamma)P(x_i) & \text{otherwise} \end{cases}$$

Determine the transition activity function in general and for the specific value $\gamma = 3/4$.

e.   With the expression

$$\overline{x_1 \vee x_2} = \overline{\overline{x_1}\,\overline{x_2}}$$

the results for a NAND2 are to be transferred to a NOR2.

f.   Formulate basic dependencies for the transition activity factor.

3 Implementation of Fundamental Operations

The sampling of digital signal processing algorithms in Chapter 1 showed that addition and multiplication are very frequently required operations. Furthermore, the inverse operations subtraction and division need to be regarded. In addition, elementary functions also occur, of which the trigonometric functions play a particular role.

This chapter presents circuit techniques and architectures for implementing the fundamental operations and some elementary functions. The circuit techniques are limited to CMOS technology. Various architectures will be elaborated on. The goals are, on the one hand, the extraction of architectures with minimal circuit delay and, on the other hand, circuits with minimal silicon area. Since the representation of numbers is an important aspect of architectures, their description will be handled first.

3.1 Numeric Representations

The basis of most numeric representations is a radix number system [29], [30]. Numbers are represented by a series of digits, and the number's value is given by the sum of the product of each digit and its weighting. This implies that the number's value depends on the position of the digits. In a number system with fixed radix value for all digits, the digit weight is an exponential function with base r and exponent i, where i is the digit's position relative to the (decimal) point. Natural numbers with n digits are represented by

$$A = a_{n-1}\, a_{n-2}\, ...\, a_1\, a_0 \qquad (3.1.1)$$

whereby the digit position is counted from the right and the digits belong to the set

$$a_i \in \{0, 1, \ldots r - 1\} \tag{3.1.2}$$

Thus the value of the number is given by

$$V(A) = \sum_{i=0}^{n-1} a_i \, r^i \tag{3.1.3}$$

The most important number systems used in arithmetic processors are listed in Table 3.1.1. Numeric representations can be extended for fractions. The fractional part is indicated with a (decimal) point. Thus the representation of fractional numbers is

$$A = a_{n-1} \ldots a_1 \, a_0 \, .a_{-1} \, a_{-2} \ldots a_{-m} \tag{3.1.4}$$

The value of the number is given by

$$V(A) = \sum_{i=-m}^{n-1} a_i \, r^i \tag{3.1.5}$$

Table 3.1.1: Examples of number systems

Name	Radix	Digit set
Binary	2	$\{0,1\}$
Octal	8	$\{0,1, \ldots 7\}$
Decimal	10	$\{0,1, \ldots 9\}$
Hexadecimal	16	$\{0,1, \ldots 9, A, \ldots F\}$

In most arithmetic expressions and functions, the representations (3.1.1) and (3.1.4) are used as synonyms for the value of the number. The actual computation of the number's value via a function $V(\cdot)$ is only specified where absolutely necessary to prevent misunderstandings.

Negative numbers are characterized by a minus sign. A useful relationship between positive and negative numbers can be derived. Given a positive number with n digits

$$A = a_{n-1} \, a_{n-2} \ldots a_1 \, a_0 \tag{3.1.6}$$

the negative number $-A$ has a negative value, i.e.

$$V(-A) = - \sum_{i=0}^{n-1} a_i \, r^i \tag{3.1.7}$$

The sum of the geometric series with the elements r^i is given by

$$\sum_{i=0}^{n-1} r^i = \frac{r^n - 1}{r - 1} \qquad (3.1.8)$$

Arithmetic manipulation leads to

$$r^n = (r - 1) \sum_{i=0}^{n-1} r^i + 1 \qquad (3.1.9)$$

Substitution of (3.1.9) into (3.1.7) results in the new equation

$$V(-A) = -r^n + \sum_{i=0}^{n-1} (r - 1 - a_i) \, r^i + 1 \qquad (3.1.10)$$

The variable $\bar{a}_i$ is said to be the complement of a_i relative to the radix r.

$$\bar{a}_i = r - 1 - a_i \qquad (3.1.11)$$

Using the complement, the negative number's value can be expressed as

$$V(-A) = -r^n + \sum_{i=0}^{n-1} \bar{a}_i \, r^i + 1 \qquad (3.1.12)$$

In digital processors, it is desirable to process positive and negative numbers together [29], [30], [31]. In this case, the digit to the far left of the representation is used as a sign indicator. The digit 0 indicates a positive number, and the digit r–1 indicates a negative number. Thus, the positive number representation of n-digit whole numbers is

$$A = 0 \, a_{n-2} \, ... \, a_1 \, a_0 \qquad (3.1.13)$$

where

$$0 \le V(A) < r^{n-1} \qquad (3.1.14)$$

There are several possibilities for representing negative numbers.

In sign magnitude representation, negative numbers are differentiated from positive numbers only through their sign.

$$\bar{A}_{SM} = (r-1) \, a_{n-2} \, ... \, a_1 \, a_0 \qquad (3.1.15)$$

In $(r-1)$'s complement (diminished-radix complement) representation, negative numbers are expressed by their digit's complements:

$$\overline{A}_{r-1} = (r-1)\,\overline{a}_{n-2}\ldots\overline{a}_1\,\overline{a}_0 \qquad (3.1.16)$$

Thus, an $(r-1)$'s complement representation follows with (3.1.12):

$$\overline{A}_{r-1} = r^n - 1 - V(A) \qquad (3.1.17)$$

A peculiarity of the two numeric representations above is their double representation of zero. A positive and a negative zero exist.

The most important form of numeric representation is r's complement (radix complement) representation, in which each digit is converted to its complement and a 1 added to the number.

$$\overline{A}_r = (r-1)\,\overline{a}_{n-2}\ldots\overline{a}_1\,\overline{a}_0 + 1 \qquad (3.1.18)$$

Using (3.1.12), it can be shown that

$$\overline{A}_r = r^n - V(A) \qquad (3.1.19)$$

A positive number A always has a 0 in its $(n-1)$th digit, whereas its negative always has the digit $(r-1)$. Thus the first two terms of the sum in (3.1.12) can be combined to

$$-r^n + (r-1)\,r^{n-1} = -r^{n-1} \qquad (3.1.20)$$

It follows that r's complements are a numeric representation where the digits get the following value:

Digit	Value		
$i = n-1$	$-r^i$	sign digit	(3.1.21)
$0 \le i \le n-2$	$a_i r^i$	other digits	

The particular advantage of r's complements is that positive and negative numbers can be added without regard to their sign. All carries into the digit with value r^n are suppressed. For the addition of two signed numbers A and B, three cases must be observed.

1. $A + B$ result correct as long as $A + B < r^{n-1}$
2. $A + (-B)$ $\rightarrow$ $A + (r^n - B)$
        $B > A$ $r^n - (B - A)$ r's complement of $B - A$
        $B \le A$ $(r^n) + A - B$ suppression of the carry to digit r^n
3. $(-A) + (-B)$ $\rightarrow$ $(r^n - A) + (r^n - B)$
                     $(r^n) + r^n - (A + B)$ r's complement of $A + B$
                     result correct as long as $A + B < r^{n-1}$

For $r = 2$, the r's complement is called a two's complement. Since two's complements have a unique representation of zero, an asymmetric range for positive and negative numbers is possible. A two's complement $\overline{A}_2$ having n bits including its sign fits

$$- 2^{n-1} \leq V(\overline{A}_2) < 2^{n-1} \tag{3.1.22}$$

In the following, two's complement numbers will not be specially marked. Whenever a two's complement number A is referred to, it is assumed that the digit to the far left indicates a sign with negative weighting $(-r^{n-1})$. A positive, n-digit number without a sign digit will be referred to as a binary number.

In addition to the usual numeric representations with positive digits, signed digit representations are also possible [31], [32]. The range of digits is then given by

$$a_i \in \{- a, \dots - 1, 0, 1, \dots a\} \tag{3.1.23}$$

whereby α must cover at least half of the desired interval $r-1$, i.e.

$$\left\lceil \frac{r-1}{2} \right\rceil \leq a \leq r - 1 \tag{3.1.24}$$

Signed digit representations are called SD numbers for short. The value of an n-digit SD number is given by (3.1.3), whereby attention must be given to the special range of α.

In the following, only SD numbers with base $r = 2$ will be treated. Since each digit of an SD number with base $r = 2$ can assume one of three values, 3^n possible representations exist for an n-digit number. However, the corresponding computation of its value according to (3.1.3) yields only $2^{n+1} - 1$ values. This means that several representations of a particular numeric value are allowed for $n > 1$. For this reason, SD representation is classified as redundant. This redundancy will be shown for a five-digit representation of the number $(-9)_{10}$. With $\overline{1} = - 1$, it follows that

$$
\begin{array}{rclcrrrr}
A & = & 0\ \overline{1}\ 0\ 0\ \overline{1} & = & -8 & -1 & & \\
  & = & 0\ \overline{1}\ 0\ \overline{1}\ 1 & = & -8 & -2 & +1 & \\
  & = & 0\ \overline{1}\ \overline{1}\ 1\ 1 & = & -8 & -4 & +2 & +1 \\
  & = & \overline{1}\ 0\ 1\ 1\ 1 & = & -16 & +4 & +2 & +1 \\
  & = & \overline{1}\ 1\ 0\ 0\ \overline{1} & = & -16 & +8 & -1 &
\end{array}
$$

$$= \bar{1} \quad 1 \quad 0 \quad \bar{1} \quad 1 \quad = \quad -16 \quad +8 \quad -2 \quad +1$$
$$= \bar{1} \quad 1 \quad \bar{1} \quad 1 \quad 1 \quad = \quad -16 \quad +8 \quad -4 \quad +2 \quad +1$$

The advantage of SD numbers is that for each number in the range of values, a representation with the least number of non-zero digits exists [32]. Binary numbers can be converted to this type of SD number by observing their so-called string property. A string of k consecutive ones within a binary number

$$...0 \quad 0 \quad 1 \quad 1 \quad 1 \quad ... \quad 1 \quad 1 \quad 0 \quad ...$$
$$\quad\quad i+k \quad\quad\quad\quad\quad\quad\quad\quad i$$

can be replaced by

$$...0 \quad 1 \quad 0 \quad 0 \quad 0 \quad ... \quad 0 \quad \bar{1} \quad 0 \quad ...$$
$$\quad\quad i+k \quad\quad\quad\quad\quad\quad\quad\quad i$$

This basically means that adding a 1 to the ith digit in the binary number creates a carry bit of 1 in the $(i+k)$th digit and that all digits in between turn to zeros. To retain the original value, a 1 must be subtracted from the ith digit. Seen mathematically, a string of ones represents a geometric series of the factor 2. The sum of the series results in

$$2^{i+k-1} + 2^{i+k-2} + \cdots + 2^{i+1} + 2^i = 2^{i+k} - 2^i \quad (3.1.25)$$

This corresponds to the statement made above. A string of ones that includes the sign bit in a negative two's complement number can also be simplified in a similar manner. The sign digit of a two's complement number is weighted negatively. A string of ones of the form

$$\bar{1} \quad 1 \quad 1 \quad ... \quad 1 \quad 1 \quad 0 \quad ...$$
$$n-1 \quad\quad\quad\quad\quad\quad\quad i$$

can be replaced by

$$0 \quad 0 \quad 0 \quad ... \quad 0 \quad \bar{1} \quad 0 \quad ...$$
$$n-1 \quad\quad\quad\quad\quad\quad\quad i$$

As above, this result can be represented by a geometric series.

$$-2^{n-1} + 2^{n-2} + \cdots + 2^{i+1} + 2^i = -2^i \qquad (3.1.26)$$

From the above it follows that for two's complement negative numbers the leading ones can be filled to match a specified fixed word length n. This is called sign extension. The sign extension has to be applied if two numbers of different word lengths are to be added. The sign of the operand with smaller word length will be extended to the word length of the other operand.

Example 3.1.1 Two's complement numbers with a different number of digits are to be added. The word width is to be matched by sign extension.

$$\begin{array}{ll} \bar{1}0110100 & \bar{1}11110110100 \\ +\ 001100111011 \quad \Rightarrow & +\ 001100111011 \\ \hline & \\ \hline & 001011101111 \end{array}$$

Signed digit numbers with the minimal number of non-zero digits and no two consecutive non-zero digits are called canonical signed digit (CSD) numbers [31], [32]. A binary number can be converted to a CSD number by consecutively converting all strings of ones from right to left according to the string properties mentioned above. Any newly occurring strings of ones must also be dealt with during the conversion process. In negative numbers, the sign digit is treated just like all other digits. Carries into the digit before the sign digit are suppressed.

Example 3.1.2 The two numbers 366 and −213 are to be represented as 12-digit binary numbers and then converted to CSD form. The binary number is created via iterative division by 2, in which the binary digits are given by the remainders from low to high. The two's complement representation of the negative number is computed according to (3.1.18).

a) $(366)_{10}$

$$\begin{array}{llll} 0\,0\,0\,1 & 0\,1\,1\,0 & 1\,1\,1\,0 \\ 0\,0\,0\,1 & 0\,1\,1\,1 & 0\,0\,\bar{1}\,0 \\ 0\,0\,0\,1 & 1\,0\,0\,\bar{1} & 0\,0\,\bar{1}\,0 \\ 0\,0\,1\,0 & \bar{1}\,0\,0\,\bar{1} & 0\,0\,\bar{1}\,0 \end{array}$$

b) $(-213)_{10}$

$$\begin{array}{llll} \bar{1}\,1\,1\,1 & 0\,0\,1\,0 & 1\,0\,1\,1 \\ \bar{1}\,1\,1\,1 & 0\,0\,1\,0 & 1\,1\,0\,\bar{1} \\ \bar{1}\,1\,1\,1 & 0\,0\,1\,1 & 0\,\bar{1}\,0\,\bar{1} \\ \bar{1}\,1\,1\,1 & 0\,1\,0\,\bar{1} & 0\,\bar{1}\,0\,\bar{1} \\ 0\,0\,0\,\bar{1} & 0\,1\,0\,\bar{1} & 0\,\bar{1}\,0\,\bar{1} \end{array}$$

The algorithm for the conversion of a binary number to the CSD form can also be described formally.

CSD conversion algorithm
Given $A = a_{n-1}a_{n-2}...a_1a_0$, an n-digit two's complement number and $D = d_{n-1}d_{n-2}...d_1d_0$, its corresponding n-digit CSD representation for radix 2:

1. Initial values: $i=0$; $c_0=0$; $a_n=a_{n-1}$
2. $c_{i+1}=a_{i+1}a_i \vee a_i c_i \vee a_{i+1}c_i$
3. $d_i=a_i+c_i-2c_{i+1}$
4. $i=i+1$; IF $i<n$ GOTO 2.

In the previous algorithm, the c_i's specify if the previous string of ones induces a carry bit in digit i. Using $a_n=a_{n-1}$, the sign is extended to the imaginary digit n. The logic function given in line 2 creates a 1 if at least two of the three arguments a_{i+1}, a_i, c_i are equal to 1. This is the condition that induces a carry bit. The arithmetic function given in line 3 specifies:

1 is created if a singular one appeared or a string of ones ended.
0 is created within a string of ones or when no string of ones appears.
$\bar{1}$ is created whenever a string of ones begins.

The use of this algorithm on the two examples in Example 3.1.2 yields the results shown. The conversion of an SD number to a two's complement number can be carried out by splitting the SD number D into two binary numbers D^+ and D^-, in which D^+ only has 1's where D has positive digits and D^- only has 1's where D has negative digits. The difference between D^+ and D^- yields the two's complement number A.

$$A = D^+ - D^- \qquad (3.1.27)$$

Example 3.1.3 Given $D = 00\bar{1}0\ 010\bar{1}\ 0010$ the splitting up of D yields $D^+ = 0000\ 0100\ 0010$ and $D^- = 0010\ 0001\ 0000$. Thus the two's complement number A is given by

D^+	0 0 0 0	0 1 0 0	0 0 1 0
$- D^-$	$-$ 0 0 1 0	0 0 0 1	0 0 0 0
A	1 1 1 0	0 0 1 1	0 0 1 0

CSD numbers have the peculiar characteristic that no two neighbouring digits can be simultaneously non-zero.

$$d_i \cdot d_{i-1} = 0 \qquad \forall i \qquad (3.1.28)$$

This means that of the nine possible pairs $d_i d_{i-1}$, only five occur.

$$d_i\, d_{i-1} \in \left\{\overline{1}0, 0\overline{1}, 00, 01, 10\right\} \qquad (3.1.29)$$

By combining groups of two consecutive digits of a CSD number, its corresponding radix 4 SD number with digit range

$$\left\{\overline{2}, \overline{1}, 0, 1, 2\right\} \qquad (3.1.30)$$

can be extracted.

A conversion of binary numbers into radix 4 SD numbers can be carried out directly using bit-pairs without the intermediate CSD representation. Using a modified Booth algorithm, sequential, 3-digit segments of the binary number are analysed and converted according to its string properties [33], [34]. Example 3.1.4 shows the conversion of the number $(366)_{10}$.

Example 3.1.4 The binary number A is to be converted into the radix 2 SD number D_2 and the radix 4 SD number D_4 by application of the Booth algorithm specified in Table 3.1.2
$A = 000101101110$
$D_2 = 0001\ 100\overline{1}\ 00\overline{1}0$
$D_4 = 01\ 2\overline{1}\ 0\overline{2}$

Table 3.1.2: Modified Booth algorithm

Binary number			SD number			Explanation
			r = 2		r = 4	
i+1	i	i−1	i+1	i		
0	0	0	0	0	0	no string of ones
0	0	1	0	1	1	end of a string of ones
0	1	0	0	1	1	simple one
0	1	1	1	0	2	end of a string of ones
1	0	0	$\overline{1}$	0	$\overline{2}$	beginning of a string of ones
1	0	1	0	$\overline{1}$	$\overline{1}$	beginning / end of a string of ones
1	1	0	0	$\overline{1}$	$\overline{1}$	beginning of a string of ones
1	1	1	0	0	0	string of ones

3.2 Adders and Subtractors

Addition is a core operation of many digital signal processing tasks. Thus circuits for implementing this operation are of particular importance. The basic circuits for addition will subsequently be derived. To achieve high throughput, circuits with negligible delay are sought. Basic principles and circuits for the implementation of adders with the desired slightest delay will be presented.

3.2.1 Basic adder circuits

One possible hardware implementation is the direct conversion of the method of adding as done by hand. In this case, the operands are added one bit at a time starting at the lowest bit level and allowing for a possible carry from the previous bit level. A circuit that sums two operand bits and a carry bit into a sum bit and a carry bit for the next higher level is called a full adder.

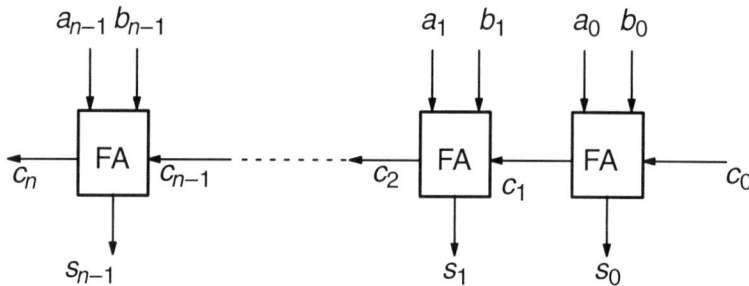

Figure 3.2.1: Adder for two operands $A+B$

Figure 3.2.1 shows an adder circuit that uses full adders to sum two n-bit numbers A and B. When two positive binary numbers are added, the carry bit c_n is used in the representation of the sum. The complete truth table for a full adder is shown in Figure 3.2.2. The behaviour of a full adder can be described as follows. The sum bit is always one when the number of ones among the three operands a_i, b_i, c_i is odd (1 or 3). The carry bit is always one when at least two of the three operands are a one. The logic function for a full adder can be derived from the ones in the truth table. A logic term that leads to a one is a minterm [18], [19]. An OR function of all minterms yields the function. The minterm representation for a full adder is:

$$
\begin{aligned}
s_i &= a_i \overline{b}_i \overline{c}_i \lor \overline{a}_i b_i \overline{c}_i \lor \overline{a}_i \overline{b}_i c_i \lor a_i b_i c_i \\
c_{i+1} &= a_i b_i \overline{c}_i \lor \overline{a}_i b_i c_i \lor a_i \overline{b}_i c_i \lor a_i b_i c_i
\end{aligned}
\tag{3.2.1}
$$

The terms can also be interpreted through observation. The first three terms of s_i check if one of the three variables is a one; the last term checks if all three variables are ones. The first three terms of c_{i+1} check if two of the three variables are ones; the last term checks if all three variables are ones.

Through the definition of an XOR function [19], [20], the function for the sum bit can be written more concisely.

$$
s_i = a_i \oplus b_i \oplus c_i
\tag{3.2.2}
$$

Through minimization [18], [19], it follows that the carry function is:

$$
c_{i+1} = a_i b_i \lor a_i c_i \lor b_i c_i
\tag{3.2.3}
$$

Any disjunctive function can be converted to a two-level NAND–NAND function with the aid of De Morgan's theorem [18], [19]. The corresponding implementation of a full adder is shown in Figure 3.2.3.

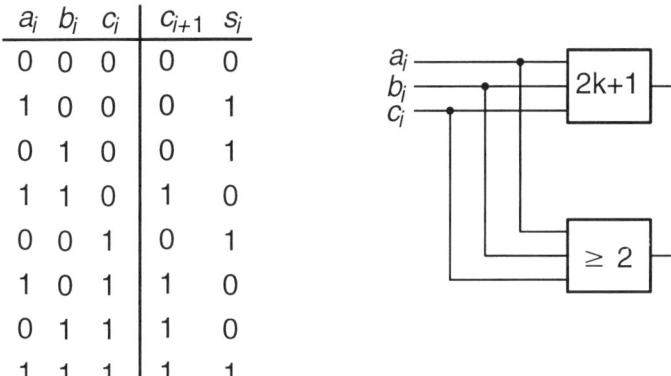

a_i	b_i	c_i	c_{i+1}	s_i
0	0	0	0	0
1	0	0	0	1
0	1	0	0	1
1	1	0	1	0
0	0	1	0	1
1	0	1	1	0
0	1	1	1	0
1	1	1	1	1

Figure 3.2.2: Truth table and logic function of a full adder

The circuit shown in Figure 3.2.1 can be used without modification for the addition of two's complement numbers. In this case, the bit level $n - 1$ indicates the sign. The carry bit c_n is not used. On account of the set word lengths, the addition of two's complement numbers is prone to overflow into the sign bit. Overflow can be identified when the sign of the two operands is identical, yet different from the sign of the sum.

$$\text{Overflow} = a_{n-1}b_{n-1}\overline{s}_{n-1} \lor \overline{a}_{n-1}\overline{b}_{n-1}s_{n-1} \qquad (3.2.4)$$

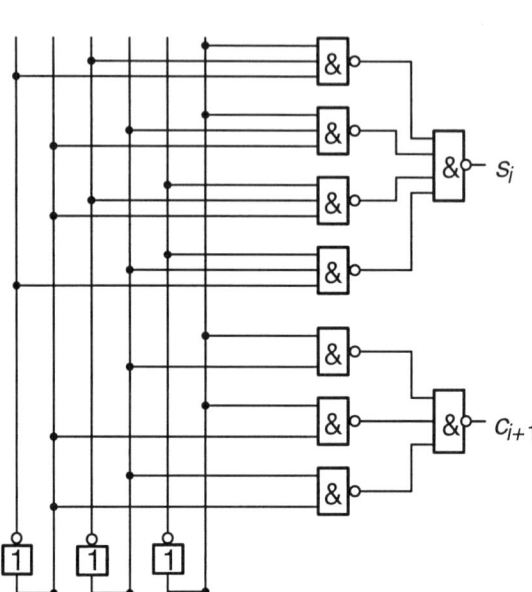

Figure 3.2.3: Gate implementation of a full adder

Based on Figure 3.2.3, the costs of implementing a full adder are one NAND4, five NAND3, three NAND2, and three INV. This leads to a CMOS circuit with a total of 56 transistors. Thus according to relation (2.3.1), an n-bit adder using minimally sized transistors has a circuit size of

$$A_{Si,\,ADD} = a_{Tr}\,56n \qquad (3.2.5)$$

Under the assumption that the circuit size is proportional to the transistor count, cost minimization requires a reduction in the number of transistors. Chapter 2 presented several alternatives to two-level NAND–NAND implementations. One possible alternative is a complex gate implementation. A special complex gate implementation of a full adder will be derived that features symmetric structures for the paths to ground and the supply voltage. From the truth table in Figure 3.2.2 it can be seen that the sum and carry functions are complemented when all three input variables a_i, b_i, c_i are complemented. Together with (3.2.1) and (3.2.3), it thus follows that

$$s_i = f_s(a_i, b_i, c_i)$$
$$\overline{s}_i = f_s(\overline{a}_i, \overline{b}_i, \overline{c}_i) \qquad (3.2.6)$$
$$= \overline{a}_i b_i c_i \vee a_i \overline{b}_i c_i \vee a_i b_i \overline{c}_i \vee \overline{a}_i \overline{b}_i \overline{c}_i$$

$$c_{i+1} = f_c(a_i, b_i, c_i)$$
$$\overline{c}_{i+1} = f_c(\overline{a}_i, \overline{b}_i, \overline{c}_i) \qquad (3.2.7)$$
$$= \overline{a}_i \overline{b}_i \vee \overline{a}_i \overline{c}_i \vee \overline{b}_i \overline{c}_i$$

Through the use of the function f_c, the function f_s can be further simplified to

$$s_i = a_i b_i c_i \vee (a_i \vee b_i \vee c_i) \overline{c}_{i+1}$$
$$\overline{s}_i = \overline{a}_i \overline{b}_i \overline{c}_i \vee \overline{(a_i \vee b_i \vee c_i)} c_{i+1} \qquad (3.2.8)$$

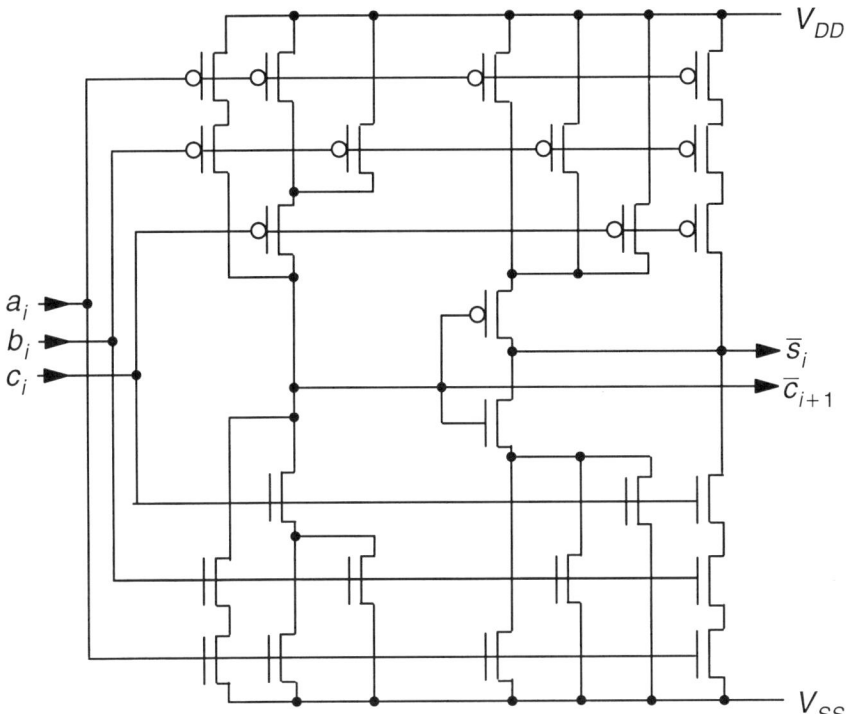

Figure 3.2.4: Symmetric full adder

The validity of this statement can be proven by substituting (3.2.3) and (3.2.7) into the previous relation and simplifying with the AND operation of complementary variables

$$x\bar{x} = 0 \qquad (3.2.9)$$

Complex gate implementations require the circuit function and its complement. According to (3.2.7) and (3.2.8), these functions are structurally identical. Figure 3.2.4 shows the transistor circuit diagram of the resulting full adder. The number of transistors in the new circuit is half as large as in the original NAND–NAND implementation. Instead of 56 transistors, only 26 are required, counting in addition the two output inverters.

As expected, the circuit size of an n-bit adder is proportional to the word width n. The adder structure shown in Figure 3.2.1 has identical circuits at each bit level. Such a circuit structure is called a bit-slice architecture. Since an adder consists of identical structural elements (FA), and the carries are determined sequentially from the lowest to the highest bit level, a particularly cost-efficient, bit-serial implementation as in Figure 3.2.5 is possible.

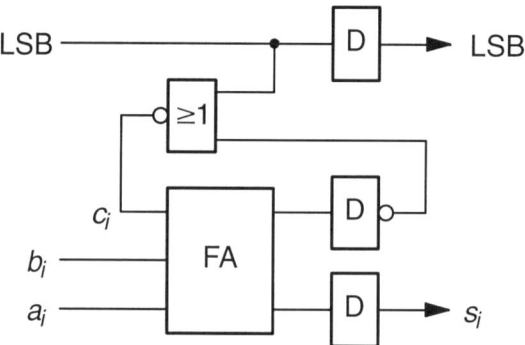

Figure 3.2.5: Bit-serial addition of two binary numbers

In this case, the least significant bits (LSB) are processed first, then the next more significant bits are clocked in. The carry bits are fed back for processing with the next higher bit level. The D-FFs are for storage in a clocked system. The control line LSB signifies the start bits. For LSB = 1, the initial carry $c_0 = 0$. For the following clock cycles, LSB = 0 making the NOR gate invisible to the carry bit feedback loop. Due to the additional D-FFs and the LSB controlled NOR gate, the total count of 110 transistors is nearly double that of a pure full adder NAND–NAND. The highest attainable throughput is dictated by the maximal delay time between input and output, i.e. through the delay of the feedback loop. The clock period is given by

$$T_{CLK} > T_{D,FA,as} + T_{D,DFF}$$
$$T_{CLK} > T_{D,FA,cc} + T_{D,DFF} + T_{D,NOR}$$

(3.2.10)

The additional index for $T_{D,FA}$ signifies the path (*as*: operand bit $\rightarrow$ sum output, *cc*: carry input $\rightarrow$ carry output). Based on the circuit delay model in section 2.4,

$$T_{CLK} > 52\ \tau_L$$

For a 1 μm CMOS process, this means that a clock frequency > 300 MHz can be reached. It should be noticed, however, that the computation of an *n*-bit result also requires *n* clock cycles.

In general, bit-parallel processing is assumed within a signal processing unit. Thus the delay characteristics of a parallel adder as in Figure 3.2.1 will be investigated. When dealing with full adders, one must differentiate between the delay between input and sum output $T_{D,FA,as}$ and the delay between input and carry output $T_{D,FA,cc}$. According to (2.4.9), NAND gates with a larger number of inputs have a larger delay. Accordingly, $T_{D,FA,as} > T_{D,FA,cc}$. The worst case for an *n*-bit adder in terms of the total delay occurs when the carry information must ripple through each bit level to determine the most significant sum bit. In this case, the delay for an *n*-bit adder is

$$T_{D,ADD} = (n-1)T_{D,FA,cc} + T_{D,FA,cs}$$

(3.2.11)

With the aid of the circuit delay model in section 2.4, the delay time can be further specified. For an adder based on full adders as in Figure 3.2.3

$$T_{D,ADD} = \tau_L(24n - 1)$$

(3.2.12)

whereby a fan-out of $F = 1$ is assumed for the sum bits. This result shows that, under the assumption of a 1 μm CMOS process, a 16-bit addition requires approximately 20 ns. For higher performance signal processing circuits, fast adders are desired. One disadvantage of ripple carry adders is the ripple delay of the carry bits from the LSB to the MSB. Various methods for shortening the addition time are give in the literature [35], [36], [37], [38], [39], [40]. Many of the original papers on fast adder implementations are compiled in an IEEE Press book [41]. Treatment of several methods can be found in the textbooks by Hwang [42] and Cavanagh [43]. The methods for shortening addition times can be roughly grouped as follows:

- fast carry propagation
- fast carry generation
- hierarchical adder structures

Examples of such adders follow.

3.2.2 Adders with fast carry chain

To accelerate the carry signals, the path logic is simplified and auxiliary signals for controlling the path logic are computed in parallel. Useful auxiliary signals for the carry function can be extracted from the Karnaugh diagram in Figure 3.2.6. These auxiliary signals derived for the carry function are

$$
\begin{aligned}
k_i &= \overline{a}_i \wedge \overline{b}_i \quad \textit{carry kill} \\
p_i &= a_i \oplus b_i \quad \textit{carry propagate} \\
g_i &= a_i \wedge b_i \quad \textit{carry generate}
\end{aligned}
\qquad (3.2.13)
$$

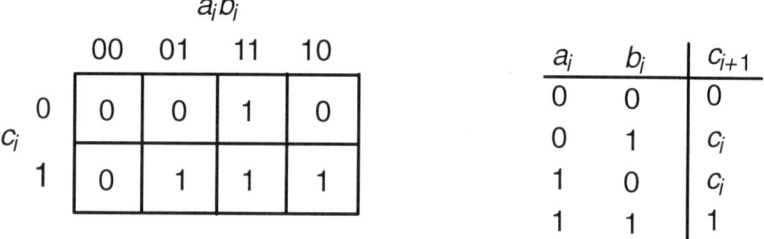

Figure 3.2.6: Karnaugh diagram and truth table for the carry function

The meaning of the auxiliary signals is clear directly from their designations. In addition to the carry propagate p_i defined above, a modified carry propagate

$$
p_i' = a_i \vee b_i \qquad (3.2.14)
$$

is sometimes used that follows directly from Equation (3.2.3). In the implementation of the logic functions, it must be noted that the use of p_i' leads to a multiple cover of 1's. Through the various, multiple cover of 1's in the carry function and through use of the auxiliary functions previously defined, several different logic functions for evaluating the carry can be derived as listed below.

$$
\begin{aligned}
c_{i+1} &= a_i b_i \lor c_i\,(a_i \lor b_i) = g_i \lor c_i\,p_i' \\
&= a_i b_i \lor c_i\,(a_i \oplus b_i) = g_i \lor c_i\,p_i \\
&= a_i\,(a_i \odot b_i) \lor c_i\,(a_i \oplus b_i) = a_i\,\bar{p}_i \lor c_i\,p_i \\
&= b_i\,(a_i \odot b_i) \lor c_i\,(a_i \oplus b_i) = b_i\,\bar{p}_i \lor c_i\,p_i
\end{aligned}
\tag{3.2.15}
$$

Using the method shown in section 2.2, switch logic for CMOS technology can be determined. Figure 3.2.7 shows an implementation of the carry function that follows directly from the truth table in Figure 3.2.6. The circuit implements the second line of (3.2.15). The parallel circuit is the OR, the term $c_i\,p_i$ is created by the transmission gate and the p-channel transistor implements the expression g_i. The third path controlled by k_i is necessary since one path must always be closed in order to retain a defined output signal level. Figure 3.2.8 shows an implementation that follows directly from the third line. The circuits shown in Figures 3.2.7 and 3.2.8 are significantly less costly than the two-stage gate logic in Figure 3.2.3 and are generally faster.

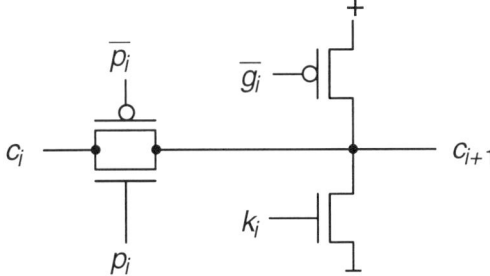

Figure 3.2.7: Logic circuit for the carry block using auxiliary signals g_i, k_i, p_i

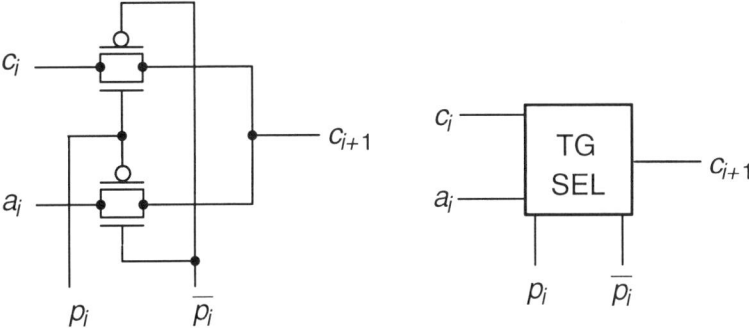

Figure 3.2.8: Logic circuit for the carry block based on a transmission gate selector

In addition to the carry block, a full adder requires an implementation of the XOR function. This XOR function is necessary for creating the propagate signal p_i and the sum signal s_i. From (3.2.2), it can be easily derived that

$$s_i = p_i \oplus c_i \qquad (3.2.16)$$

Besides the structure shown in Figure 2.2.11, several further XOR implementations are possible. The structure shown in Figure 3.2.9a based on transmission gate selectors follows directly from the definition of XOR. It is a 2:1 multiplexer that transmits the signal a for $b = 0$ and the signal $\bar{a}$ for $b = 1$. Due to the symmetry of XOR, an exchange of a and b is also possible. Including the inverters, eight transistors are required. Figure 3.2.9b shows an additional structure that requires only six transistors including the inverters.

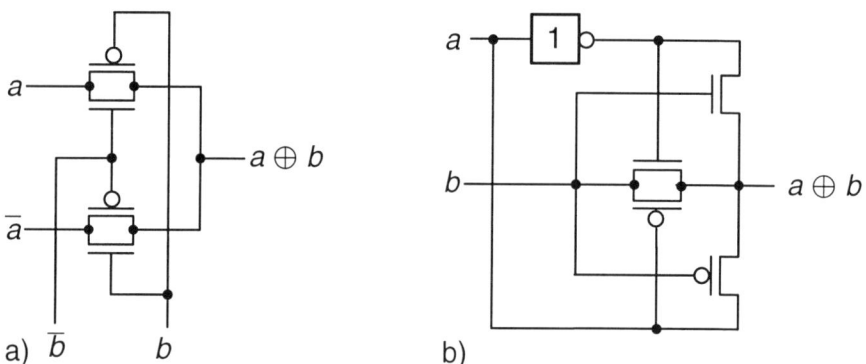

Figure 3.2.9: Implementations for the XOR function

From the definition of XOR and of the equivalence function, it can be shown that

$$a \oplus b = \overline{a \odot b} = \bar{a} \odot b = a \odot \bar{b}$$
$$a \odot b = \overline{a \oplus b} = \bar{a} \oplus b = a \oplus \bar{b} \qquad (3.2.17)$$

From the equations above, it follows that the exchange of a and $\bar{a}$ or b and $\bar{b}$ in the circuit of Figure 3.2.9a creates an equivalence function XNOR from the antivalence function XOR and vice versa. For a corresponding change in the function of the circuit of Figure 3.2.9b, the inverter must be transferred from the upper path to the lower path. Combining equation (3.2.16) with the circuits of Figures 3.2.8 and 3.2.9a, a full adder can be constructed from three transmission gate selectors and six inverters. The circuit shown in Figure 3.2.10 has a carry output with inverted value. Thus, a second type of full adder is necessary that yields a normal output carry given an inverted input carry. These two types must then be alternately connected in sequence.

The inverter at the output of the transmission gate is necessary to reduce the delay. As shown in Figure 2.4.3, transmission gates can be modelled as RC circuits. The carry blocks of m adder stages, connected in series and without the inverters in between, can be modelled as $m + 1$ RC circuits. Under the simplifying assumption that all resistances and capacitances have the same value and using the Elmore time constants from (2.4.11), the delay becomes

$$T_D \approx (ln2)RC\frac{(m + 2)(m + 1)}{2} \tag{3.2.18}$$

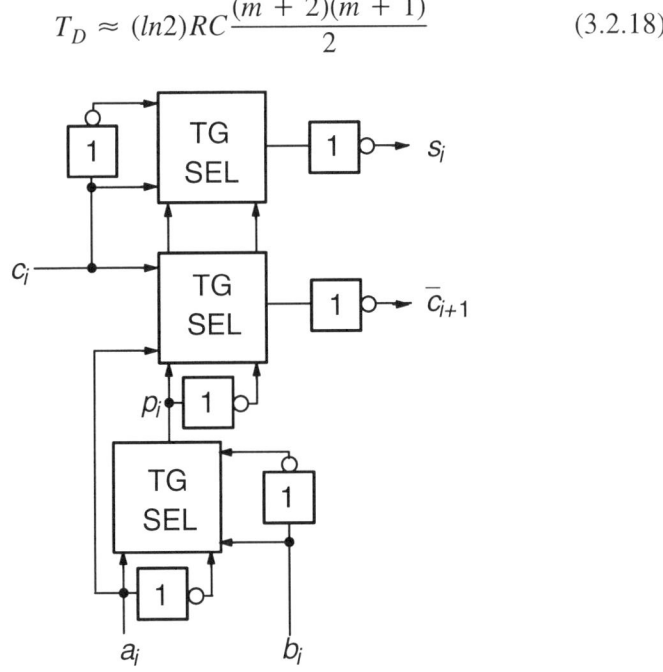

Figure 3.2.10: Transmission gate adder

This means that the delay time is proportional to m^2. For an n-bit adder, n/m clusters of m adders plus one inverter are switched in series. Thus the delay time from (3.2.18) should be multiplied by n/m for an n-bit adder. A minimization of the delay time leads to a value for m between 1 and 2. Since m must be a natural number, $m = 1$ is normally chosen.

In comparison to adder implementations based on NAND gates (Figure 3.2.3), transmission gate adders have practically half as many transistors–24 instead of 56 transistors per full adder. Adder implementations with symmetric full adders such as in Figure 3.2.4 have approximately the same transistor count, but the delay characteristics of transmission gate adders are significantly better. The time-critical path of a transmission gate adder goes from $a_0 \rightarrow p_0 \rightarrow c_1$ and then along the carry path to $\overline{c}_{n-1}$ and then to the sum output s_{n-1}. The delay model yields:

$$T_{D,ADD} = \tau_L(13n + 20.4) \tag{3.2.19}$$

Figure 3.2.11 shows a graphical comparison with the adders based on full adders as in Figure 3.2.3. The slope of the transmission gate adder is lower due to the faster carry path, and its higher minimal value results from the parallel computation of the propagate signal. For a 16-bit addition, the delay is reduced by approximately 8 ns. Thus the transmission gate adder features better delay characteristics at lower cost. Further variants of adder structures with a carry path based on transmission gates can be found in the literature. These are called fast carry chain or Manchester adders [16], [17], [39].

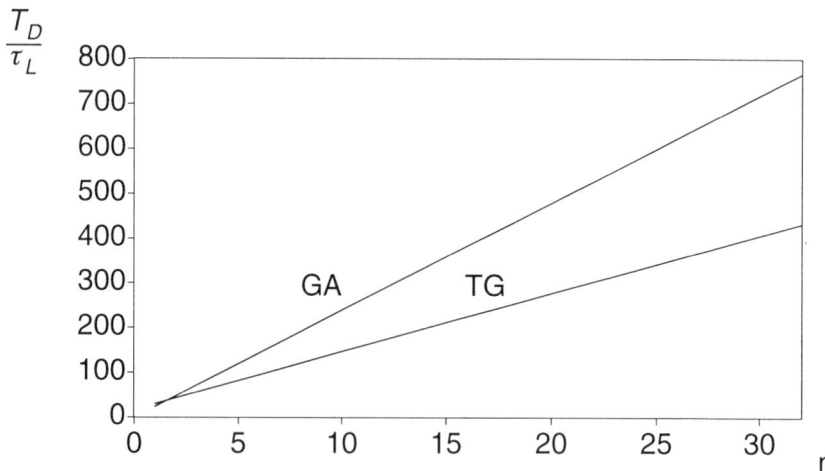

Figure 3.2.11: Delay time T_D over the word width n of adders with bit slice structure (GA - gate implementation of the full adder, TG - transmission gate adder)

3.2.3 Adders with parallel carry computation

The sequential computation of the carry bits in the adders shown above is disadvantageous for the delay characteristics. From the inputs a_i and b_i, g_i and p_i are computed in parallel. The carry functions are then evaluated sequentially with increasing i in accordance with

$$c_{i+1} = g_i \lor p_i\, c_i \tag{3.2.20}$$

A parallel evaluation of the carry bits requires a function that incorporates solutions from previous bit levels. The substitution of c_1 into c_2 results in

$$
\begin{aligned}
c_2 &= g_1 \lor p_1\, c_1 \\
&= g_1 \lor p_1\, (g_0 \lor p_0\, c_0) \\
&= g_1 \lor p_1\, g_0 \lor p_1\, p_0\, c_0
\end{aligned} \tag{3.2.21}
$$

Accordingly for c_3

$$
\begin{aligned}
c_3 &= g_2 \lor p_2\, c_2 \\
&= g_2 \lor p_2\, (g_1 \lor p_1\, g_0 \lor p_1\, p_0\, c_0) \\
&= g_2 \lor p_2\, g_1 \lor p_2\, p_1\, g_0 \lor p_2\, p_1\, p_0\, c_0
\end{aligned} \tag{3.2.22}
$$

Continuing by this method, the following equation for the carry function c_{i+1} results:

$$
\begin{aligned}
c_{i+1} &= g_i \lor p_i\, g_{i-1} \lor p_i\, p_{i-1}\, g_{i-2} \lor \ldots \lor p_i\, p_{i-1} \cdots p_1\, p_0\, c_0 \\
&= g_i \lor \sum_{j=0}^{i-1} \left(\prod_{k=j+1}^{i} p_k \right) g_j \lor \prod_{k=0}^{i} p_k\, c_0
\end{aligned} \tag{3.2.23}
$$

The product symbol $\prod$ signifies multiple AND operations and the summation symbol $\sum$ stands for multiple OR operations. Adder implementations with parallel carry bit computation based on (3.2.23) are called carry lookahead.

The disjunctive functions in (3.2.23) can be implemented as two-stage NAND–NAND structures (Figure 3.2.12). Through the increase in the number of terms at higher bit levels i, the corresponding gate configuration becomes ever more complex. The delay characteristics of gates with multiple inputs also become less economical. According to (2.4.9), the input count m directly affects the delay characteristics. If an n-bit adder were to be implemented as a carry block as in Figure 3.2.12, each bit level i would require $(i+2)(i+5)$ transistors. Summing over all bit planes results in a total transistor count for the carry block of

$$N_{Tr} = (n^3 + 9n^2 + 20n)/3 \qquad (3.2.24)$$

The entire adder including the computation of generate and propagate signals and the sum bits would require

$$N_{Tr} = (n^3 + 9n^2 + 86n)/3 \qquad (3.2.25)$$

transistors. The circuit area is significantly larger in comparison to a simple ripple carry adder as in Figure 3.2.3. In respect to the delay characteristics, one must observe that, according to (2.4.9), the delay of a NAND gate increases with the number of inputs. Thus the time-critical path contains the gate with the largest number of inputs for the computation of c_{n-1}. Of the auxiliary signals, the propagate signal $p_{n/2}$ with its $n(n+2)/4$ inverter capacitances has the largest capacitive load. Thus the time-critical path goes along $(a_{n/2}, b_{n/2}) \rightarrow p_{n/2} \rightarrow c_{n-1} \rightarrow s_{n-1}$. An approximation for the total delay yields

$$T_{D,ADD} = \tau_L(0.25n^2 + 4.4n + 33.1) \qquad (3.2.26)$$

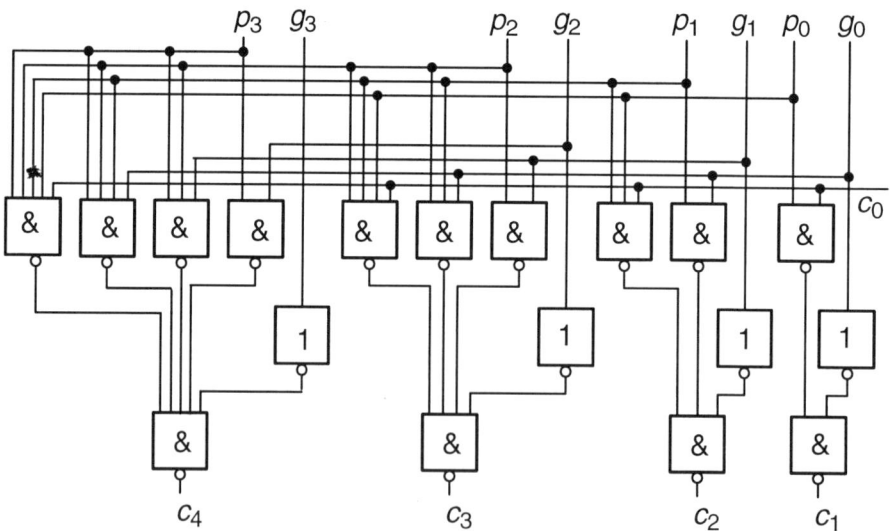

Figure 3.2.12: Gate logic for evaluating the carry bits in a carry lookahead adder

This approximation shows that the delay grows with the square of n. Up to $n = 32$, this circuit yields improved delay characteristics. For higher values, it is worse than a transmission gate adder. However, it should be noted that the delay characteristics for a NAND gate with more than four inputs are badly modelled by (2.4.9). Correspondingly large errors also occur for large values of n in (3.2.26).

The carry block can be implemented by a one-level complex gate instead of a two-level NAND–NAND structure. Figure 3.2.13 shows a complex gate for the determination of c_4. The function for c_4 is given by

$$c_4 = g_3 \vee p_3 (g_2 \vee p_2 (g_1 \vee p_1 (g_0 \vee p_0 c_0))) \qquad (3.2.27)$$

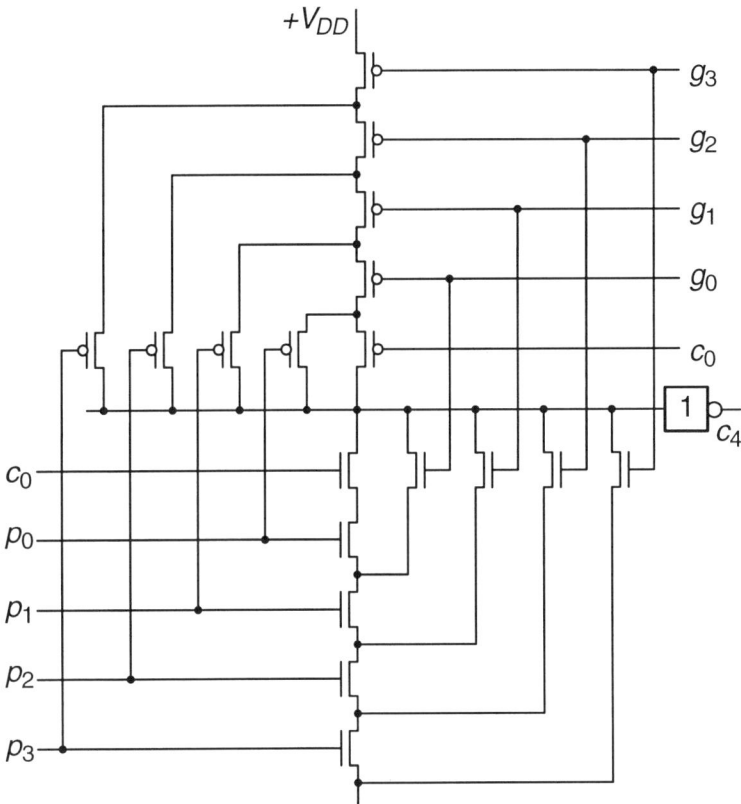

Figure 3.2.13: Complex gate for the carry function c_4

The statements made in section 2.2 about parallel and series transistor circuits can be directly transferred to the structure in Figure 3.2.13. The complementary structures between the n-circuit and the p-circuit are also clearly visible. A complex gate implementation reduces, in particular, the number of transistors required. The transistor count now only increases in proportion to n^2. The delay characteristics are not improved since a larger number of inner capacitances must be charged and discharged within the complex gate. In this case, characteristic data of an n-bit adder are given by

$$
\begin{aligned}
N_{Tr} &= 2n^2 + 28n \\
T_{D,ADDn} &= \tau_L(n^2 + 2n + 36)
\end{aligned}
\qquad (3.2.28)
$$

3.2.4 Hierarchical adder structures

In particular for large word widths n, the carry lookahead technique presented is not very efficient. Through a hierarchical adder structure, a further reduction of the delay can be achieved. In this case, the total word width n is split into smaller bit groups m, and the addition of the bit groups is carried out in parallel. The results of the bit groups are then combined to give the total result. This splitting up into bit groups can also be applied exclusively to the carry block. In the following, three important hierarchical adder structures are presented.

Binary lookahead carry adder
For the description of hierarchically structured carry blocks, new auxiliary generate and propagate functions are required. So-called block carry functions are defined for the block comprising the lower i bits. With G_i as the block generate and P_i as the block propagate signal, the carry function is

$$
c_{i+1} = G_i \lor P_i\, c_0 \qquad\qquad 0 \le i \le n - 1 \qquad (3.2.29)
$$

Using the block carry function for c_i and substituting into (3.2.20), it can be shown that

$$
c_{i+1} = g_i \lor p_i\, (G_{i-1} \lor P_{i-1}\, c_0) \qquad (3.2.30)
$$

From comparison with (3.2.29), it follows that

$$
\begin{aligned}
G_i &= g_i \lor p_i\, G_{i-1} \\
P_i &= p_i\, P_{i-1}
\end{aligned}
\qquad (3.2.31)
$$

The equations above can also be described using an operator $\circ$. Then

$$
(G_i, P_i) = (g_i, p_i) \circ (G_{i-1}, P_{i-1}) \qquad (3.2.32)
$$

With $G_0 = g_0$ and $P_0 = p_0$ and through repeated application of the $\circ$ operator, the block carry function becomes

$$(G_i, P_i) = (g_i, p_i) \circ ... \circ (g_1, p_1) \circ (g_0, p_0) \qquad (3.2.33)$$

It can now be shown that the $\circ$ operator is associative, i.e. the order in which the operations are carried out is irrelevant. This can be proved by combining three generate and propagate signals in various orders. The resulting two possibilities are listed below.

$$
\begin{aligned}
[(g_3, p_3) \circ (g_2, p_2)] \circ (g_1, p_1) &= [(g_3 \vee p_3 g_2, p_3 p_2)] \circ (g_1, p_1) \\
&= [g_3 \vee p_3 g_2 \vee p_3 p_2 g_1, p_3 p_2 p_1] \\
(g_3, p_3) \circ [(g_2, p_2) \circ (g_1, p_1)] &= (g_3, p_3) \circ [g_2 \vee p_2 g_1, p_2 p_1] \\
&= [g_3 \vee p_3(g_2 \vee p_2 g_1), p_3 p_2 p_1]
\end{aligned}
\qquad (3.2.34)
$$

Using the distributivity of $\wedge$ and $\vee$, it can be shown that the two expressions are identical.

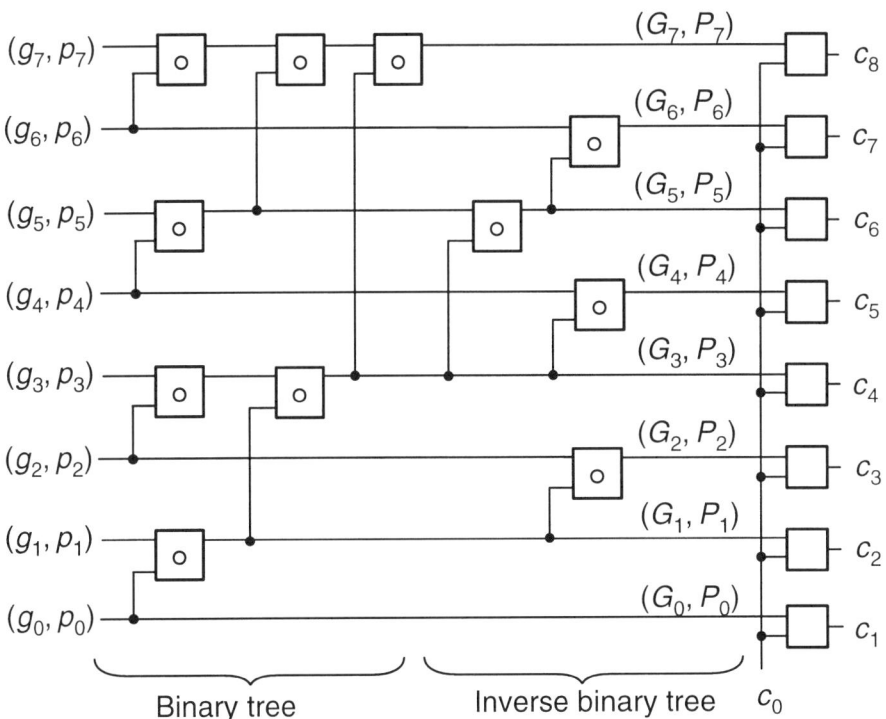

Figure 3.2.14: Carry block of a binary lookahead adder ($n = 8$)

An obvious way of determining all the block carry functions is through sequential evaluation in increasing order from bit level 0 to bit level $n-1$. As with ripple carry addition, this leads to processing time that is proportional to the word width n. Due to the associativity of the $\circ$ operator, it is possible to reduce the processing time via more parallel processing in a tree structure. First, sequential bit levels are connected in parallel sets of two. The results of this first hierarchical stage are then combined in a similar manner, as is repeated by each following stage. This tree structure produces $k = 0, 1, \dots \log n$ carry block functions for all powers of two, $2^k - 1 < n$. (Note: Whenever no other base is specified for the logarithm, base 2 is assumed, i.e. $\log x = \log_2 x$.) The block carry functions for bit levels that lie between the powers of two are evaluated with an inverse tree.

Figure 3.2.14 shows a carry block constructed in this manner, described as binary lookahead carry by Brent and Kung [40]. The number of operators for a binary tree is $n-1$. The inverse binary tree is not a complete binary tree. Thus it has fewer operators. It contains $n-1-\log n$ operators. The time-critical path runs from (g_0, p_0) via $(G_{n/2-1}, P_{n/2-1})$ and (G_{n-2}, P_{n-2}) to the output c_{n-1}. This path contains the most operators connected in series. A total of $2\log(n/2)$ operators are encountered along this path. When computing the circuit delay, it should be noted that the capacitive loads are not identical, but rather increase towards the middle. The delay induced along the time-critical path can be expressed as

$$T_{D,g_0-G_{n-2}} = \sum_{j=1}^{\log\frac{n}{2}} [T_{D,OP}(j+2) + T_{D,OP}(j)] \qquad (3.2.35)$$

where $T_{D,OP}$ is the operator delay. The two summands represent the two binary trees and the arguments of $T_{D,OP}$ specify the variable load capacitances. The operator delay is given by

$$T_{D,OP}(j) = (7 + 2j)\tau_L \qquad (3.2.36)$$

Through substitution in (3.2.35) and using the relation $\log\frac{n}{2} = \log n - 1$, it follows that

$$T_{D,g_0-G_{n-2}} = \tau_L (2\log_2^2 n + 16\log_2 n - 18) \qquad (3.2.37)$$

Taking into consideration the logic for the evaluation of the auxiliary signals (g_i, p_i), the logic for determining the carry signals and the XOR operation for the creation of the sum bits s_i, the following characteristic transistor counts and delay times can be given for binary lookahead carry adders.

$$N_{Tr} = 64n - 16 \log_2 n - 32 \quad \text{with } n = 2^k$$
$$T_{D,ADD} = \tau_L[2 \log_2^2 n + 16 \log_2 n + 17.8] \qquad (3.2.38)$$

For large n, the transistor count is proportional to n and the delay proportional to $\log^2 n$. This shows the clearly improved circuit delay over a non-hierarchical fast carry chain adder.

Carry select adders

In addition to parallel processing based on division of the carry block into bit groups, this method can also be applied to other types of adders. The basic concept consists of computing alternative results in parallel and subsequently selecting the correct result with single or multiple stage hierarchical techniques [37], [38]. A carry select adder with single stage selection is shown in Figure 3.2.15. Bit groups of length $m = 4$ are formed, and both sum bits and output carry bits are calculated for the two alternatives: input carry 0 or 1. The multiplexers for choosing the correct result are controlled by carry select (CS) logic. The CS logic function for the $(k+1)$th group is given by

$$\text{CS } (k+1): \qquad c_{(k+1)m} = c_{(k+1)m}(0) \vee c_{(k+1)m}(1) \; c_{km} \quad (3.2.39)$$

where $c_{(k+1)m}(0)$ specifies the resulting carry function for $c_{km} = 0$ and $c_{(k+1)m}(1)$ is the corresponding result for $c_{km} = 1$. For the derivation of the preceding relation, group carry functions are defined as with the block carry functions. Thus

$$c_{(k+1)m} = GG_k \vee GP_k \; c_{km} \qquad (3.2.40)$$

where GG_k represents the group generate and GP_k the group propagate of the kth group. The group carry functions are calculated through m applications of the $\circ$ operator on the corresponding generate and propagate signals. The relation (3.2.39) is derived through insertion of the two alternatives $c_{km} = 0$ and $c_{km} = 1$ into (3.2.40) and subsequent logical reduction. The use of the CS logic functions of the next lowest bit level leads to a logic function that depends only on the output carries of the adder.

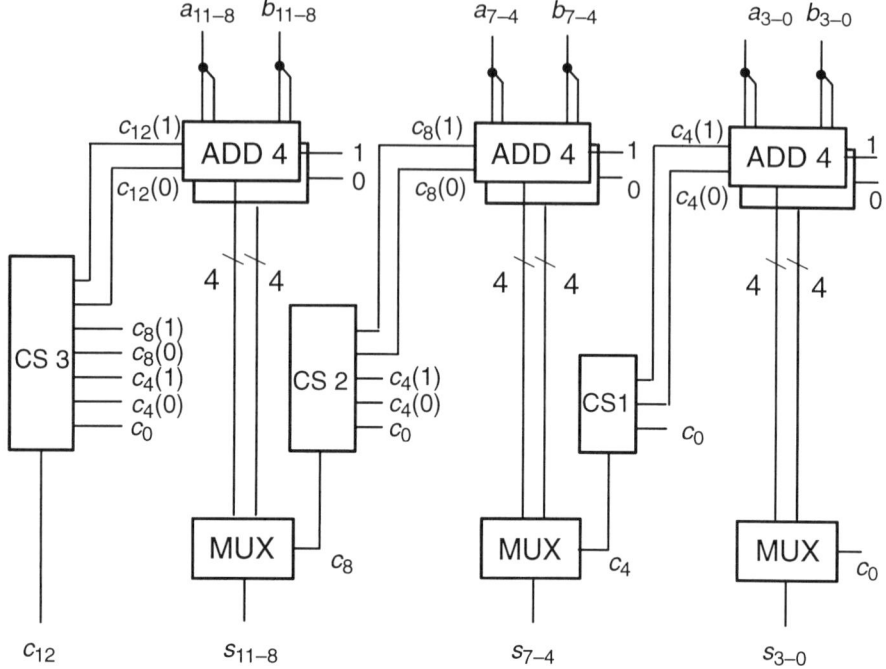

Figure 3.2.15: Carry select adder

The time-critical path runs through the adder of the lowest bit group, the carry select logic of the highest bit group and its corresponding multiplexer. For bit groups with m bits, the path $a_0 \rightarrow c_m(1) \rightarrow c_{n-m} \rightarrow s_{n-1}$ results in the largest delay. When determining the delay characteristics, it should be noted that the output $c_m(1)$ drives n/m CS blocks and that the load capacitance is likewise proportional to n/m. Assuming the adder is implemented as a transmission gate adder (Figure 3.2.10), the CS logic is created of complex gates and the multiplexers consist of transmission gates, the transistor count can be given as

$$N_{Tr} = 58n + 8\frac{n}{m} + 2\left(\frac{n}{m}\right)^2 \qquad (3.2.41)$$

With attention given to the diverse capacitive loads, the circuit delay model of section 2.4 specifies the delay to be

$$T_{D,ADD} = \tau_L[4\frac{n}{m} + 15m + 34.4] \qquad (3.2.42)$$

In Figure 3.2.15, 4-bit groups were assumed. The optimal number of bits per group m_{opt} can be computed from the zero of the first derivative of (3.2.42) with respect to m. The optimal value is

$$m_{opt} = \sqrt{\frac{4n}{15}} \qquad (3.2.43)$$

Substituting into (3.2.42) leads to

$$T_{D,ADD} = \tau_L[4\sqrt{15n} + 34.4] \qquad (3.2.44)$$

This means that the delay is proportional to $\sqrt{n}$. The substitution of real values for n in relation (3.2.43) results in an m_{opt} of 2 or 3. These rather low values for adder groups result from the rather minor delay induced by CS logic and multiplexers.

In the literature an alternative realization of a carry select adder is reported where the CS logic blocks are replaced by multiplexers for the carry selection [38], [44]. Each multiplexer is extended by an additional plane having $c_{(k+1)m}(0), c_{(k+1)m}(1)$ as input and $c_{(k+1)m}$ as output. Thus, relation (3.2.29) is realized by a multiplexer controlled by c_{km}. The disadvantage of this structure is that the multiplexer selection operates sequentially since the carry controlling the multiplexer operation must be ready before the carry for the next bit group gets its correct value. The implementation according to Figure 3.2.15 displays a parallel evaluation of the group carries.

Conditional sum adders

An adder implementation with hierarchical selection and high resolution through bit groups of size $m = 1$ is termed a conditional sum adder [38]. For each bit level i, sum and carry bits for both possible input carry bits are evaluated. The logic block that carries out these operations is called a condition cell, CC. The logic function of the CC is derived by substituting both cases $c_i = 0$ and $c_i = 1$ into Equations (3.2.2) and (3.2.3).

$$
\begin{aligned}
s_i(0) &= a_i \oplus b_i & c_{i+1}(0) &= a_i\, b_i \\
s_i(1) &= a_i \odot b_i & c_{i+1}(1) &= a_i \vee b_i
\end{aligned}
\qquad (3.2.45)
$$

A circuit diagram of a CC is shown in Figure 3.2.16.

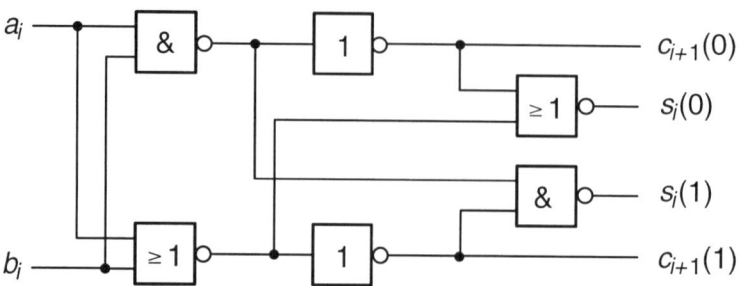

Figure 3.2.16: Conditional cell in gate logic

The correct results of the addition are chosen through hierarchical multiplexing of the outputs of the CC. The first stage merges the results of two consecutive bit groups, i.e. R_1'

$$
\begin{array}{cc}
c_{i+1} & s_i \\
\underline{c_{i+2} \quad s_{i+1}} \\
c_{i+2}' & s_{i+1}' \quad s_i
\end{array}
\qquad i = 1, 3, 5, \ldots
$$

The results c_{i+2}' s_{i+1}' are chosen by the value of c_{i+1}. If $c_{i+1} = 0$, the results corresponding to 0 are selected by its successor cell, otherwise the results corresponding to 1 are selected. In bit level 0, the results c_1 can be chosen directly by c_0. The second level merges the results of the first level, i.e.

$$
\begin{array}{ccc}
c_{i+2} & s_{i+1} & s_i \\
\underline{c_{i+4} \quad s_{i+3} \quad s_{i+2}} \\
c_{i+4}' & s_{i+3}' \quad s_{i+2}' \quad s_{i+1} \quad s_i
\end{array}
\qquad i = 3, 7, \ldots
$$

As before, the value of c_{i+2} is used to choose the results c_{i+4}' c_{i+3}' c_{i+2}'. In the first bit level, the results c_3 s_2 s_1 are selected by c_1. The third level is merged as follows.

$$
\begin{array}{ccccc}
c_{i+4} & s_{i+3} & s_{i+2} & s_{i+1} & s_i \\
\underline{c_{i+8} \quad s_{i+7} \quad s_{i+6} \quad s_{i+5} \quad s_{i+4}} \\
c_{i+8}' & s_{i+7}' \quad s_{i+6}' \quad s_{i+5}' \quad s_{i+4}' \quad s_{i+3} \quad s_{i+2} \quad s_{i+1} \quad s_i
\end{array}
\qquad i = 7, 15, \ldots
$$

Here again the selection is controlled by c_{i+4}. In the third bit level, the results c_7 s_6 s_5 s_4 s_3 are selected by c_3. An example of the general principle is shown in Table 3.2.1. The arrows signify selection based on the value of the carry bit.

Table 3.2.1: Example of the principles of a conditional sum adder. The decimal addition 45 + 52 = 97 is shown

i	6	5	4	3	2	1	0	Carry in	Level
a_i	0	1	0	1	1	0	1		
b_i	0	1	1	0	1	0	0		
$s_i(0)$	0	0	1	1	0	0	1	0	
$c_{i+1}(0)$	0	1	0	0	1	0	0		1
$s_i(1)$	1	1	0	0	1	1	0	1	
$c_{i+1}(1)$	0	1	1	1	1	0	1		
$s_i(0)$	1	0	1	1	0	0	1		
$c_{i+1}(0)$	0		0		1		0		2
$s_i(1)$	1	1	0	0	0	1			
$c_{i+1}(1)$	0		1		1				
$s_i(0)$	1	0	1	1	0	0	1		
$c_{i+1}(0)$	0				1				3
$s_i(1)$	1	1	0	0					
$c_{i+1}(1)$	0								

The entire circuit for a conditional sum adder is shown in Figure 3.2.17. If the conditional cell is implemented in gate logic and the multiplexers as transmission gates, the circuit requires

$$N_{Tr} = (10n - 2)\log_2(n + 1) + 34n \quad \text{with } n = 2^k - 1 \quad (3.2.46)$$

transistors. According to the circuit delay model of section 2.4, the delay for $n > 7$ is

$$T_{D,ADD} = \tau_L[2n + 13\log_2(n + 1) + 11]$$
$$\text{with} \quad n = 2^k - 1 \quad (3.2.47)$$

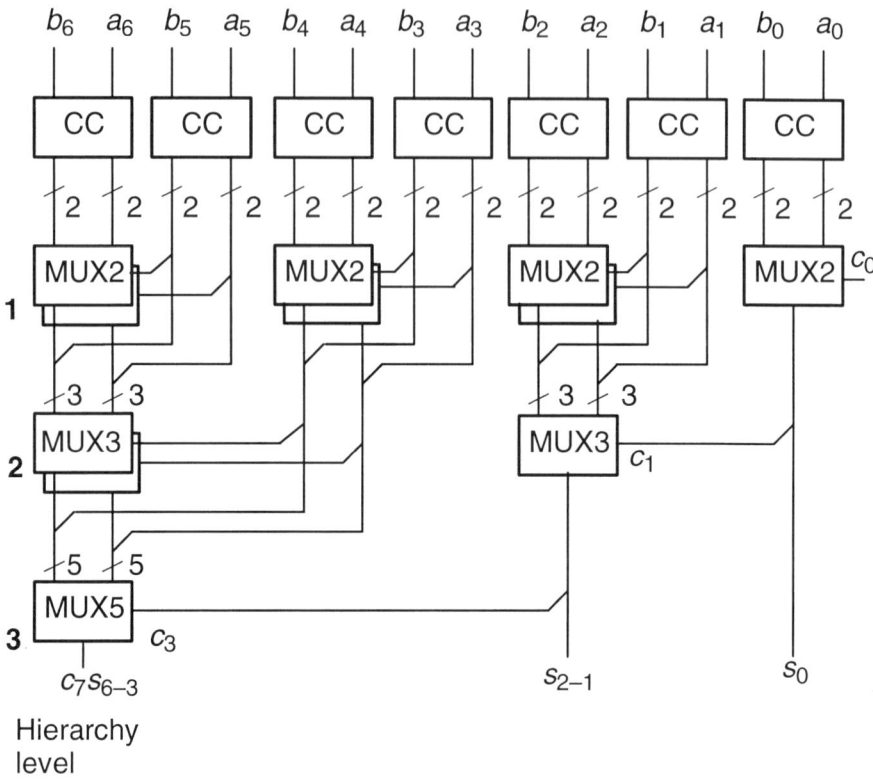

Figure 3.2.17: Conditional sum adder

Due to the hierarchical structure of the multiplexers, a logarithmic dependency might be expected in the delay characteristics. However, for large n the time-critical path runs through the outer multiplexers for the final evaluation of the sum and carry bits that control the multiplexers in the next higher level of the hierarchy. Thus the time-critical path contains the CC cells for a_0, b_0 and the chain of multiplexers MUX2, MUX3, MUX5, MUX9, etc. The number of capacitive loads increases with each level. The capacitive loads for the carry bits are given by

c_1:	$F = 4$
c_3:	$F = 6$
c_7:	$F = 10$
c_{15}:	$F = 18$

Thus the delay characteristics of the multiplexer tree are described by

$$T_{D,MUX-Tree} = T_{D,MUX}(2,4)$$
$$+ \sum_{i=1}^{\log(n+1)-2} T_{D,MUX}(2^i + 1, 2^{i+2} + 2) \quad (3.2.48)$$
$$+ T_{D,MUX}\left(\frac{n+1}{2} + 1, F\right)$$

The first argument of $T_{D,MUX}$ specifies the number of inputs controlled; the second argument represents the load driven by the carry outputs of the particular multiplexer. The delay of a single multiplexer from its control line input to its output is

$$T_{D,MUX}(j, k) = \tau_L [10 + j + k] \quad (3.2.49)$$

Substituting this into (3.2.48) with consideration of the delay characteristics of the CC cell yields the value given by (3.2.47). A linear component in the delay characteristics even of logarithmic trees results from the exponential increase in the capacitive loads. Yet it should be noted that, for typical n, conditional sum adders have the best delay characteristics. A complete conditional sum adder structure is not defined for powers of two, but rather for $n = 2^k - 1$. For word lengths that are a power of two, a CC and an MUX2 multiplexer must be added for the missing bit level. The delay given by (3.2.47) must be increased by 12.6 τ_L for powers of two.

Adder structures in comparison
Summarizing, it can be said that bit-slice structures such as the transmission gate adder in Figure 3.2.10 have a transistor count and delay that is proportional to n. They are said to be of the order $O(n)$.

The order $O(\cdot)$ of a function is an asymptotic upper limit. Given

$$|f(x)| \le c \, |g(x)| \quad \forall \, x > x_0, \; c > 0 \quad (3.2.50)$$

then

$$f(x) = O(g(x)) \quad (3.2.51)$$

Thus, for bit-slice structures,

$$N_{Tr} = O(n)$$
$$T_D = O(n) \quad (3.2.52)$$

Hierarchical structures were shown to have a delay of logarithmic order. For the binary lookahead carry adder

$$N_{Tr} = O(n)$$
$$T_D = O(\log^2 n) \tag{3.2.53}$$

The corresponding result for a conditional sum adder is

$$N_{Tr} = O(n \log n)$$
$$T_D = O(n) \tag{3.2.54}$$

The order of a function correctly describes only its asymptotic behaviour. For non-infinite values, the actual function values must be observed. Figure 3.2.18 shows the exact curves for the functions discussed, from which it can be seen that the use of hierarchical structures for reducing the delay pays off even with small word widths n.

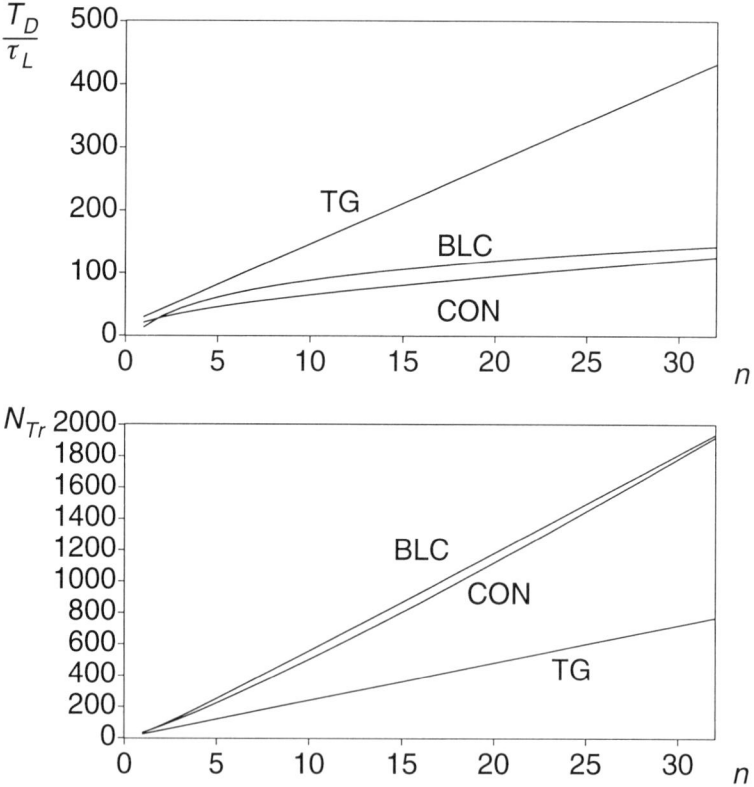

Figure 3.2.18: Delay and transistor count over the word width for three adder structures (TG–transmission gate adder, BLC–binary lookahead carry adder, CON–conditional sum adder)

The calculation of delay characteristics using the circuit delay model presented is quite complicated. Ordinarily, simple and plausible correlations are desired. For bit-slice structures such as transmission gate adders, they are easily found since the delay is primarily determined by the carry path. Along this path, only one type of cell occurs, and all cells possess an identical load capacitance and thus have set delay characteristics. Evaluation of the delay characteristics of the other adder structures is significantly more complex since diverse cell types occur and even identical cell types can differ in the number of inputs they possess. Furthermore, the delay characteristics are influenced by the various capacitive loads. Delay computations that do not consider these influences will lead to other descriptive equations.

3.2.5 Subtractors

As a first approach, subtraction can be implemented by a similar process to the way in which it is done by hand. Starting at the lowest bit level, the subtraction is carried out one bit level at a time, borrowing whenever the previous level produced a negative result. Figure 3.2.19 shows a subtractor circuit. As with adders (see section 3.2.1), the subtractor units are designated as full subtractors. The black dots signify negative evaluation, not complementing. The bits of the operand B and the borrow values are taken to be negative.

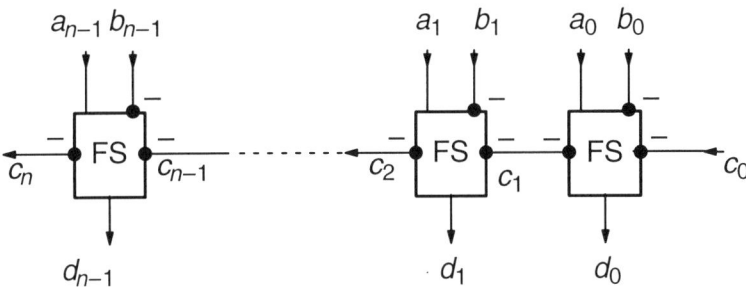

Figure 3.2.19: Subtractor for two operands A and B

The logic function of a full subtractor can be specified by a complete truth table. Through function minimization the following relations can be derived:

$$d_i = a_i \oplus b_i \oplus c_i$$
$$c_{i+1} = \bar{a}_i b_i \lor c_i (\bar{a}_i \lor b_i) \tag{3.2.55}$$

These logic functions are practically identical to those of a full adder (compare (3.2.2) and (3.2.3)). Just the borrow function uses $\bar{a}_i$ instead of a_i.

Since subtraction can be carried out as an addition with a two's complement subtrahend, the results found for addition can be reused. For the conversion to two's complement, the bits of the second operand must be complemented and a one added. For $c_0 = 0$, the addition of 1 can be carried out via the carry input c_0. In detail, given

$$A - B = A + (-B) \qquad (3.2.56)$$

the following holds for each bit level:

$$
\begin{array}{ll}
\textit{Addition} & \textit{Subtraction} \\
c_0 = 0 & c_0 = 1 \\
g_i = a_i\, b_i & g_i = a_i\, \bar{b}_i \\
p_i = a_i \oplus b_i & p_i = a_i \odot b_i \\
s_i = p_i \oplus c_i & s_i = p_i \oplus c_i \\
c_{i+1} = g_i \vee p_i\, c_i & c_{i+1} = g_i \vee p_i\, c_i
\end{array}
\qquad (3.2.57)
$$

The statements made above can be used for a switchable adder/subtractor (Figure 3.2.20). Here, m specifies the mode signal, and the XOR gates provide m-controlled complementing.

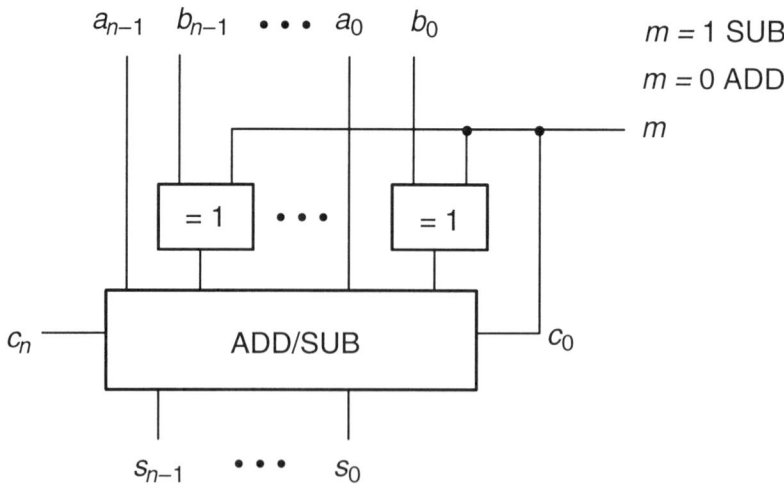

Figure 3.2.20: Switchable addition/subtraction

3.2.6 Carry save adders

The bit-parallel adders handled up to this point are grouped under the term carry propagate adders (CPA). Characteristic of such adders is that the carry signal is evaluated to determine the sum. This means that depending on the particular structure, the carry signal either travels sequentially through the logic blocks from the lowest to the highest level or is used throughout a tree structure for virtually parallel evaluation of the sum bits. In particular for the summation of a larger number of operands, carry save adders (CSA) are a viable alternative.

Here the carry signals are not used for the current addition, but rather for its successor. Figure 3.2.21 shows an n-bit carry save adder that uses n full adders to merge the three operands U, V and W to n sum bits s_i and n carry bits c_{i+1}. The sum bits constitute the sum vector S; the carry bits form the carry vector C, whereby the carry vector is shifted by one position. The representation of the intermediate results using sum and carry vectors is redundant since only $n+2$ bits are necessary to represent the sum of three operands but $2n$ bits are employed. Not until the sum vector S and the carry vector C have been added again in a carry propagate adder are the results complete. Such an adder is called a vector merging adder.

$$U + V + W = S + 2C \qquad (3.2.58)$$

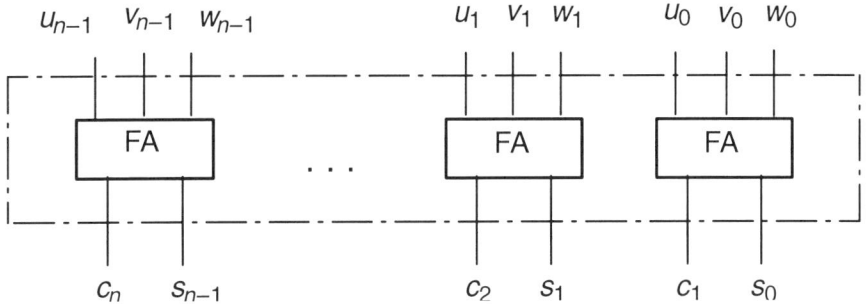

Figure 3.2.21: Carry save adder

In a multi-operand adder, carry save adders are used as long as possible until the final step when just a sum and a carry vector remain. These two are then merged in a vector merging adder. Figure 3.2.22 shows a carry save adder tree for eight operands. The particular shift by one bit of intermediate result vectors must be observed at each stage. This is signified by arrows in

Figure 3.2.22. When using two's complement negative numbers, the sign digits must be expanded accordingly to achieve correct results. According to (3.1.26), the sign digits can be extended without changing the value.

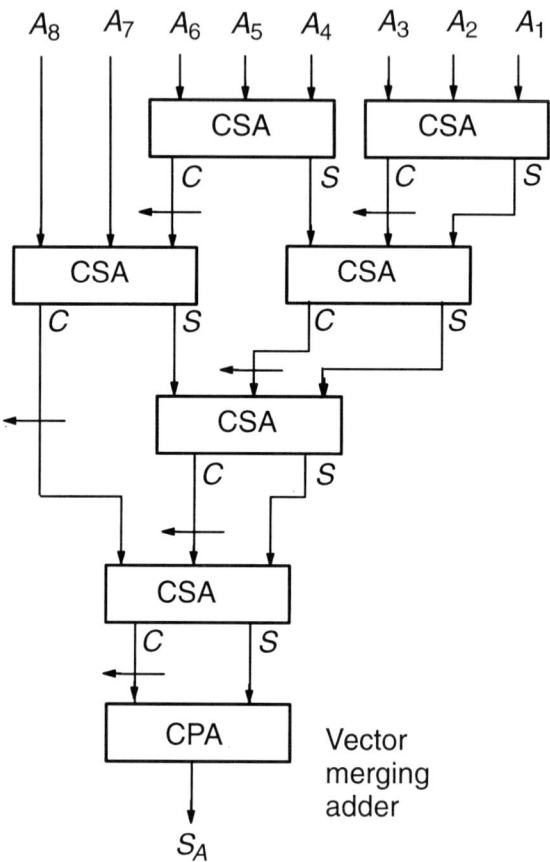

Figure 3.2.22: Carry save adder tree for eight operands

For space saving implementations, accumulators can be used. In this case, the operands are read in sequentially and added to the previous results.

$$S_A = \sum_{k=1}^{N} A_k \qquad\qquad (3.2.59)$$

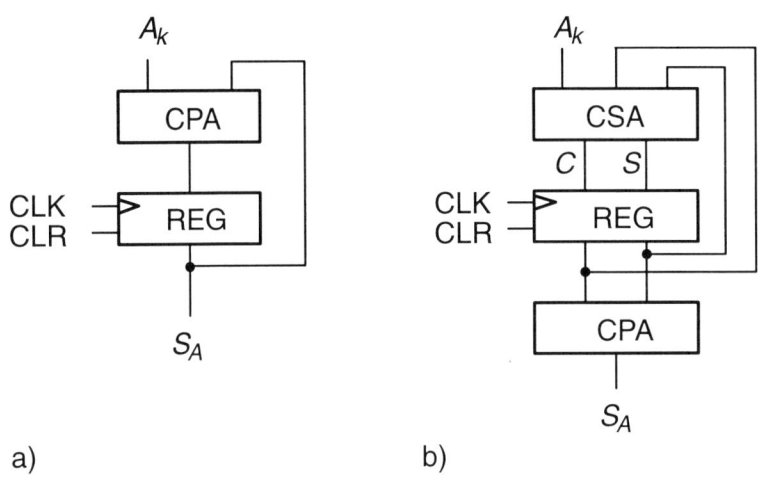

Figure 3.2.23: Accumulator implementations

An accumulator based on a carry propagate adder is shown in Figure 3.2.23a. At the beginning of the addition cycle, the register is cleared. Then, in N clock cycles, the operands A_k are read in and added to the intermediate sum stored in the register. After all N clock cycles have been carried out, the register contains the desired sum. Alternatively, such an accumulator can be implemented according to the carry save method shown in Figure 3.2.23b. In this case, the intermediate results are represented by both a sum and a carry vector. Both vectors must be stored in a register and fed back for the following addition. After N clock cycles, a redundant representation of the sum is stored by the two vectors in the registers. These two vectors must subsequently be merged in a carry propagate adder.

The particular advantage of a CSA accumulator is the high clock frequency for accumulation, which is dictated by the delay characteristics of just one full adder and a D-FF. In comparison, the delay that dictates the clock frequency of accumulation as in Figure 3.2.23a results from the delay of an n-bit CPA and a D-FF. However, due to the additional CPA at the output and the higher number of D-FFs required for the registers, a CSA accumulator requires more transistors. In addition, in many implementations, the additional CPA must be uncoupled via additional transfer registers.

3.3 Multipliers

Besides addition, multiplication is a very heavily used core operation for signal processing. To achieve high throughput, fast multipliers are required. Although such particular multiplier configurations have higher hardware costs, the goal of achieving high signal processing performance usually has priority. In the following, various possibilities for implementing multipliers with short computation times will be presented. In the first section, multipliers for positive numbers will be presented, and in the second section, multipliers for the multiplication of two's complement signed numbers will be explored.

3.3.1 Multipliers for positive numbers

The multiplication of two unsigned numbers A and B creates the product

$$P = A \cdot B \tag{3.3.1}$$

where A is called the multiplicand and B the multiplier. Given that A is an m-bit positive whole number and B is an n-bit positive whole number, then the numeric representation of the product P requires $(m+n)$ bits. The value of the product P results from the product of the values of the two operands A and B, i.e.

$$V(P) = V(A) \cdot V(B) \tag{3.3.2}$$

Substituting the value of B by the relation (3.1.3), the product can be represented as a function of the individual bits b_j of the multiplier.

$$\begin{aligned} V(P) &= V(A) \sum_{j=0}^{n-1} b_j 2^j \\ &= \sum_{j=0}^{n-1} V(A) \, b_j 2^j \end{aligned} \tag{3.3.3}$$

According to (3.3.3), partial products are formed by the product between A and the bits b_j of B and shifted versions of the partial products are added to get the product P. The partial product $V(A)b_j$ is 0 for $b_j = 0$ and $V(A)$ otherwise. This method is used for computation by hand and can also be similarly used for hardware implementations. Sequential multiplication using ALUs is implemented in this fashion, for example. The product P can be represented as a sum of partial products at bit level by splitting the number A into its bits a_i:

$$V(P) = \sum_{i=0}^{m-1} \sum_{j=0}^{n-1} (a_i \, b_j) \, 2^{i+j} \tag{3.3.4}$$

The partial products that occur and their summation for the creation of the product bits p_k are illustrated in Figure 3.3.1 for $m = n = 4$.

				a_3	a_2	a_1	a_0		
				$a_3 b_0$	$a_2 b_0$	$a_1 b_0$	$a_0 b_0$	←	b_0
			$a_3 b_1$	$a_2 b_1$	$a_1 b_1$	$a_0 b_1$		←	b_1
		$a_3 b_2$	$a_2 b_2$	$a_1 b_2$	$a_0 b_2$			←	b_2
	$a_3 b_3$	$a_2 b_3$	$a_1 b_3$	$a_0 b_3$				←	b_3
p_7	p_6	p_5	p_4	p_3	p_2	p_1	p_0		

Figure 3.3.1: Partial product matrix for representing the operations during the binary multiplication of positive numbers

Ripple carry array multipliers
A multiplier implementation can be derived directly from equation (3.3.4). The evaluation of the partial products is implemented by a logical AND gate, since the result can only be 1 when both bits are 1. Through a regular array configuration of AND gates and adder cells, the computation of the product can be implemented as given by (3.3.4) [45]. Figure 3.3.2 shows an array multiplier based on ripple carry adders. The function of each multiplier cell includes both the evaluation of the partial products and their summation. Based on (3.2.2) and (3.2.3), their function is given by

$$\begin{aligned} s &= (ab) \oplus s' \oplus c' \\ c &= abs' \vee abc' \vee s'c' \end{aligned} \tag{3.3.5}$$

where a,b are bits of the operands A and B, s' is the sum bit of the preceding cell, and c' the carry bit of the preceding cell. The sum bit s and the carry bit c are passed to the next cell as results. In the representation in Figure 3.3.2, the carry bits are transported horizontally and the sum bits vertically. In real implementations, array multipliers are rectangular, not diamond shaped. This rectangular form is achieved by shifting the sides of the rhombus such that the operand bits a_i run vertically and the sum bits run at $-45°$.

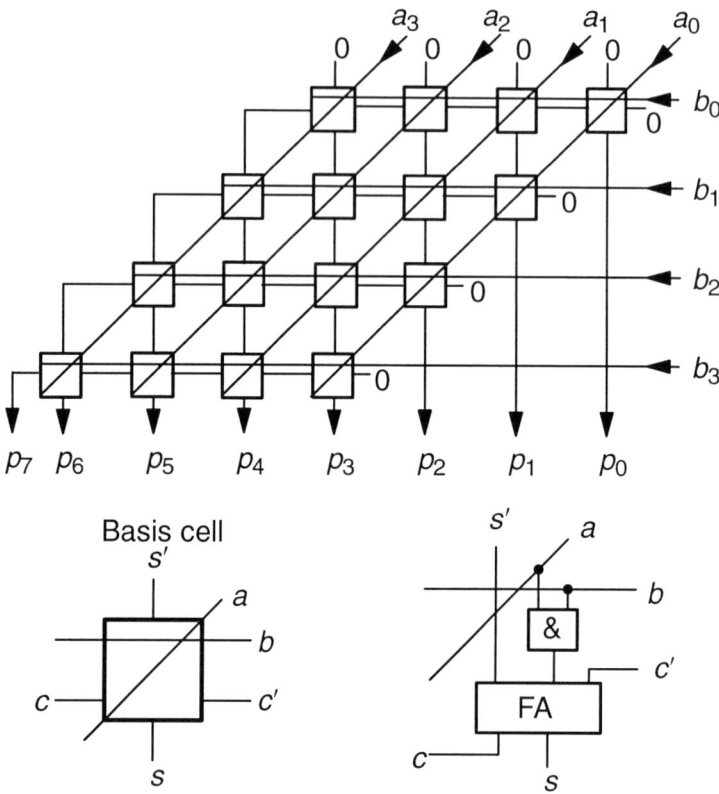

Figure 3.3.2: Ripple carry array multiplier

Figure 3.3.3 shows such a configuration. In the following, both structures (diamond shaped and rectangular) will be used. Wherever the digit valency is to be emphasized by the structural representation, the diamond shape is chosen.

For some special cases along the edges of the array, the multiplier cells can be simplified. If $s' = 0$ or $c' = 0$, the full adder FA simplifies to a half adder HA. A half adder is an adder cell with two instead of three input bits. If both $s' = 0$ and $c' = 0$ ($s' \vee c' = 0$), the adder cell disappears entirely. Counting the special cases, the total implementation costs of an array multiplier consist of

AND: nm
HA: n
FA: $nm - n - m$

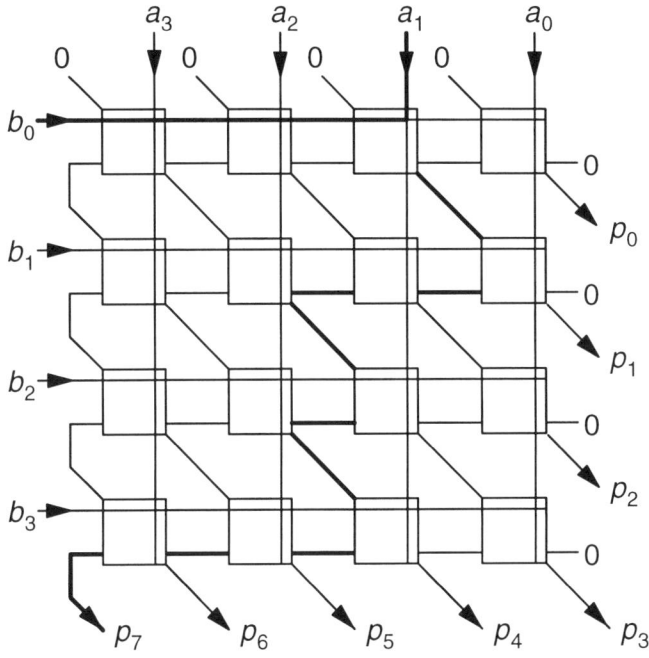

Figure 3.3.3: Ripple carry array multiplier in rectangular configuration

The AND gates can be implemented by NAND gates followed by an inverter. A great number of circuit variations for implementing full and half adders are given in the literature [15], [16], [17]. In addition to implementations in gate logic (Figure 3.2.3), implementations with symmetric p- and n-circuits (Figure 3.2.4) and implementations with transmission gates (Figure 3.2.10), many diverse mixtures exist. All these implementations have varying characteristics in terms of circuit size, delay and power consumption.

For the comparison of the various multiplier structures, transmission gate configurations as in Figure 3.3.4 will be used, in which a FA requires 26 transistors and a HA 16 transistors. One of their characteristics is that the delay for the transition $c' \rightarrow c$ ($T_{D,c'c}$) is significantly shorter than for $s' \rightarrow c$ ($T_{D,s'c}$). However, for this to hold, s' must be applied early enough before c' such that the internal propagate signal reaches its final stable state. In this architecture, the outputs s and c have identical delay characteristics. Using the model from section 2.4,

$$\text{FA}: \quad T_{D,c'c} = T_{D,c's} = (12 + F)\tau_L$$
$$T_{D,s'c} = T_{D,s's} = (26.4 + F)\tau_L$$
$$\text{HA}: \quad T_{D,c'c} = T_{D,c's} = (8 + F)\tau_L \qquad (3.3.6)$$
$$T_{D,s'c} = T_{D,s's} = (12 + F)\tau_L$$

can be derived.

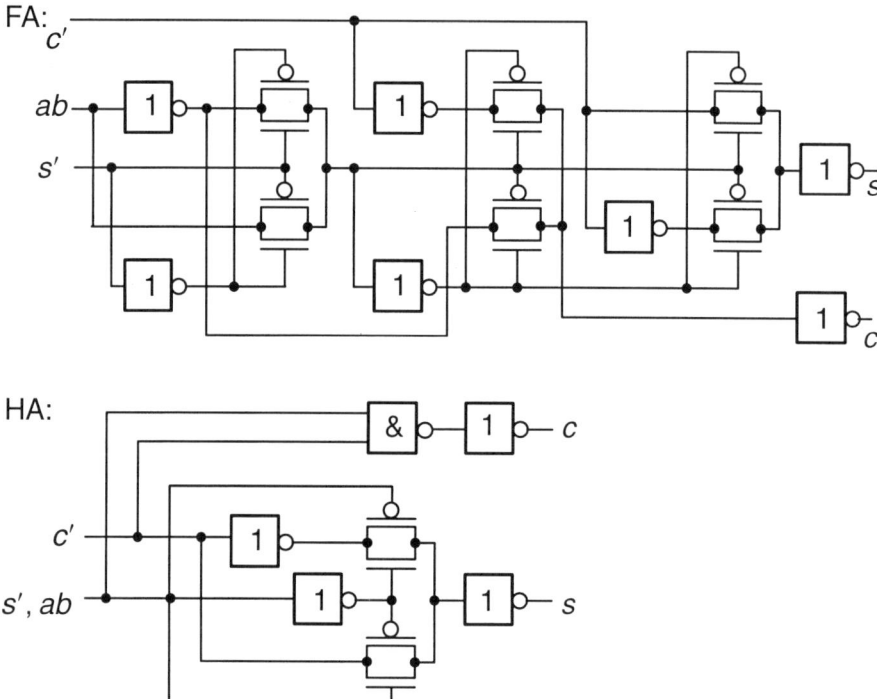

Figure 3.3.4:   Full adder FA and half adder HA for multiplier implementation

The fan-in capacitances of the adders for 3 and 4 must be considered when calculating the delay of the entire array. In addition, it must be observed that n or rather m AND gates must drive the input signals a_i and b_j. The enlarged input capacitance leads to an increase in the transient time for the signals a_i and b_j. This delay can be reduced through appropriate dimensioning of the drivers (see also section 2.4). For the delay times given in the following, the influence of the size of the input capacitances was taken into consideration. Nonetheless, no special driver circuits were assumed. To simplify the comparison, $n = m$ is assumed. Based on these assumptions, the following holds for a ripple carry array multiplier:

$$N_{Tr} = 32n^2 - 36n$$
$$T_{D,MUL} = (62.4n - 72.4)\tau_L \qquad (3.3.7)$$

The transitions $c'c$ and $c's$ have approximately half the delay of $s'c$ and $s's$. Thus, the time-critical path zigzags through the array between the cells for $i = 1$ and $i = 2$. After reaching $j = n - 1$, it then goes horizontally along the carry path. The time-critical path is emphasized in Figure 3.3.3 by a thick line. The delay characteristics of the time-critical path are given by

$$
\begin{aligned}
T_{D,MUL} &= T_{D,AND} + (n-1)\,T_{D,FA,\,s'c} + (n-2)\,T_{D,FA,\,c's} \\
&\quad + (n-2)\,T_{D,FA,c'c} \qquad (3.3.8) \\
&\approx 4T_{D,ADDn}
\end{aligned}
$$

Approximately $3n$ adder cells lie along the time-critical path, of which approximately $2n$ carry out the fast carry of c'. Thus the total delay of the multiplier can be simply given as $4n\ T_{D,c'c}$. Although the array multiplier consists of n n-bit adders, the delay is not n times the delay of one n-bit adder, but only about four times. The largest delay only occurs during transitions from very small numbers $(A = 0, B = 0)$ to very large numbers $(A = 2^n - 1, B = 2^n - 1)$ or vice versa.

Carry save array multipliers
An array multiplier is implemented as an n-fold addition. For the addition of several operands, the carry save technique introduced in section 3.2.6 is practical. Figure 3.3.5 illustrates a carry save array multiplier [45]. The carry signals are transferred to the adders in the next bit level respectively. A final m-bit adder adopts the function of a vector merging adder. In particular for large word lengths m, this final adder can be implemented hierarchically according to the techniques of section 3.2.4. The term Braun array multiplier is used to describe multipliers whose final adder is implemented as a ripple carry adder [46]. Figure 3.3.6 depicts such a multiplier and its simplified special components. The letter M denotes a pure multiplication cell, the letter A a pure addition cell and the letter H a half adder. In this figure, the cells are configured rectangularly as in Figure 3.3.3.

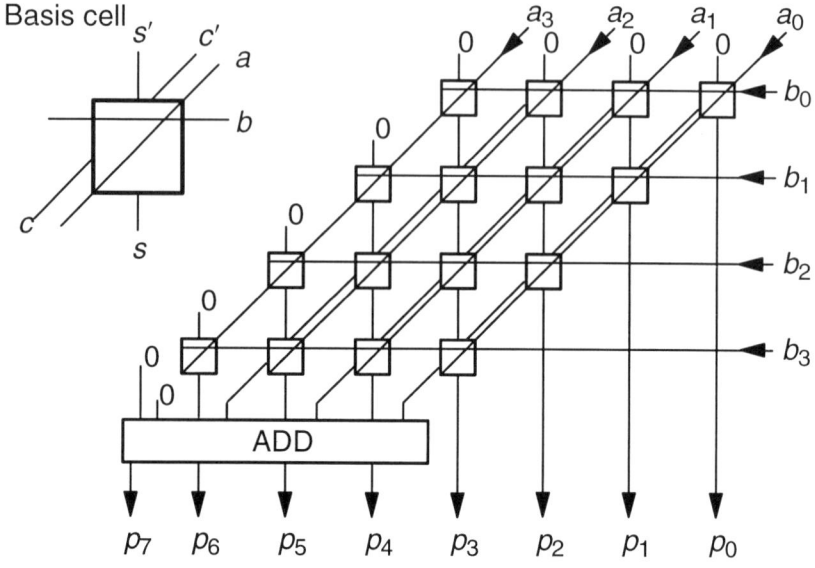

Figure 3.3.5: Carry save array multiplier

The total costs of a Braun array multiplier are identical to those of a ripple carry array multiplier since the number and types of logic elements are identical. The individual logic elements are simply configured and wired differently. Under the assumptions previously made, the delay characteristics of a Braun array multiplier are approximated by

$$T_{D,MUL} = (46.4n - 54.8)\tau_L \qquad (3.3.9)$$

The time-critical path runs down diagonally, starting at $(i,j)=(n-1,0)$, and then entering the next adder along the carry path. The time-critical path is emphasized in Figure 3.3.6 by a thick line. The delay induced by the time-critical path is given by

$$T_{D, MUL} \approx T_{D, AND} + (n-2) T_{D,FA,s's} + T_{D,FA, s'c}$$
$$+ (n-1) T_{D,FA,c'c} \qquad (3.3.10)$$
$$\approx 3 T_{D, ADDn}$$

Approximately $2n$ adder cells lie along the time-critical path, of which approximately n serve the fast carry of c'. The delay is reduced in comparison to a ripple carry array multiplier; it is approximately three times that of an n-bit adder. Braun array multipliers have particular advantages when used in pipelines. This will be shown later in Chapter 4.

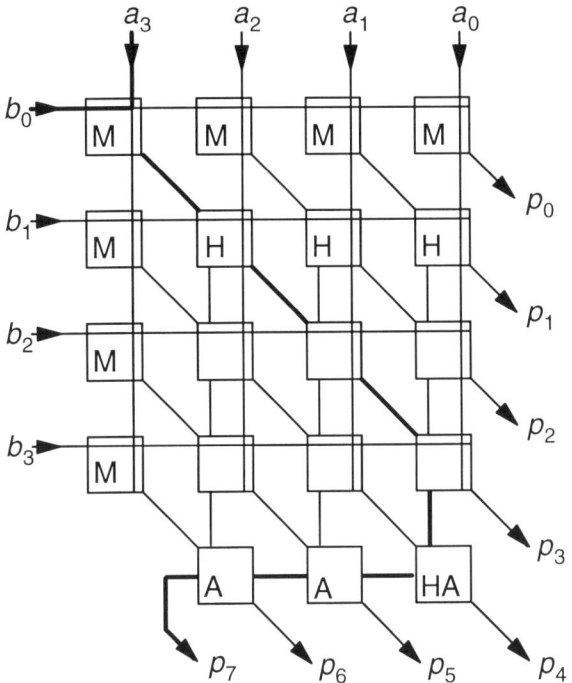

Figure 3.3.6: Braun array multiplier

Array multipliers with bit-pair evaluation

A multiplier implementation with low hardware costs but simultaneously improved delay characteristics can be derived through the evaluation of bit-pairs [42], [46], [47]. If two consecutive multiplier bits are simultaneously evaluated for the creation of partial products, equation (3.3.3) changes to

$$V(P) = \sum_{j=0}^{n/2-1} V(A) \, (2b_{2j+1} + b_{2j}) \, 2^{2j} \qquad (3.3.11)$$

For each bit-pair $b_{2j+1} \, b_{2j}$, the following values can occur:

$$2b_{2j+1} + b_{2j} \in \{0, 1, 2, 3\} \qquad (3.3.12)$$

Thus, in addition to 0 and A, $2A$ and $3A$ must also be available. The value $2A$ is simply A shifted by one position. The value of $3A$ must be computed in a separate adder from $A + 2A$. An array multiplier based on bit-pairs is shown in Figure 3.3.7. The required partial products for (3.3.11) are created using bit-pair controlled multiplexers. The digit shifting corresponding to 2^{2j}

is achieved by hard-wire shifting of the bit levels. All partial products are combined to the final product in a CSA adder tree.

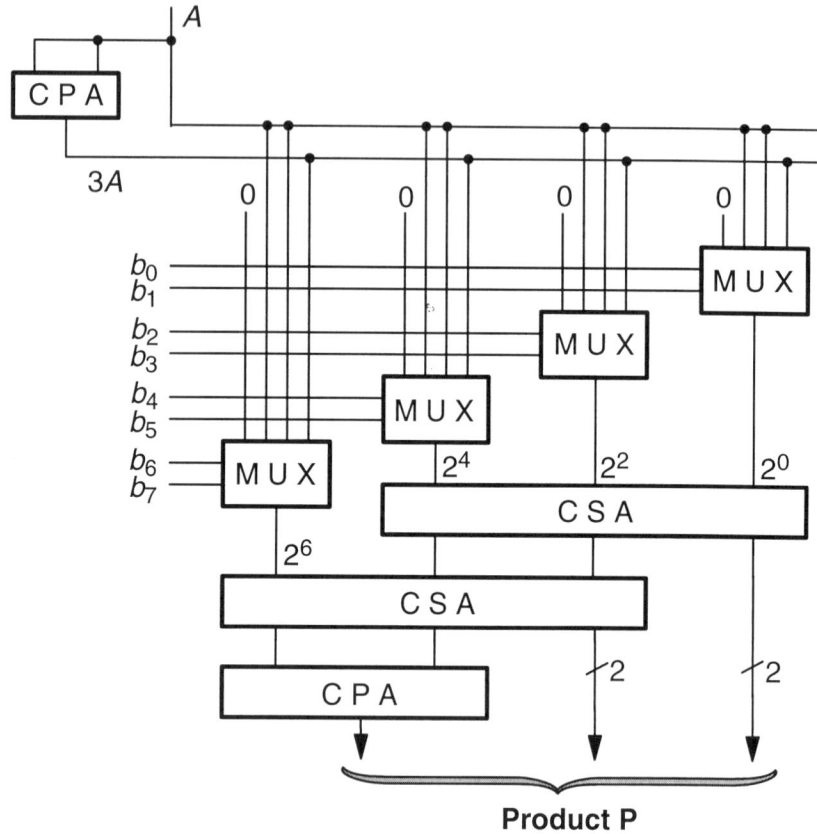

Figure 3.3.7: Array multiplier with bit-pair evaluation

Due to the operands that have been shifted by two bits relative to one another in the CSA adder, not all FAs have three operand bits. Those with two operands reduce to a half adder; those with one operand are simply hard-wired. If the multiplexers are implemented based on transmission gates, the following characteristic data hold:

$$N_{Tr} = 25n^2 + 26n - 56$$
$$T_{D,MUL} = (48,7n - 34,2)\tau_L \tag{3.3.13}$$

This does not result in a delay improvement in comparison to Braun multipliers. The reason for this lies in the lengthy computation of $3A$ and the high load capacitance of the wiring for $3A$. The time-critical path consists of the complete carry path of the CPA adder for the computation of $3A$, a run through the CSA tree along bit level n with its $n/2 - 2$ FA and the carry path of the following CPA adder with the top n FA. A total of $2.5n - 2$ FA are encountered, of which $2n$ carry out the fast carry. This corresponds to approximately three times the delay of an n-bit adder. On account of the increased capacitive loads of A and $3A$ and the delay of the multiplexer, there is even a slight worsening of the delay characteristics in comparison to Braun multipliers. Yet the use of hierarchical CPA adders and special drivers for A and $3A$ enables a decrease in the delay. The low hardware costs remain an essential advantage of this adder, albeit with less structural regularity.

Wallace tree multipliers
A further alternative for multiplier implementation is Wallace tree multipliers [48], in which all partial products are calculated in parallel in the first stage. These partial products are then added in an adder tree according to the carry–save principle. The final resulting carry and sum vectors are merged in a CPA. The basic structure is shown in Figure 3.3.8. This structure can be classified to the carry–save array multipliers if the AND gates used to create the partial products are separated from the array. However, a tree structure is used instead of a linear configuration such as in Figure 3.3.5 for the fast addition of the partial products.

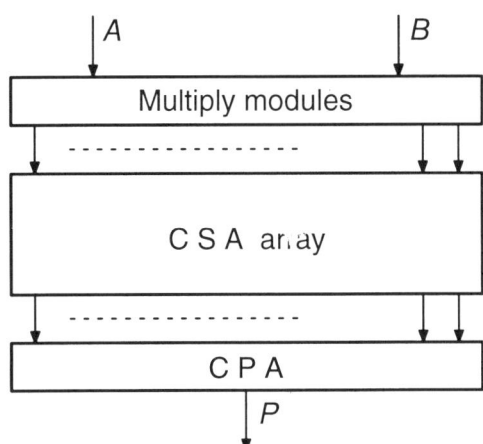

Figure 3.3.8: Multiplier structure based on Wallace trees

Wallace suggested a tree structure with a minimal number of full adders along the time-critical path. Configurations such as in Figure 3.2.22 are used. It must be noted that, in the adder trees of multipliers, the numbers to be added are shifted versions of Ab_j. As can be seen in the partial product matrix in Figure 3.3.1, this means that the numbers to be added contain many zeros. The knowledge of these zeros is used to simplify the adder tree. Depending on the number of zeros, the full adder is reduced to either a half adder, a hard-wired connection or a non-existent connection. In carry–save adders, three operand bits are compressed to a sum and a carry bit in each full adder. Full adders can be interpreted as counters that add three equal valued operand bits (counting the ones) and compress them to a 2-bit binary number. This implies a reduction of the operands available in parallel by at most 3/2 in each hierarchical level of the adder tree. Thus, the number k of hierarchical levels in an adder tree necessary to combine n operands to a sum and carry vector is

$$k \geq \frac{\log(n/2)}{\log(3/2)} \qquad (3.3.14)$$

The $\geq$ symbol follows from the criterion that the number of operands in each level of hierarchy must be a whole number. The series {2, 3, 4, 6, 9, 13, 19, 28, 42 ...} gives the number of operands when reduced at each level of hierarchy. Computed from low to high, the next member results from multiplication by 1.5, whereby non-whole numbers are replaced by the next lower whole number.

It follows from equation (3.3.14) that the delay time of an adder tree can be approximated by a log function. Since the hierarchical implementation of a two-operand adder (vector merging adders) also has a logarithmic delay dependency, a logarithmic delay characteristic for multipliers with Wallace trees can be achieved.

Dadda multipliers
The use of adder trees has been more closely analysed by Dadda [49], [50]. He was able to deduce implementations with fewer full adders through specific observations at the bit level. His methods will be demonstrated on an 8×8 bit multiplier. As was deduced from Figure 3.2.22, an operand reduction of 8, 6, 4, 3, 2 is carried out. In a Wallace tree multiplier, the maximum possible number of FAs and HAs are incorporated into each level of hierarchy. Dadda suggests the incorporation of just as many adders as are actually needed by the operand reduction striven for. Figure 3.3.9 illustrates this principle. In the figure, dots represent the bits in each level of hierarchy. The ellipses exemplify the FAs and HAs used to group two or three bits. The connected dots in the following level illustrate the sum and carry bit-pairs. From

the illustration, it follows that this adder tree requires 35 FAs and seven HAs. Figure 3.3.10 shows an adder tree for bit level p_7 of an 8×8 multiplier as an example. In comparison to the structure in Figure 3.3.5, only four as opposed to seven full adders are encountered in the sequence. Using ripple carry adders for the vector merging adders, the following characteristic data result for 8×8 bit and 16×16 bit Dadda multipliers:

$$
\begin{aligned}
8 \times 8 \text{ bit}: \quad N_{Tr} &= 1760 \\
T_{D,MUL} &= 280.2\ \tau_L \\
16 \times 16 \text{ bit}: \quad N_{Tr} &= 7616 \\
T_{D,MUL} &= 544.2\ \tau_L
\end{aligned}
\tag{3.3.15}
$$

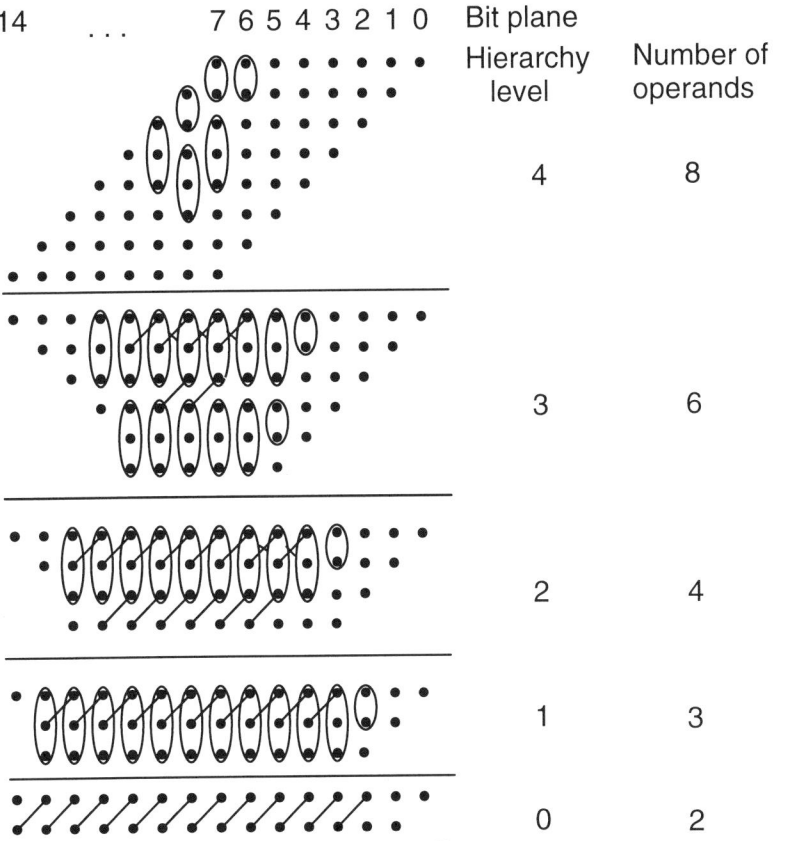

Figure 3.3.9: Hierarchical operand reduction based on Dadda for a 8×8 bit multiplier

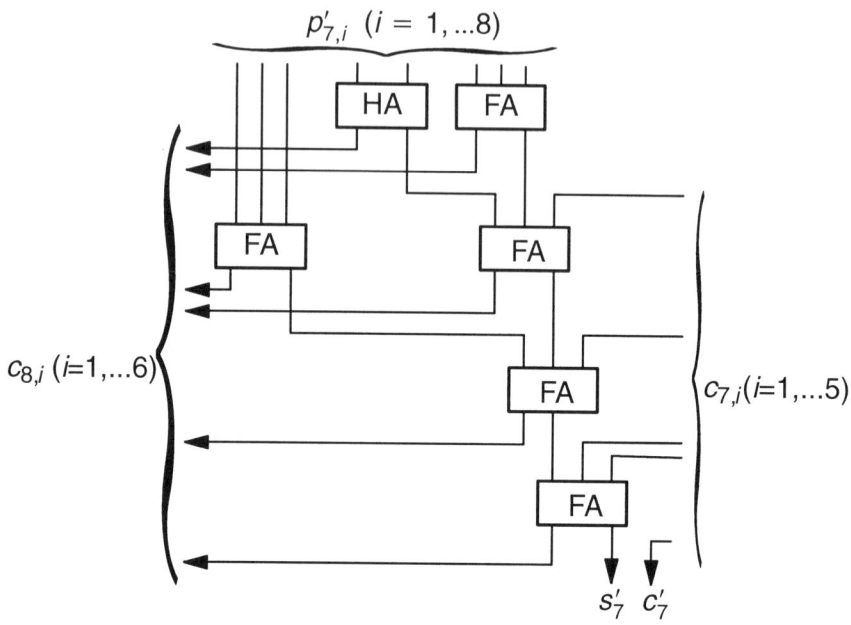

Figure 3.3.10: Adder tree for the computation of the sum and carry bits in bit level p_7 of an 8×8 bit Dadda multiplier

These are the same hardware costs as for a Braun multiplier, yet with reduced delay. Using hierarchical adders for the vector merging adders, the delay can be reduced even further. In comparison to Dadda multipliers, a Wallace tree multiplier has a slightly increased delay and higher hardware costs.

The Dadda multiplier consists of a CSA adder tree with levels as given by (3.3.14) and a CPA for $2n - 2$ bit. If the CPA is implemented in full adders according to the ripple carry principle, then the major contribution to the delay of the total multiplier comes from this adder. The time-critical path does not run through the middle of the CSA adder tree with the maximal serial number of adders, but rather through the lower bit planes. In a 16×16 bit multiplier, the time-critical path runs along bit level 4 with its half adder and two full adders. For this reason, the delay characteristics of an $n \times n$ bit multiplier can be roughly compared with the delay characteristics of a $2n$ bit wide CPA. Thus, it has approximately twice the delay of an n-bit adder.

Wallace tree and Dadda multipliers are particularly advantageous for large word widths. In this case, as long as a hierarchical adder implementation is used for the CPA, the total delay characteristics show a roughly logarithmic dependence on the word width. A disadvantage of these multipliers is their complex wiring. This leads to slightly less efficient use of the chip area. The advantages in the delay characteristics are further diminished through the additional wiring capacitances.

In addition to the use of partial product modules at the bit level, they can also be used with a large number of bits, for instance in modules that implement 4×4 bit multiplication [42]. In such a case, the n-bit word to be multiplied must be split up into 4-bit groups. The intermultiplication of the 4-bit groups leads to partial products that must be combined using Wallace adder trees for the total product.

Standard Add–Shift Multipliers

Similar to the way in which extremely cost-efficient implementations can be achieved using sequential configurations in adders, multipliers can also profit from sequential configurations. Based on equation (3.3.3), the summation of the partial products Ab_j can be carried out sequentially. The corresponding hardware configuration is shown in Figure 3.3.11. Its operation is described by the following pseudo-program.

$$
\begin{aligned}
AC &\leftarrow 0 \\
MR &\leftarrow B \\
AX &\leftarrow A
\end{aligned}
$$

repeat n times
$$
\begin{aligned}
\{AC \quad &\leftarrow \quad (AC) + (AX \wedge MR_0) \\
\text{SHR} \quad &AC/MR\}
\end{aligned}
$$

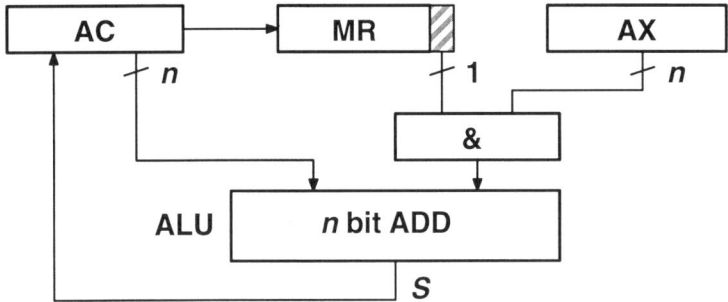

Figure 3.3.11: Multiplication implementation using sequential add/shift technique

First the starting values are loaded into all three registers. The accumulation of the partial products between the multiplicand (AX) and the multiplier bit (MR_0) follows. The serial change in the bits' valency is accounted for through use of the shift operation SHR. After n cycles, the product P is contained in the doubly long register AC/MR. The primary disadvantage of the add/shift technique shown in Figure 3.3.11 is the multiplication time of n cycles, whereby the clock rate is limited by the delay of an AND gate, an n-bit ADD and one shift operation. In comparison to an array multiplier, the total processing time is roughly $n/3$ times larger. The hardware costs are significantly lower since only three registers and some control circuitry are needed in addition to the n-bit adder and the logical AND gates.

3.3.2   Signed number multipliers

Multipliers with pre- and post-complement

If two two's complement numbers are to be multiplied, the most obvious solution seems to be to multiply the absolute values of the two numbers and then convert the result into a two's complement number according to the usual sign rules [42]. Yet taking into account that two's complement numbers usually have an asymmetric range (see (3.1.22)), the representation of their absolute values requires as many bits as their two's complement representation due to the smallest negative number.

Consider are a multiplicand A and a multiplier B as n-bit two's complement numbers. The conversion to an absolute value representation is regulated by the sign bit a_{n-1} or b_{n-1}. If $a_{n-1} = 1$ or $b_{n-1} = 1$, a bitwise complement must be carried out and a 1 added. The product of the absolute values $|A| \cdot |B|$ is given the sign

$$a_{n-1} \oplus b_{n-1} \qquad (3.3.16)$$

and must be converted to a two's complement number accordingly. A configuration for the multiplication of two's complement numbers in illustrated in Figure 3.3.12. The COMPL blocks carry out the complementing controlled by a mode signal. In order to get also a correct result if one of the operands is zero and the other is negative, the complementer at the output has to be extended by an additional bitplane.

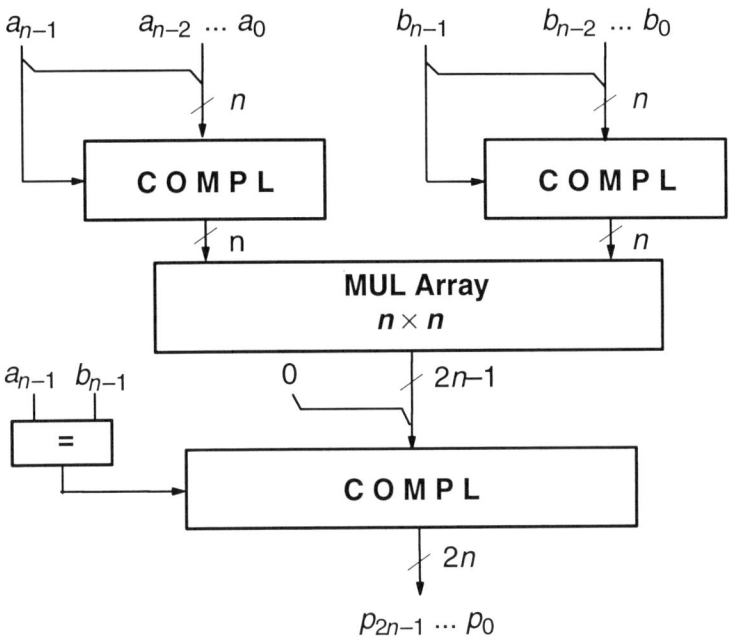

Figure 3.3.12: Multiplier with pre- and post-complementing

So as to be able to estimate the hardware costs of such a multiplier implementation, a circuit for the creation of two's complements will be derived. A mode signal m for controlling is assumed, whereby $m = 0$ leaves the number unchanged and $m = 1$ complements. Figure 3.3.13 shows a bit-slice structure that uses modified half adders MHA. From the desired functionality, the following behaviour for the MHA block of bit level i can be derived:

$$a_i' = \begin{cases} a_i & m = 0 \\ \overline{a}_i \oplus c_i & m = 1 \end{cases} \tag{3.3.17}$$

$$c_{i+1} = \overline{a}_i \, c_i \tag{3.3.18}$$

Using the equations given under (3.2.17), the two lines in (3.3.17) can be combined.

$$a_i' = a_i \oplus (m\overline{c}_i) = a_i \odot (\overline{m} \lor c_i) \tag{3.3.19}$$

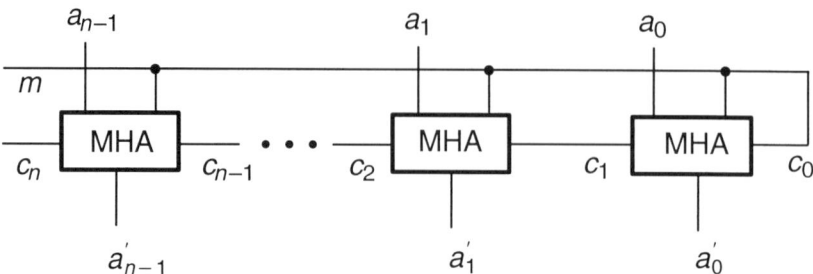

Figure 3.3.13: Circuit for the creation of two's complements

The carry path of the two's complement circuit consists of logical ANDs, as specified by (3.3.18). Similar to an adder, equation (3.3.18) represents a carry transfer with $\bar{a}_i$ as its propagate signal. For implementations with fast carry path, two cell types are used, as in transmission gate adders. The one type creates an inverted carry, the other has an inverted carry input. Figure 3.3.14 shows a segment of such a carry path. In addition to the carry path, (3.3.19) implies the necessity of an XOR and an AND gate for the implementation of an MHA. Using the XOR implementation of Figure 3.2.9a, the following characteristic data can be derived after simplification of the first and last bit levels:

$$T_{D,COMPL} = (9.8n - 4.9 + F)\, \tau_L$$
$$N_{Tr,COMPL} = 19n - 24 \tag{3.3.20}$$

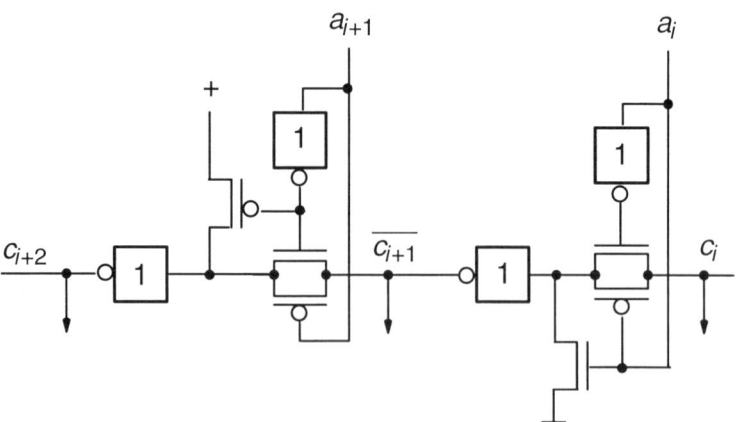

Figure 3.3.14: Segment of the carry path of a two's complementer

The given delay is the maximum value that occurs when the input carry c_0 is passed through to the highest bit level. This is true of the complement of the largest negative number -2^{n-1}.

If a Braun array multiplier is used for the multiplier block in Figure 3.3.12, the following transistor count for the total configuration results:

$$N_{Tr, MUL} = 32n^2 + 40n - 64 \qquad (3.3.21)$$

The array multiplier contributes most significantly to the hardware costs. In evaluating the delay characteristics of the configuration in Figure 3.3.12, it must be observed that the maximal delay in the individual blocks (COMPL, MUL) only occurs for special bit patterns among the input operands. Bit patterns that lead to maximal total delay in the array multiplier do not necessarily lead to maximal delay in the complementers. The maximal delay has been found to be

$$T_{D, MUL} = (56.2n - 89.1)\, \tau_L \qquad (3.3.22)$$

The increase in the delay results primarily from the complementing of the multiplier. In the worse case, the carry path extends over n levels. Due to its lower capacitive loads, the carry path of the complementer (Figure 3.3.14) has only two-thirds the delay of a normal adder. Thus the delay of the configuration in Figure 3.3.12 is roughly 3.6 times that of an n-bit adder. For this reason, alternative multiplier configurations are sought, that have improved delay characteristics by not requiring these additional complementers.

Pezaris array multipliers

According to (3.1.21), the sign digit of two's complement numbers is weighted negatively. This knowledge can be exploited for the construction of special multiplier arrays. Consider two two's complement numbers A and B, whose signs are in parentheses to illustrate their negative weighting.

$$\begin{aligned} A &= (a_{n-1})\ a_{n-2}...a_1\ a_0 \\ B &= (b_{n-1})\ b_{n-2}...b_1\ b_0 \end{aligned} \qquad (3.3.23)$$

p_9	p_8	p_7	p_6	p_5	p_4	p_3	p_2	p_1	p_0	
					(a_4)	a_3	a_2	a_1	a_0	
					$(a_4 b_0)$	$a_3 b_0$	$a_2 b_0$	$a_1 b_0$	$a_0 b_0$	$\leftarrow b_0$
				$(a_4 b_1)$	$a_3 b_1$	$a_2 b_1$	$a_1 b_1$	$a_0 b_1$		$\leftarrow b_1$
			$(a_4 b_2)$	$a_3 b_2$	$a_2 b_2$	$a_1 b_2$	$a_0 b_2$			$\leftarrow b_2$
		$(a_4 b_3)$	$a_3 b_3$	$a_2 b_3$	$a_1 b_3$	$a_0 b_3$				$\leftarrow b_3$
	$a_4 b_4$	$(a_3 b_4)$	$(a_2 b_4)$	$(a_1 b_4)$	$(a_0 b_4)$					$\leftarrow (b_4)$

Figure 3.3.15: Partial product matrix for two's complement multiplication. The parentheses indicate negative weighting

Figure 3.3.15 shows the partial product matrix for 5-bit two's complement numbers, using the sign notation specified above. All partial products in parentheses (all products with a sign) are weighted negatively; all without parentheses are positive. The product of both signs is also positive. The product P is obtained through a mixture of addition and subtraction.

Pezaris generalized full adders for such structures [51]. In a normal adder, all three input bits x, y, z are weighted positively, and the result lies in the range $[0, 3]$, which can be represented by a 2-bit binary number cs, where c and s are also weighted positively. Pezaris classifies such full adders to be of type 0, since none of their input bits are weighted negatively. A full adder, at whose input one bit, for instance z, is weighted negatively are called type 1. The range of type 1 full adders in which z is weighted negatively is $[-1, 2]$. This range can be expressed by a binary number with negative weighting of the sum output, i.e. $c(-s)$. Similarly, type 2 full adders have the range $[-2, 1]$ and a result representation through $(-c)s$. Type 3 full adders, it follows, have the range $[-3, 0]$ and the representation $(-c)(-s)$. The four types of full adders are summarized in Figure 3.3.16. The dots on the blocks signify negative weighting. They are not inverters. Using a complete truth table, it can be easily verified that all four full adders represent similar logic functions. Given the input signals x, y, z and the output signals c, s for a generalized full adder, its logic function can be given as

$$s = x \oplus y \oplus z \qquad \text{Types 0–3} \qquad (3.3.24)$$

$$c = \begin{cases} xy \vee z\,(x \vee y) & \text{Types 0, 3} \\ xy \vee \bar{z}\,(x \vee y) & \text{Types 1, 2} \end{cases} \qquad (3.3.25)$$

In spite of the negative weighting of several signals, only the logic function of the carry differs slightly between types 0, 3 and types 1, 2. Thus the implementation costs for all types of full adders are practically identical.

Operation

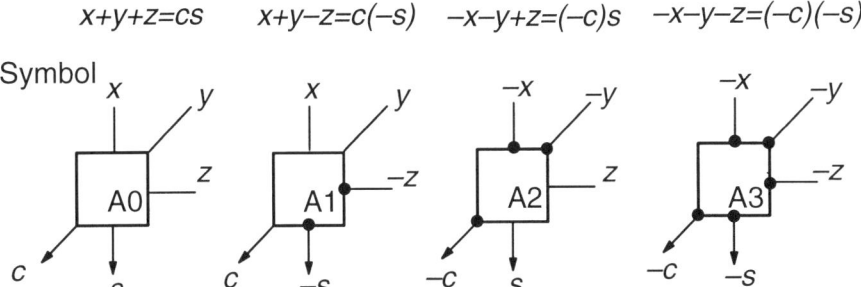

Figure 3.3.16: The four types of generalized full adders

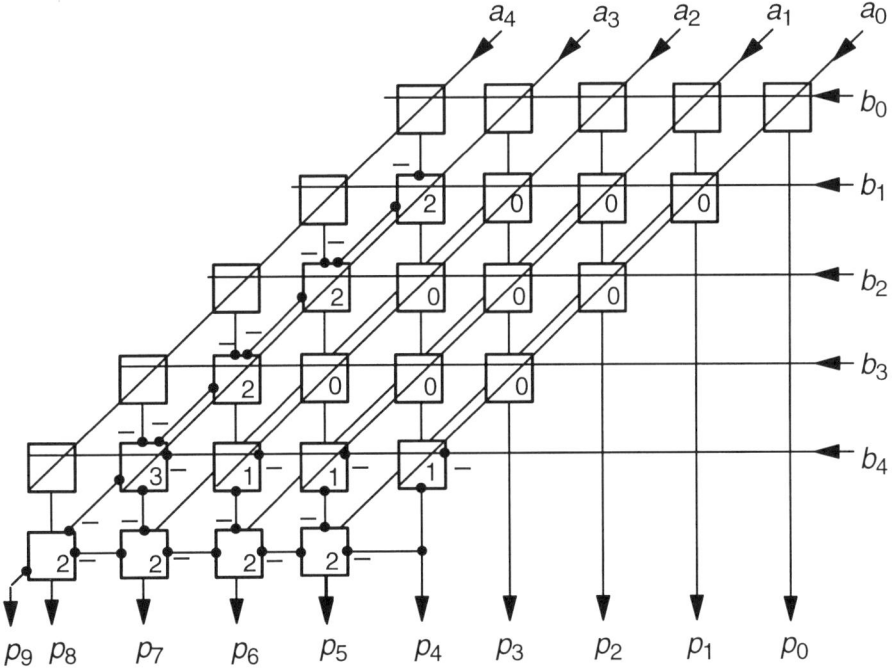

Figure 3.3.17: Pezaris array multiplier

If the full adder types shown in Figure 3.3.16 are incorporated in an array multiplier, a structure such as that in Figure 3.3.17 results. This is a modified Braun multiplier, in which the negative weighting of partial products is given by a sign bit. In order for the partial products to be able to be summed in normal full adders without a sign bit, a type 2 full adder must be used for the addition of partial products with sign bit a_{n-1}. An additional peculiarity occurs in bit level p_{n-1}. Due to the result of the type 1 full adder ($a_0 b_4$), p_{n-1} must be weighted negatively. A positive weighting is achieved as with sign extension by adding a negatively weighted 1 at the next higher bit level.

Pezaris array multipliers have costs and delay characteristics that closely correspond to those of a Braun array multiplier. Their characteristics are given by

$$\begin{aligned} N_{Tr} &= 32n^2 - 26n \\ T_{D,MUL} &= (46.4n - 38.4)\tau_L \end{aligned} \tag{3.3.26}$$

Baugh–Wooley array multipliers

A further alternative for two's complement array multipliers was suggested by Baugh and Wooley [52].

			(b_4)	b_3	b_2	b_1	b_0	
			(a_4)	a_3	a_2	a_1	a_0	
				$a_3 b_0$	$a_2 b_0$	$a_1 b_0$	$a_0 b_0$	
				$a_3 b_1$	$a_2 b_1$	$a_1 b_1$	$a_0 b_1$	
			$a_3 b_2$	$a_2 b_2$	$a_1 b_2$	$a_0 b_2$		positive
	$a_4 b_4$	0	$a_3 b_3$	$a_2 b_3$	$a_1 b_3$	$a_0 b_3$		summands
		$(a_4 b_3)$	$(a_4 b_2)$	$(a_4 b_1)$	$(a_4 b_0)$			negative
		$(a_3 b_4)$	$(a_2 b_4)$	$(a_1 b_4)$	$(a_0 b_4)$			summands
(p_9)	p_8	p_7	p_6	p_5	p_4	p_3	p_2	p_1 p_0

Figure 3.3.18: Partial product matrix separated into positive and negative summands

For the derivation of this method, the partial product matrix must be split into positive and negative summands (Figure 3.3.18). The negatively weighted summands can be interpreted as two numbers that must be subtracted. Instead of subtracting, the two's complement of these two numbers can be added.

For $a_4 = 1$ in the example given in Figure 3.3.18, the number

$$0 \qquad 0 \qquad a_4\,b_3 \qquad a_4\,b_2 \qquad a_4\,b_1 \qquad a_4\,b_0$$

must be complemented and incremented by 1:

$$1 \qquad 1 \qquad \overline{a_4\,b_3} \qquad \overline{a_4\,b_2} \qquad \overline{a_4\,b_1} \qquad \overline{a_4\,b_0}$$
$$+\,1$$

A logic function is sought that produces the values above for $a_4 = 1$ and zero for $a_4 = 0$. The partial products

$$q_{4+j} = \begin{cases} 0 & a_4 = 0 \\ \overline{a_4\,b_j} & a_4 = 1 \end{cases}$$

are given by the logic function

$$q_{4+j} = a_4\,\overline{b_j}$$

For the ones, the variable a_4 yields the desired result. For the leading two positions

$$a_4 \qquad a_4 \,\ldots$$

the function

$$1 \qquad \overline{a}_4 \,\ldots$$
$$+\,1$$

produces the same results. If the method described is also used for the negative summands containing b_4 and the leading ones of a_4 and b_4 are added, the result is a partial product matrix of positive summands, such as Figure 3.3.19, whose sum produces the desired product.

p9	p8	p7	p6	p5	p4	p3	p2	p1	p0
				(b_4)	b_3	b_2	b_1	b_0	
				(a_4)	a_3	a_2	a_1	a_0	
						$a_3 b_0$	$a_2 b_0$	$a_1 b_0$	$a_0 b_0$
					$a_3 b_1$	$a_2 b_1$	$a_1 b_1$	$a_0 b_1$	
				$a_3 b_2$	$a_2 b_2$	$a_1 b_2$	$a_0 b_2$		
	$a_4 b_4$	0	$a_3 b_3$	$a_2 b_3$	$a_1 b_3$	$a_0 b_3$			
	$\bar{a}_4$	$a_4 \bar{b}_3$	$a_4 \bar{b}_2$	$a_4 \bar{b}_1$	$a_4 \bar{b}_0$				
	$\bar{b}_4$	$\bar{a}_3 b_4$	$\bar{a}_2 b_4$	$\bar{a}_1 b_4$	$\bar{a}_0 b_4$				
1					a_4				
					b_4				
(p_9)	p_8	p_7	p_6	p_5	p_4	p_3	p_2	p_1	p_0

Figure 3.3.19: Partial product framework for the Baugh–Wooley algorithm

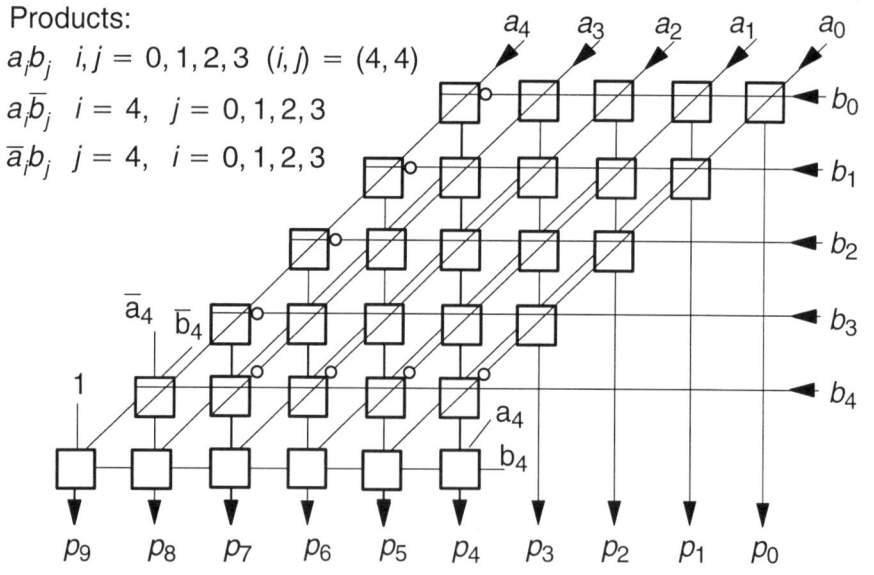

Products:
$a_i b_j$ $i, j = 0, 1, 2, 3$ $(i, j) = (4, 4)$
$a_i \bar{b}_j$ $i = 4$, $j = 0, 1, 2, 3$
$\bar{a}_i b_j$ $j = 4$, $i = 0, 1, 2, 3$

Figure 3.3.20: Array multiplier after Baugh–Wooley

The structure of the corresponding array multiplier is shown in Figure 3.3.20. Naturally, the structure shown for $n = m = 5$ can be applied to any n or m. The costs and delay of a Baugh–Wooley are comparable to those of a Braun array multiplier. They are given by

$$N_{Tr} = 32n^2 - 32n + 88$$
$$T_{D,MUL} = (46.4n - 14.8)\tau_L \qquad (3.3.27)$$

The essential differences between Pezaris and Baugh–Wooley array multipliers can be generalized as follows. In the former, the partial products are created as with positive numbers, excepting that two different full adder implementations are employed. In the latter, only one full adder implementation is necessary; however, the deviant partial products must be created using complementers.

Booth array multiplier
Array multipliers for the multiplication of two's complement numbers that evaluate bit-pairs are also possible. A modified Booth algorithm (section 3.1) with radix 4 builds the foundation for the necessary recoding.
The following holds for the multiplier B:

$$V(B) = \sum_{j=0}^{n/2-1} d_j \, 2^{2j} \quad d_j \in \{-2, -1, 0, 1, 2\} \qquad (3.3.28)$$

The product is defined as

$$V(P) = \sum_{j=0}^{n/2-1} V(A) \, d_j \, 2^{2j} \qquad (3.3.29)$$

This means that the partial products $-2A, -A, 0, A, 2A$ must be available. The two's complement number $-A$ is extracted from A with the aid of a complementer. By means of hard-wired shifting by one position, the numbers $2A$ and $-2A$ are created.
Figure 3.3.21 shows a Booth array multiplier. The digits for a SD representation are decoded through serial analysis of the multiplier B three bits at a time, whereby one bit overlaps the previous three bits. The decoder output controls multiplexers for the selection of the necessary partial products in (3.3.29). The valency shift of 2^{2j} is carried out through hard-wired shifting of the bit slices. All partial products are combined to the total product in a CSA. One notable peculiarity of the adder tree is that it adds two's complement numbers that have been shifted by two bits relative to one another. In

order to achieve correct results, the sign bits must be extended accordingly. This implies that, for the addition of three numbers that have each been shifted by two bits, the sign of the number to the far right must be extended by four bits, the number in the middle by two bits. For all further additions in the adder tree, the sign must be extended such that the allowable range is observed and an overflow of the number into the sign digits is avoided.

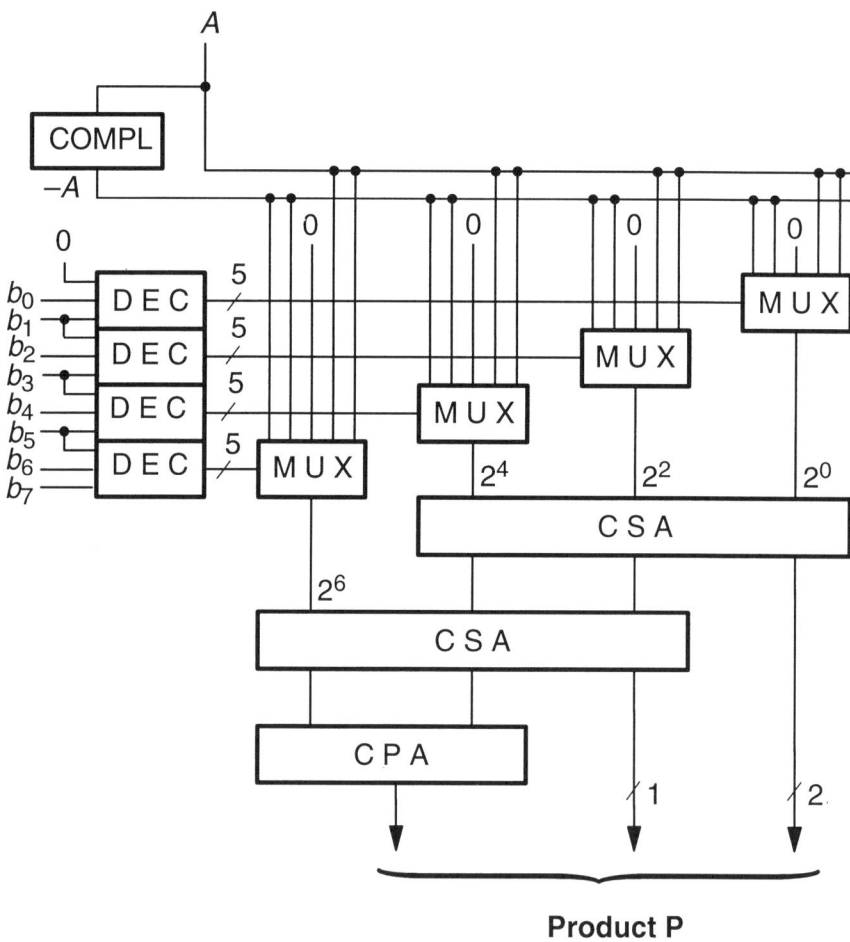

Figure 3.3.21: Booth array multiplier

Due to its bit-pair evaluation, the adder tree of Booth multipliers is significantly less cost intensive. In spite of the rather complex decoder, the multiplexers and the additional complementers, the total costs are noticeably smaller than with the other two's complement multipliers discussed.

$$N_{Tr} = 23n^2 + 56n - 58 \qquad (3.3.30)$$

In addition to the adder tree, the contributions from the complementers and the multiplexers must be taken into consideration when evaluating the delay. The primary contributions to the delay come from the complementer and the final adder in the adder tree. If a carry ripple adder is used here, the delay characteristics can be approximated by

$$T_{D,MUL} = (41.5n + 1.9)\tau_L \qquad (3.3.31)$$

This is the slightest delay of any of the two's complement multipliers considered. This is enabled by the favourable delay characteristics of the complementer in addition to the reduced number of adder stages in the adder tree. The time-critical path consists roughly of n stages along the carry path of the complementer, $(n/2 - 2)$ FA in the CSA tree and n stages along the carry path of the final CPA adder. Since the delay of the complementer is two-thirds that of a normal adder, a total delay of approximately 2.6 times that of an n-bit adder results.

The use of the circuit delay model of section 2.4 is quite cumbersome for complex configurations such as multipliers. Therefore, in addition simplified approximations were given. In the case of multipliers, this is valid since most adder cells have identical capacitive loads. And since only general architectural comparisons were to be carried out here, the accuracy of results is sufficient. Exact delay computations require simulation in SPICE that takes the wiring capacitance into account. Adder elements were used in these multiplier architectures that have significantly faster carry paths in comparison to their sum paths. Other full adder implementations, in comparison to those used here, have a faster sum path and a slower carry path. Since in most cases the time-critical path has more carry transitions than sum transitions, the use of other adder cells normally does not lead to any improvement.

3.4 Dividers

Division occurs in signal processing algorithms less frequently than multiplication. Divisions are primarily required for matrix operations and for normalizing intermediate results. In order for the divisions not to be the time-limiting factor in such algorithms, special hardware structures for the implementation of fast divisions are desired. In the following section, array structures for the implementation of fixed-point division are introduced. Standard division algorithms are explained thereafter.

3.4.1 Binary division algorithms

Division is the inverse operation of multiplication. It is defined as the evaluation of a quotient that, when multiplied by the divisor, produces the dividend. It is rare that the quotient of an integer divisor and an integer dividend is also an integer. Thus, for fixed-point division, a remainder is introduced. Given a dividend A, a divisor D, a quotient Q and a remainder R, then

$$\frac{A}{D} = Q + \frac{R}{D} \qquad (3.4.1)$$

Through multiplication with D, the equivalent equation

$$A = Q \cdot D + R \qquad (3.4.2)$$

results.

For fixed-point division, the range of the operands must be predetermined. If Q and D are n-bit binary numbers, then A requires $2n$ bits and R requires n bits since $R < D$ must hold. It will first be assumed that the numbers are positive integers. The quotient is then given by

$$Q = \sum_{i=0}^{n-1} q_i \, 2^i \qquad (3.4.3)$$

Substitution into (3.4.2) leads to

$$A = q_{n-1}D \, 2^{n-1} + q_{n-2}D \, 2^{n-2} + \ldots + q_0 D \, 2^0 + R \quad (3.4.4)$$

The evaluation of the quotient bits q_i is carried out in sequence from the highest valued bit q_{n-1} to the lowest valued bit q_0. Once a particular quotient bit has been calculated, its corresponding term is subtracted from the value of A and a sequence of positive remainders R_j is created.

Starting with

$$R_0 \, 2^n = A \qquad (3.4.5)$$

and using (3.4.4), a recursive equation for the remainder can be formulated.

$$R_j \, 2^{n-j} = q_{n-(j+1)} D \, 2^{n-(j+1)} + R_{j+1} \, 2^{n-(j+1)}$$

$$j = 0, 1, ...n-1 \qquad (3.4.6)$$

Through modification of (3.4.5) and (3.4.6), a recursive algorithm for the evaluation of the quotient bits can now be given. The first step is

$$R_0 = A \, 2^{-n} \qquad (3.4.7)$$

This means that the point in A is shifted by n digits. The new remainder results from the previous remainder in accordance with

$$R_{j+1} = 2R_j - q_{n-(j+1)} \, D \qquad j = 0, 1, ... \; n-1 \quad (3.4.8)$$

The goal is to create a sequence of smallest possible remainders. Thus

$$q_{n-(j+1)} = \begin{cases} 0 & 2R_j < D \\ 1 & 2R_j \geq D \end{cases} \qquad (3.4.9)$$

must hold for $q_{n-(j+1)}$.

The quotient bit can also be evaluated by complementing the sign of a preliminary remainder

$$\underline{R}_{j+1} = 2R_j - D \qquad (3.4.10)$$

The actual remainder is given by

$$R_{j+1} = \begin{cases} \underline{R}_{j+1} & q_{n-(j+1)} = 1 \\ \underline{R}_{j+1} + D = 2R_j & q_{n-(j+1)} = 0 \end{cases} \qquad (3.4.11)$$

This addition of D is called restoring since it wins back the original remainder.

This binary division algorithm was derived for positive integers. It can be shown that it also holds for non-integers in a similar way. By multiplying both sides of Equation (3.4.2) by 2^{-k}, the point can be arbitrarily shifted. Yet the sequence of the quotient bits is not affected. For true rational numbers, the (decimal) points of A, D, Q and R occur directly before the highest bit, for example:

$$Q = \sum_{i=-n}^{-1} q_i \, 2^i \qquad (3.4.12)$$

In this case, (3.4.5) and (3.4.6) must be adjusted accordingly. Now

$$R_0 = A \tag{3.4.13}$$

and

$$R_j \, 2^{-j} = q_{-(j+1)} \, D \, 2^{-(j+1)} + R_{j+1} \, 2^{-(j+1)} \tag{3.4.14}$$

The equivalent equation to (3.4.8) is

$$R_{j+1} = 2R_j - q_{-(j+1)}D \quad j = 0,\dots n-1 \tag{3.4.15}$$

In summary, the result of these observations states that the division algorithm can be carried out independently of the particular position of the (decimal) point and that the position of the (decimal) point needs only to be evaluated for the final result.

A set number of bits is used for the representation of Q. If Q requires more than the allowable n bits, the quotient is said to overflow. The corresponding condition is

$$R_0 \geq D \tag{3.4.16}$$

Quotient overflow can be detected by a cycle of pre-testing similar to (3.4.10) and (3.4.11).

As with multiplication, the division of signed numbers can be carried out using their absolute values and the sign then determined separately. Given the signs a_n, d_n, q_n and r_n corresponding to the numbers A, D, Q and R, the standard multiplicative sign rules lead to

$$\begin{aligned} q_n &= a_n \oplus d_n \\ r_n &= a_n \end{aligned} \tag{3.4.17}$$

The division algorithm in this section is based on binary numbers. For the sake of completeness, it is worth noting that algorithms exist that evaluate the quotient Q in SD notation [42], [55].

Example 3.4.1

$n = 3$

$A =$	101001	$(41)_{10}$
$D =$	111	$(\ 7)_{10}$

R_0	101.001	Test overflow
$-D\quad -$	111.	$\bar{1} = -1$
R_0	$\bar{1}110.001$	$R_0 < 0$, no overflow

R_0	101.001	Restore

$2R_0$	1010.01	Start division
$-D\quad -$	0111.	
R_1	0011.01	$R_1 \geq 0 \quad q_2 = 1$

$2R_1$	0110.1	
$-D\quad -$	0111.	
R_2	$\bar{1}1111.1$	$R_2 < 0 \quad q_1 = 0$

R_2	110.1	Restore

$2R_2$	1101	
$-D\quad -$	0111	
R_3	0110	$R_3 \geq 0 \quad q_0 = 1$

$Q =$	101	$(\ 5\ )_{10}$
$R =$	110	$(\ 6\ )_{10}$

Note: In place of the subtraction of D, the two's complement representation of D could have been added. However, an additional sign digit must then be handled.

3.4.2 Array dividers

Division is principally a sequence of specially controlled subtractions. As with multipliers, array configurations can be used in the implementation of dividers for more time-efficient execution of its operations. Time-efficient execution is the primary goal of array configurations, not low-cost hardware. The elementary structures of restoring array dividers [53] and non-restoring array dividers [54] will be presented. More complex and faster array dividers can be found in the literature [42].

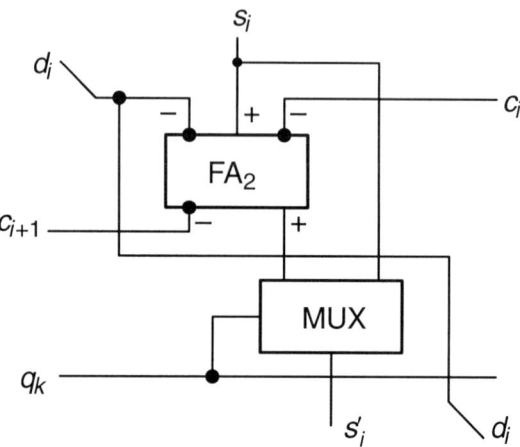

Figure 3.4.1: Controlled subtractor cell CS

The binary division presented in the previous section requires a subtraction as in (3.4.10), the evaluation of the quotient bit from the results of the subtraction (3.4.9) and the selection of the new remainder according to (3.4.11). The restoration is implemented as a multiplexer operation and not as a re-addition of D. Figure 3.4.1 shows a subtractor cell with controlled reconstruction (CS = controlled subtract) for the implementation of this function. Using the nomenclature of Figure 3.3.16, the full adder FA_2 is of type 2 with two negative inputs and one negative carry output. From (3.3.24) and (3.3.25), the following logic function for the cell CS results:

$$s_i' = \begin{cases} s_i \oplus d_i \oplus c_i & q_k = 1 \\ s_i & q_k = 0 \end{cases} \tag{3.4.18}$$

and

$$c_{i+1} = d_i \, c_i \lor \bar{s}_i \, (d_i \lor c_i) \tag{3.4.19}$$

The sum bit s'_n need not be evaluated for the highest bit level n of the computation, and $d_n = 0$. This means that this bit level of the cell CS can be simplified considerably. Its carry path is thus

$$c_{n+1} = \bar{s}_n \, c_n \tag{3.4.20}$$

Each quotient bit q_k is identical to the complement of c_{n+1}. It follows that the quotient bit is

$$q_k = \overline{c_{n+1}} = \overline{\overline{s_n}\,\overline{c_n}} = s_n \vee \overline{c}_n \qquad (3.4.21)$$

accordingly.

Figure 3.4.2 shows a restoring array divider based on the CS cells and simplified at the highest bit level.

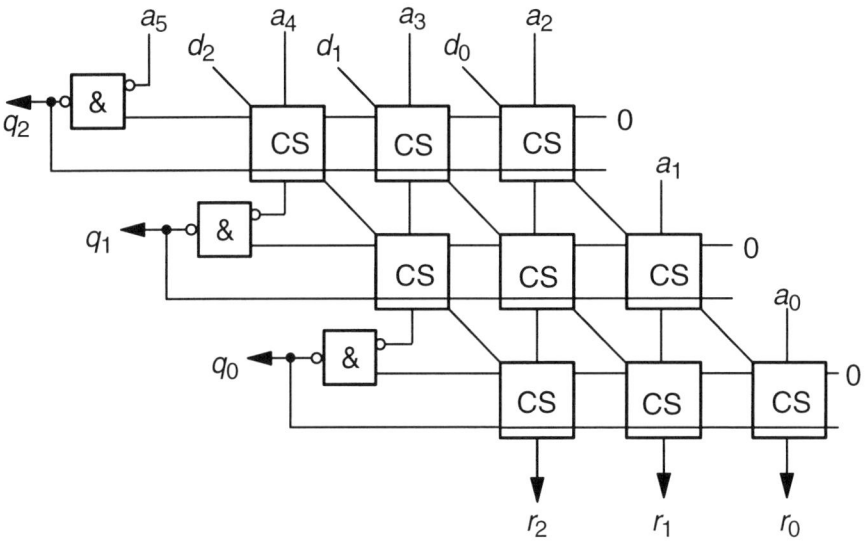

Figure 3.4.2: Restoring array divider ($n = 3$)

The array consists of n^2 CS cells, n inverters and n NAND gates. Thus the transistor count is proportional to n^2. Using a full adder based on transmission gates, the transistor count of the divider is given by

$$N_{Tr} = 40n^2 + 6n \qquad (3.4.22)$$

The time-critical path of each subtraction stage runs along the carry path until the evaluation of the quotient bit and then via the parallel multiplexers to the next level. The time-critical path of the first level passes through n CS cells and $n - 1$ CS cells in the higher levels. Thus, the delay is given by the sum

$$T_{D,\,DIV} = \Delta T + [n + (n - 1)^2]T_{D,FA} + nT_{D,\,NAND} + nT_{D,\,MUX} \qquad (3.4.23)$$

The time ΔT represents the increase in the pre-steady-state delay in the divisor line d_i due to the increased load capacitances on this wire. Taking the particular load capacitances into account, the circuit delay model of section 2.4 yields

$$T_{D,\,DIV} = (18n^2 + 23.4n + 12)\tau_L \qquad (3.4.24)$$

Due to the sequential passing through all levels, the delay is proportional to n^2. This result demonstrates that array divider implementations have no considerable advantage in terms of delay over purely sequential implementations using ALUs. The use of an array was much more efficient in multipliers since the delay characteristics were then proportional to n.

The high load capacitances in the bit quotient line contribute considerably to the delay. These high load capacitances come from the multiplexers working in parallel. Their share can be reduced using special driving circuits. However, non-restoring array dividers have also been proposed as an alternative.

In accordance with (3.4.10), the division algorithm evaluates the difference

$$\underline{R}_{j+1} = 2R_j - D \qquad (3.4.25)$$

Whenever $q_{n-j} = 0$, R_j was restored. Substituting the restoration into the previous equation yields

$$\underline{R}_{j+1} = 2(\underline{R}_j + D) - D = 2\underline{R}_j + D \qquad (3.4.26)$$

This means that the restoring process can be avoided by carrying out a controlled addition/subtraction. The computation of the remainder from (3.4.10) is to be altered as follows:

$$R_{j+1} = \begin{cases} 2R_j - D & q_{n-j} = 1 \\ 2R_j + D & q_{n-j} = 0 \end{cases} \qquad (3.4.27)$$

The remainders R_{j+1} can be positive or negative. The quotient bits are derived from the sign of the remainders as before.

$$q_{n-(j+1)} = \begin{cases} 0 & R_{j+1} < 0 \\ 1 & R_{j+1} \geq 0 \end{cases} \qquad (3.4.28)$$

Figure 3.4.3 shows a cell with controlled addition/subtraction (CAS) for non-restoring division. Depending on the state of the mode line $q_k = 0$, either the bit d_i or $\overline{d_i}$ is used as input for the FA. Thus the mode signal $q_k = 1$ induces a bit complement. The mode line continues to the carry input of the lowest bit level where it steers the requisite addition of 1 for the creation of two's complements.

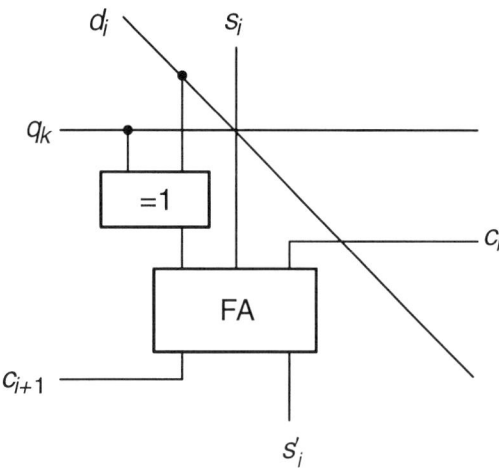

Figure 3.4.3: Subtractor cell CAS with controlled add/subtract

Due to the particular requirements of division, the number of CAS cells necessary can be reduced by 1. The term $2R_j$ of an n-bit divisor requires at most $n+1$ bits. An additional sign bit is necessary for a two's complement representation of R_{j+1}. This means that $n+2$ bits are required. Yet

$$| R_j | < D \tag{3.4.29}$$

Therefore, the sign bit and its preceding bit in R_{j+1} are always identical. Thus one digit can always be spared. In addition, (3.4.28) states that the quotient bit is the complement of the sign. Yet due to the particular requirements given above, the carry bit of level $n+1$ is the complement of the sign.

$$s_{n+2} = s_{n+1} = \overline{c}_{n+2} \tag{3.4.30}$$

It follows that the carry bit c_{n+2} can be used directly as the quotient bit. Figure 3.4.4 shows a non-restoring array divider resulting from these considerations.

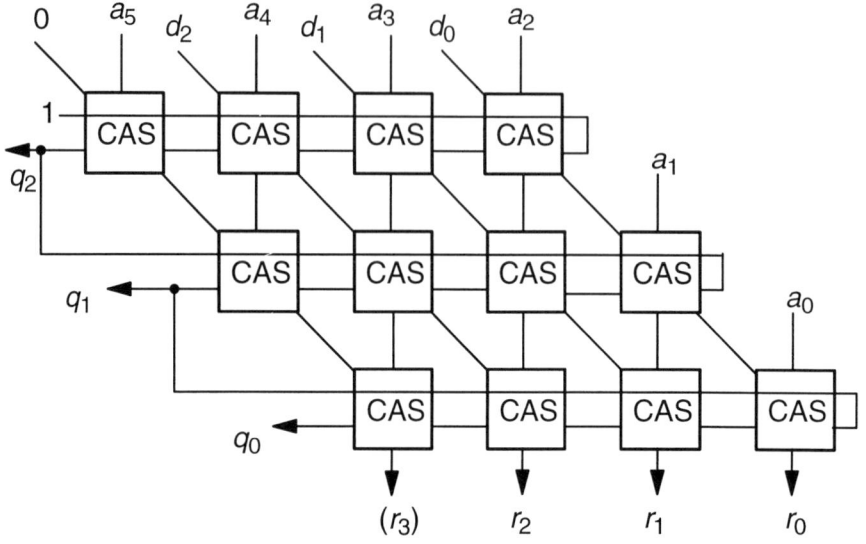

Figure 3.4.4: Non-restoring array divider ($n = 3$)

A total of $n(n+1)$ CAS cells are required. The total transistor count is thus

$$N_{Tr} = 34n^2 + 34n \qquad (3.4.31)$$

This is only a slight reduction in comparison to restoring array dividers. Due to its controlled addition/subtraction, a CAS array can output the remainder as a negative, two's complement number for special operations. A correct representation of the remainder as a positive number can be created via subsequent addition of the divisor D. Thus, whenever a correct representation of the remainder is required, an additional adder controlled by the sign of the remainder is essential. In this case, the costs are even higher than those of a restoring divider.

The time-critical path runs through the XOR gate of the lowest bits and then along the carry path of the FA. This is the same at all subsequent levels. Thus, the delay consists of the sum

$$T_{D,\,DIV} = \Delta T + (n^2 + n)T_{D,FA} + nT_{D,XOR} \qquad (3.4.32)$$

Taking the particular load capacitances into account, the delay is given by

$$T_{D,\,DIV} = (19n^2 + 46.8n - 8)\tau_L \qquad (3.4.33)$$

This shows that non-restoring array dividers have no advantages in terms of their delay characteristics.

In the array dividers shown, it was assumed that no overflow occurred. If an overflow check is to be carried out, a pre-test similar to that for regular division (see Example 3.4.1) must be included. This can be implemented by extending the arrays by one row. In principle, this row evaluates a quotient bit q_n that is 1 in the case of overflow and 0 otherwise.

The delay characteristics of an array divider can be very roughly described as n subtractions of length $n+1$. The delay characteristics are adversely influenced by the carry throughput from bit level to bit level. With aid of the techniques shown in section 3.2, the computations can be accelerated. Possible measures would be, for instance, carry lookahead or hierarchical structures such as carry select or conditional sum. However, such measures reduce the regularity of the arrays considerably.

3.5 Implementations of Elementary Functions

In addition to the basic operations of addition, subtraction, multiplication and division already treated, elementary functions are often required to carry out signal processing tasks. Examples of such elementary functions are trigonometric functions and functions such as square root, logarithm and exponentiation. In the following section, general methods for the computation of elementary functions will be briefly explained. Then special architectures for implementation of CORDIC will be treated. CORDIC was chosen since the same hardware can implement several elementary functions and CORDIC is becoming increasingly popular for signal processing.

3.5.1 Methods of computing elementary functions

The essential methods applicable for implementation in integrated circuits are to be presented here along with a discussion of their implementational aspects. Table-oriented methods, polynomial approximations and iterative methods will be described.

Table-oriented methods
An obvious method of implementing elementary functions is the use of tables. The function values are stored directly in memory. The function argu-

ment addresses a memory location, and the contents yield the function value. The disadvantage of this method is the exponential growth in the memory requirements. Thus, this method is only economical for small word widths.

Given an n-bit function argument and an m-bit function value representation, direct storage in a table would require

$$N_{Sp} = m \ 2^n \qquad (3.5.1)$$

storage locations. Typical memory structures for permanent storage are ROM (read only memory) and PLAs (programmable logic arrays). The principle configuration of such permanent memory is shown in Figure 3.5.1.

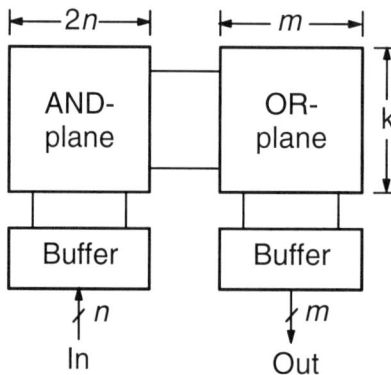

Figure 3.5.1: Schematic representation of permanent memory

The illustration implies that each logic function can be implemented as a two-stage sum of products (AND/OR) or a two-stage product of sums (OR/AND) [18], [19]. If working with minterms, the minterms are created by ANDing all variables. The actual function results from the OR of the minterms. In ROM, all 2^n minterms ($k=2^n$) are evaluated in the AND stage. The AND stage is thus also called an address decoder (n to 1 decoder), i.e. for a given input, one of the 2^n decoder outputs will be 1, all others will be 0. Complementary inputs are necessary for the evaluation of the minterms. These can be created with an input driver. By paying special attention to the locations of the ones in the logic function, the number of terms (implicants) and the number of variables per term can be minimized. The number of implicants k in several functions can be considerably smaller than 2^n. This fact is exploited for PLA table implementations. ·

PLAs and ROM can be efficiently implemented in a geometrically regular layout consisting of an array of transistors. Programming is carried out by connecting the transistors with the particular outputs. By exploiting the inverting character of drivers, both AND and OR gates can be implemented with NOR gates. NOR gates with many inputs can be efficiently implemented in pseudo-NMOS technology. A count of all transistor locations (including the non-connected ones) yields the following relation, upon which the circuit size can be approximated:

$$N_{Tr,\ PLA} = 2nk + mk + 4n + k + 3m \qquad (3.5.2)$$

For a complete address decoder (ROM), $k = 2^n$. In ROMs, however, the hardware costs in the AND stage can be further reduced through the use of a hierarchical decoder. Yet it should be noted that table implementations in ROMs and even as PLAs are quite inefficient in terms of size. Whenever several functions are to be implemented as tables, the memory requirements grow accordingly.

Polynomial approximations
Polynomial approximations can be used as an alternative to table implementations. A function $f(x)$ is approximated by a polynomial $P_n(x)$ of degree n, whereby the approximation error within the defining interval should be smaller than a given tolerance value.

$$f(x) \approx P_n(x) = \sum_{i=0}^{n} a_i\ x^i \qquad (3.5.3)$$

The polynomial coefficients are usually calculated by truncating an infinite power series, for instance a Taylor series [56]. In this case, the convergence characteristics of the Taylor series are exploited. A disadvantage of a truncated Taylor series is its error. This is small near the point of expansion, yet it increases considerable the farther the argument strays from the point of expansion. A more constant approximation error in the defining interval and a significant reduction of the total error within an interval can be achieved through Chebyshev approximation [56]. In order to limit the degree of the polynomial for a given error tolerance, approximations using rational polynomials are applied in many cases.

$$f(x) \approx \frac{P_n(x)}{Q_m(x)} \qquad (3.5.4)$$

The rational Chebyshev approximation can be carried out with the aid of the Remez exchange algorithm [56]. The implementation of functions through polynomial approximation requires efficient polynomial arithmetic.

An expansion of the polynomial using the Horner scheme shows that a total of n multiplications and n additions are necessary to evaluate an n-degree polynomial.

$$P_n(x) = (..(a_n\, x + a_{n-1})\, x + ...\, a_1)\, x + a_0 \qquad (3.5.5)$$

Polynomial arithmetic is generally implemented with one multiplier and one adder, i.e. the evaluation of the polynomial is carried out sequentially in n steps. In addition to the multiplier and the adder, coefficient storage (ROM, for instance) is also necessary. If the coefficient memory contains the coefficients of several functions, then the same hardware can be used for several functions. The known methods of function approximation are described in [56]. Furthermore, polynomials for the approximation of the most important functions are listed in tables. The following polynomial approximation, for example, describes the sine and cosine functions to within an absolute error of less than 10^{-5} over the range $[0, \pi/2]$.

$$
\begin{aligned}
\sin x &\approx x P_{3,s}(x^2) \\
\cos x &\approx P_{3,c}(x^2)
\end{aligned}
\qquad x \in [0, \pi/2] \qquad (3.5.6)
$$

a_i	$P_{3,s}$	$P_{3,c}$
0	0.9999966157	0.9999932946
1	–0.1666482836	–0.4999124376
2	0.00830632518	0.0414877472
3	–0.00018363653	–0.00127120948

Iterative methods

In the iterative methods, the value of the function is approximated by singular steps that are designed to converge when iterated, whereby the next step is often not determined until after the previous step has been carried out. The particular steps of an iteration should be as simple as possible, for instance a continuous computation of products or sums. In many cases, the algorithm contains two or even more recursive equations that are interconnected in such a way that one variable (the control variable) converges to a constant, 0 or 1 for instance, whereas the other variable (the production variable) converges to the result. In the algorithms used, the results gains in accuracy with each iteration.

The principle behind such algorithms will be explained using convergence division as an example [57]. The quotient

$$Q = \frac{A}{D} \tag{3.5.7}$$

is to be calculated. In each iteration, both the numerator and the denominator are to be multiplied by a factor R_k, where $k = 0, 1, \ldots, n$.

$$\frac{A}{D} = \frac{A}{D} \cdot \frac{R_0}{R_0} \cdot \frac{R_1}{R_1} \cdot \ldots \cdot \frac{R_n}{R_n} \tag{3.5.8}$$

The sequence of factors is chosen such that, for significantly large n,

$$D \cdot R_0 \cdot R_1 \cdot \ldots R_n \rightarrow 1 \tag{3.5.9}$$

and thus

$$A \cdot R_0 \cdot R_1 \cdot \ldots R_n \rightarrow Q \tag{3.5.10}$$

It is assumed that both A and D are positive, rational numbers normalized to the range

$$\frac{1}{2} \leq A, D < 1 \tag{3.5.11}$$

By taking their absolute values and shifting, this is possible for all numbers. From the range of D, it follows that

$$D = 1 - \delta \qquad \text{with } 0 < \delta \leq 1/2 \tag{3.5.12}$$

Taking

$$R_i = 1 + \delta^{2^i} \tag{3.5.13}$$

the desired behaviour can be achieved. D_i is designated to be an intermediate partial product of (3.5.9). Given

$$D_0 = D \cdot R_0 \tag{3.5.14}$$

it follows that for each iteration

$$D_i = D_{i-1} R_i \tag{3.5.15}$$

Given the inital condition (3.5.12) and substituting (3.5.13), it can be shown that

$$D_i = 1 - \delta^{2^{i+1}} \qquad (3.5.16)$$

For large enough values of $i = n$, D_n is practically 1 and its deviation from 1 is negligible due to the limited accuracy of the representation. For the implementation, it is important that R_i can be evaluated in two's complement form from D_{i-1}.

$$\begin{aligned} R_i &= 2 - D_{i-1} \\ &= 2 - (1 - \delta^{2^i}) \\ &= 1 + \delta^{2^i} \end{aligned} \qquad (3.5.17)$$

Iterative algorithms consist of the calculation of the initial value and the sequential iterations.

Division algorithm

Initial value: $R_0 =$ two's complement of D
$$D_0 = D\,R_0$$
$$A_0 = A\,R_0$$

For $i = 1$ to n
$$R_i = \text{two's complement of } D_{i-1}$$
$$D_i = D_{i-1}\,R_i$$
$$A_i = A_{i-1}\,R_i$$

The algorithm is finished when D_i first reaches the exact value $1.0 \ldots 0$ ($D_n = 1$) within the accuracy of the representation. The quotient sought is then

$$Q = A_n$$

Since this method converges quadratically in accordance with (3.5.16), it requires fewer iterations than the division algorithms shown in section 3.4. Even if the intermediate results are limited to a set number of bits, the method converges quickly enough to the value sought.

The convergence methods presented have been generalized by Chen [58]. Even functions such as $1/x$, $\sqrt{x}$, e^x and $\ln x$ can be computed with the generalized convergence methods. A large number of alternative, iterative algorithms are presented in the literature. The CORDIC method will be treated in detail in the following section as an important example. The CORDIC method provides a large range of functions and can be implemented in modular hardware. A particular characteristic is that the iteration time is independent of the argument of the function.

3.5.2 The CORDIC method

The name CORDIC stands for <u>Co</u>ordinate <u>R</u>otation <u>Di</u>gital <u>C</u>omputer. The underlying method for computing the rotation of a vector in a Cartesian coordinate system and evaluating the length and angle of a vector was developed by Volder [59]. The CORDIC method was later expanded for multiplication, division and the hyperbolic functions. The various function computations were summarized into a unified technique by Walther [60].

The CORDIC algorithm
The rotation of a vector $[x_0, y_0]^T$ by an angle θ in Cartesian coordinates leads to the vector $[x_n, y_n]^T$ (Figure 3.5.2). The resulting vector can be computed with the aid of matrix operations:

$$\begin{bmatrix} x_n \\ y_n \end{bmatrix} = \begin{bmatrix} \cos\theta & -\sin\theta \\ \sin\theta & \cos\theta \end{bmatrix} \begin{bmatrix} x_0 \\ y_0 \end{bmatrix} \tag{3.5.18}$$

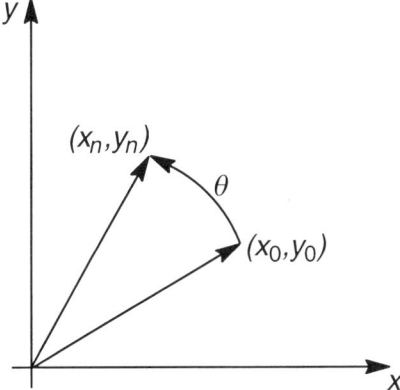

Figure 3.5.2: Rotation in a cartesian coordinate system

Using the identity

$$\cos\theta = \frac{1}{\sqrt{1 + \tan^2\theta}} \tag{3.5.19}$$

and factoring out $\cos\theta$, Equation (3.5.18) can be modified to

$$\begin{bmatrix} x_n \\ y_n \end{bmatrix} = \frac{1}{\sqrt{1 + \tan^2\theta}} \begin{bmatrix} 1 & -\tan\theta \\ \tan\theta & 1 \end{bmatrix} \begin{bmatrix} x_0 \\ y_0 \end{bmatrix} \tag{3.5.20}$$

In the CORDIC method, the rotation by an angle θ is implemented as several partial rotations by a given step angle α_i. Similar to the way in which any whole number in a finitely defined interval can be represented by n digits, any angle θ within a defined interval can be represented to a certain accuracy by a set of n partial angles α_i. Through specification of the sign σ_i, the sum of the partial angles α_i approximates the given angle θ.

$$\theta = \sum_{i=0}^{n-1} \sigma_i \alpha_i \qquad \sigma_i \in \{-1, 1\} \qquad (3.5.21)$$

In order to keep n small for a specified accuracy, the absolute value of α_i decreases for larger indices. Figure 3.5.3 shows a method for determining the sign. The initial α_i's are weighted positively until the sum exceeds the value θ. Afterwards, the α_i's are negatively weighted until the sum falls below the value θ. This is then continued accordingly. This means that the sign of the difference between θ and the sum controls the sign of the partial angles.

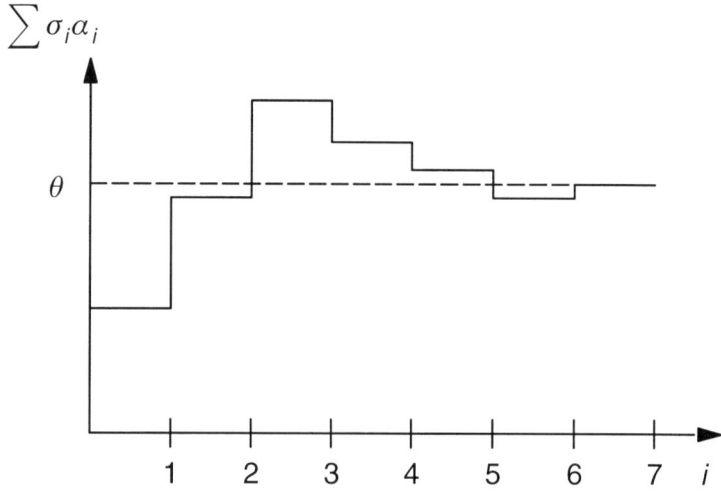

Figure 3.5.3: Iteration of the α_i for the representation of θ

To simplify the computation of the matrix product (3.5.20), the angles α_i are chosen such that $\tan \alpha_i$ represents a series of powers of 2.

$$\tan \alpha_i = 2^{-i} \qquad i = 0, 1, ...n - 1 \qquad (3.5.22)$$

Furthermore, an auxiliary variable z_i is introduced that contains the accumulated partial angles and can be used to control the sign of the partial angles. For $z_0 = \theta$

$$z_{i+1} = z_i - \sigma_i \tan^{-1}(2^{-i}) \tag{3.5.23}$$

$$\sigma_i = \begin{cases} +1 & z_i \geq 0 \\ -1 & z_i < 0 \end{cases} \tag{3.5.24}$$

The matrix product can be simplified for each step to

$$\begin{bmatrix} x_{i+1} \\ y_{i+1} \end{bmatrix} = k_i \begin{bmatrix} 1 & -\sigma_i 2^{-i} \\ \sigma_i 2^{-i} & 1 \end{bmatrix} \begin{bmatrix} x_i \\ y_i \end{bmatrix} \tag{3.5.25}$$

$$k_i = \frac{1}{\sqrt{1 + 2^{-2i}}} \tag{3.5.26}$$

The partial factors k_i can be combined for all n iterations to a total factor

$$k = \prod_{i=0}^{n-1} \frac{1}{\sqrt{1 + 2^{-2i}}} \tag{3.5.27}$$

Under the previous assumptions, the iteration equations without the partial factors k_i are

$$\begin{aligned} x_{i+1} &= x_i - \sigma_i 2^{-i} y_i \\ y_{i+1} &= y_i + \sigma_i 2^{-i} x_i \\ z_{i+1} &= z_i - \sigma_i \tan^{-1}(2^{-i}) \end{aligned} \tag{3.5.28}$$

For scaling to the correct amplitude, a final multiplication by the total factor k must also be carried out.

$$\begin{aligned} k\, x_n &\rightarrow x_n \\ k\, y_n &\rightarrow y_n \end{aligned} \tag{3.5.29}$$

The previous iteration procedure was called "rotation" since a vector $[x_0, y_0]^T$ was rotated by a given angle θ. If the resulting vector is rotated in such a way that it lies along the x-axis ($y = 0$) after the iterations, the accumulated partial angles represent the angle $\tan^{-1}(y_0/x_0)$ and the x-coordinate represents the absolute value of the vector.

$$x_n = \sqrt{x_0^2 + y_0^2}$$
$$y_n = 0$$
$$z_n = z_0 + \tan^{-1}\left(\frac{y_0}{x_0}\right)$$

(3.5.30)

Similar to how z_i previously converged to zero through deliberate altering of the sign, this must also occur for y_i. The sign σ_i is now given by

$$\sigma_i = \begin{cases} -1 & x_i\, y_i \geq 0 \\ +1 & x_i\, y_i < 0 \end{cases}$$

(3.5.31)

Since the direction of rotation depends on the quadrant in which the current vector is located, the sign of the product $x_i\, y_i$ must be used to specify the sign of σ_i. To simplify the implementation, the sign of the product is specified by the signs of x_i and y_i. This last method of operation is called "vectoring".

Square root and trigonometric functions can be calculated with the iteration process defined above. Additional elementary functions can be computed using the extensions formulated by Walther [60]. The radius R and the angle Φ of a vector $[x_0, y_0]^T$ are defined using a parameter m that specifies the coordinate system.

$$R = \sqrt{x_0^2 + m y_0^2}$$

(3.5.32)

$$\Phi = \frac{1}{\sqrt{m}} \tan^{-1}\left(\sqrt{m}\,\frac{y_0}{x_0}\right)$$

(3.5.33)

or, inversely,

$$x_0 = R \cos(\sqrt{m}\,\Phi)$$
$$y_0 = \frac{1}{\sqrt{m}} R \sin\left[\frac{\Phi}{\sqrt{m}}\right]$$

(3.5.34)

where m is in the set

$$m \in \{1, 0, -1\}$$

(3.5.35)

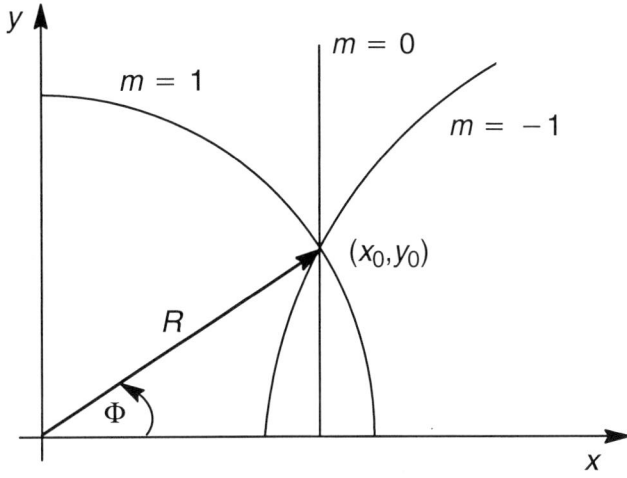

Figure 3.5.4: Plots of R = const. for various coordinate systems

From the shape of their plots for R = constant, the three coordinate systems are termed hyperbolic ($m = -1$), linear ($m = 0$) and circular ($m = 1$) (see Figure 3.5.4). The equations for the unified CORDIC algorithm are analogous to Equation (3.5.28).

$$x_{i+1} = x_i - m\,\sigma_i\delta_{m,i}\,y_i$$
$$y_{i+1} = y_i + \sigma_i\delta_{m,i}\,x_i \qquad (3.5.36)$$
$$z_{i+1} = z_i - \sigma_i\,\alpha_{m,i}$$

The individual parameters are termed as follows:

m coordinate system parameter $\{1, 0, -1\}$

σ_i direction of rotation of the ith iteration $\{-1, 1\}$

$\delta_{m,i}$ incremental change of the ith iteration $\{0 < \delta_{m,i} \leq 1\}$

$\alpha_{m,i}$ partial angle of the ith iteration

i iteration index $\{0, 1, \dots n-1\}$

The resulting partial scaling factors (radius factor) and partial angles for each step (also called microrotations) in the various coordinate systems are summarized in Table 3.5.1.

Table 3.5.1: Partial angle and partial scaling factor (radius factor) relative to m

m	$\alpha_{m,i}$	$k_{m,i}^{-1}$
1	$\tan^{-1}\delta_{1,i}$	$\sqrt{1 + \delta_{1,i}^2}$
0	$\delta_{0,i}$	1
-1	$\tanh^{-1}\delta_{-1,i}$	$\sqrt{1 - \delta_{-1,i}^2}$

As in Equations (3.5.27) and (3.5.21), the scaling factor and the total angle after n iterations are given by

$$k_m = \prod_{i=0}^{n-1} k_{m,i} = \prod_{i=0}^{n-1} \frac{1}{\sqrt{1 + m\delta_{m,i}^2}} \tag{3.5.37}$$

$$\alpha_m = \sum_{i=0}^{n-1} \sigma_i \alpha_{m,i} = \begin{cases} \displaystyle\sum_{i=0}^{n-1} \sigma_i \tan^{-1}\delta_{1,i} & m = 1 \\[2ex] \displaystyle\sum_{i=0}^{n-1} \sigma_i\, \delta_{0,i} & m = 0 \\[2ex] \displaystyle\sum_{i=0}^{n-1} \sigma_i\, \tanh^{-1}\delta_{-1,i} & m = -1 \end{cases} \tag{3.5.38}$$

The equations in Table 3.5.2 specify the results after n iterations and subsequent scaling with k_m for the modes of operation known as "rotation" ($z_n \to 0$) and "vectoring" ($y_n \to 0$). From the table it can be seen that multiplication, division and the hyperbolic functions are also possible due to the unified theory. Through special choice of the input variables, elementary functions can be selected from the CORDIC functions in Table 3.5.2. For $y_0 = 0$ in rotation mode, $x_0\cos z_0$, $x_0\sin z_0$, $z_0 x_0$, $x_0\cosh z_0$ and $x_0\sinh z_0$ are calculated. Similarly, for $z_0 = 0$ in vectoring mode, $\tan^{-1}(y_0/x_0)$, y_0/x_0 and $\tanh^{-1}(y_0/x_0)$ are computed. Further functions can be created from the following list of special equations.

$$e^z = \cosh z + \sinh z$$

$$e^{-z} = \cosh z - \sinh z$$

$$\ln w = 2\tanh^{-1}\frac{w-1}{w+1}$$

$$\sqrt{w} = \sqrt{\left(w+\frac{1}{4}\right)^2 - \left(w-\frac{1}{4}\right)^2} \qquad (3.5.39)$$

$$\tan w = \frac{\sin w}{\cos w}$$

$$\tanh w = \frac{\sinh w}{\cosh w}$$

Table 3.5.2: CORDIC functions

Mode	m	CORDIC functions
$z_n \to 0$	1	$x_n = x_0\cos z_0 - y_0\sin z_0$ $y_n = y_0\cos z_0 + x_0\sin z_0$
	0	$x_n = x_0$ $y_n = y_0 + z_0 x_0$
	−1	$x_n = x_0\cosh z_0 + y_0\sinh z_0$ $y_n = y_0\cosh z_0 + x_0\sinh z_0$
$y_n \to 0$	1	$x_n = (x_0^2 + y_0^2)^{1/2}$ $z_n = z_0 + \tan^{-1}(y_0/x_0)$
	0	$x_n = x_0$ $z_n = z_0 + y_0/x_0$
	−1	$x_n = (x_0^2 - y_0^2)^{1/2}$ $z_n = z_0 + \tanh^{-1}(y_0/x_0)$

In some cases, the evaluation of special elementary functions requires double application of the CORDIC method. For the evaluation of $\tan w$, for instance, first $\cos w$ and $\sin w$ are computed, then in a second step, their quotient is calculated. Functions such as cot, $\sin^{-1}$, $\cos^{-1}$, coth, $\sinh^{-1}$ and $\cosh^{-1}$ require double application of the CORDIC method.

To simplify the implementation of CORDIC, the incremental steps $\delta_{m,i}$ of each iteration should be a power of 2.

$$\delta_{m,i} = 2^{-S(m,i)} \qquad i \in \{0, 1, \dots n-1\} \qquad (3.5.40)$$

In that case, matrix multiplication can be implemented through shift and add. An integer series $S(m,i)$ is termed a shift sequence in CORDIC.

The shift sequence influences the convergence, the range of convergence and the scaling factor. First of all, the partial angles $a_{m,i}$ and their resulting partial sum should create a monotonic, decreasing series. Furthermore, the shift sequences should be constructed so as to guarantee to fulfil the particular objective of the iteration to a given accuracy. Thus, the partial angles must fulfil the following criterion.

$$a_{m,i} - \sum_{j=i+1}^{n-1} a_{m,j} < a_{m,n-1} \qquad i \in \{0, 1, ...n - 2\} \quad (3.5.41)$$

This means that a partial angle in iteration i can be compensated for, except for a small remaining error, by all the following partial angles through a change of direction. This criterion is fulfilled whenever the absolute value of the partial angles does not decrease faster than a geometric series. A series that fits

$$a_{i+1} > \frac{1}{2}a_i \qquad (3.5.42)$$

fulfils this desired criterion. This criterion is not fulfilled for integer shift series in a hyperbolic coordinate system ($m = -1$). Therefore more complex shift series are required in this case. For circular rotations, the criterion is fulfilled. In the linear case, the system is on the edge of the convergence range since it fits the equality.

The sum of all partial angles belonging to a shift series determines the maximum angle A_0 (convergence range) for which the goal of the iteration can be reached.

$$\max |A_0| = \sum_{i=0}^{n-1} a_i + a_{n-1} \qquad (3.5.43)$$

Table 3.5.3 lists shift series after Walther [60], their corresponding range of convergence and their scaling factor. In the hyperbolic case, the shifts $\{4, 13, ... k, 3k+1, ...\}$ are repeated to fulfil the convergence criterion. Since $\tanh^{-1}(2^0)$ is not defined, the shift sequence for $m = -1$ begins at 1. For the integer shift sequences listed in Table 3.5.3, the number of cycles for a given accuracy depends on the coordinate system parameter m. An accuracy of 16 bits, for example, requires a shift series with 17 cycles for $m = 1$, 16 cycles for $m = 0$ and 18 cycles for $m = -1$.

Table 3.5.3: Examples of positive integer shift sequences and their corresponding range of convergence [60]

m	$S(m,i)$	$\max |A_0|$	k_m^{-1}
1	0,1,2,3,4,5,...,i,...	1.743287	1.646760
0	1,2,3,4,5,6,...,i+1,...	1.000000	1.000000
−1	1,2,3,4,4,5,...12,13,13,14,...	1.118173	0.828159

The use of the CORDIC algorithm with integer shift series from Table 3.5.3 requires a final scaling step by k_m. This scaling can be implemented through CORDIC with $m = 0$. In this case the CORDIC implements a constant multiplier for the y component. Thus the total number of cycles including scaling increases significantly. As a result, for the given example of 16-bit accuracy, including scaling for one output, 33 cycles are necessary for $m = 1$ and 34 cycles for $m = -1$. The final scaling can also be carried out with an additional external multiplier. Through CSD coding of the scaling factor, the number of adders for the implementation of the multiplier can be reduced. For the particular coefficients of a 16-bit implementation, seven adders/subtractors are sufficient.

The shift sequences given in Table 3.5.3 have a limited range of convergence. With the aid of some basic identities, the computation of these functions for larger arguments can also be achieved. In the case of the trigonometric functions, the extension of the argument range is based on the periodicity of the functions. For the hyperbolic functions, Walther [60] put together more complex equations for argument reduction.

The literature provides various methods for simplifying the scaling factor compensation step. Most of the recommendations introduce iterations with double shifts in addition to the standard iterations with one shift. The scaling factor is particularly simple to implement in many of these methods. In Deprettere's method [61], the scaling factor can be implemented by a few shifts and adds.

CORDIC architectures
In principle, the CORDIC algorithm can be implemented as a cascade of N CORDIC elements (CEs). Each of these CORDIC elements carries out a microrotation in accordance with equation (3.5.36). Each element must be supplied with the following control values: the mode parameter (rotation or vectoring), the coordinate system parameter m and the iteration level i. Figure 3.5.5 shows such a configuration. More efficient use of the hardware is achieved by a recursive configuration as in Figure 3.5.6. The start values

x_0, y_0, z_0 are loaded into registers and then altered with each iteration cycle. A recursive configuration also takes into account the various lengths of the shift series. A parallel implementation of a CORDIC element for integer shift series is shown in Figure 3.5.7. The basic elements are switchable adders/subtractors and controllable shifters. The ROM in the control path is required for the evaluation of the angle $\alpha_{m,i}$ corresponding to the increment $2^{-S(m,i)}$. Due to the small range of i and m, only a few values must be stored in the ROM. In cascade implementations, the CORDIC elements at each iteration level can be further simplified since the parameter i is predefined.

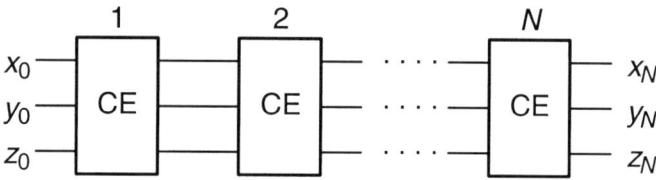

Figure 3.5.5: CORDIC implementation using cascaded CORDIC elements (CEs)

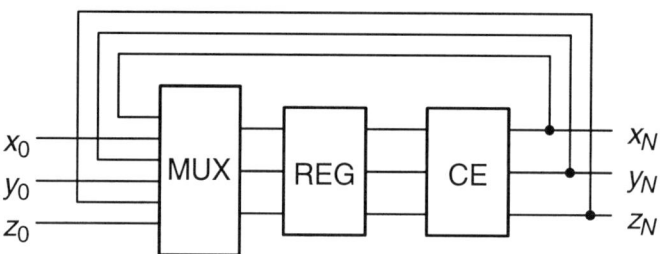

Figure 3.5.6: Recursive CORDIC implementation

A particular implementation aspect is the accuracy of the representation in terms of the number of bits for the intermediate results. Due to the shift operation, the number of bits ought to increase by $S(m,i)$ with each iteration. If no limiting effects were allowed for the intermediate results, intermediate results with a maximum of $L^2/2$ bits would have to be used for an L-bit result. Taking advantage of statistical error characteristics at each stage of iteration, the number of bits for the intermediate results can be reduced. Since the number of cycles is proportional to the number of bits required for the result,

$$L_Z = L_E + \log_2 L_E \tag{3.5.44}$$

is used as a rule of thumb, where L_E denotes the number of bits for the result and L_Z the number of bits for the intermediate results. To account for the influence of the scaling factor compensation, L_Z should be extended by an additional bit. In determining the number of bits to be used, one should note that the data in the x and y paths increase in amplitude even if they are normalized. In cyclic mode ($m = 1$), this is dictated in particular by the scaling factor k_1. The amplitude can increase by $\sqrt{2}\,k_1$. Due to the addition, a factor of 2 should be taken into account in linear mode. Due to the exponential core of the function in hyperbolic mode, overflow can become a problem. In this case, a factor of about 3 is possible. Thus to avoid overflow, an additional two bits must be used.

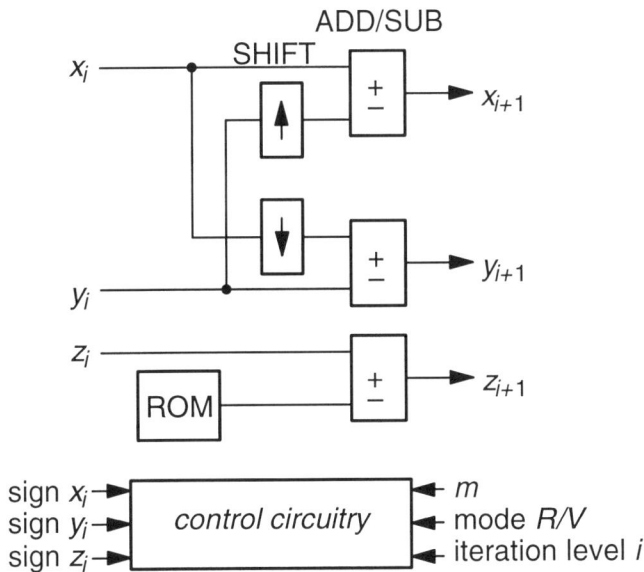

Figure 3.5.7: Implementation of a CORDIC element for singular shifts

The hardware costs of a CORDIC element consist of the transistors for implementing the registers, the shifters, the adders, the ROM and the control logic. For each bit, a register requires 16 transistors (see Figure 2.2.14). If the adder is implemented as a bit-slice, transmission gate adder, then 34 transistors are necessary per bit since an additional XOR is required to switch between addition and subtraction (Figure 3.2.20). The shifters are generally implemented as barrel shifters [15]. A barrel shifter is a special, hardwired, rectangular array of transistor switches. The number of transistor switches is the product of the word width and the maximal shift. The previous observations show that a CORDIC element requires a significant number of transistors for its implementation.

In a normal barrel shifter, the input signal must pass only one transistor in order to arrive at the desired output. In the present case of CORDIC elements, the numbers must be made smaller via a shift to the right. In order for the subsequent adder/subtractor to yield the correct result, the sign must be sufficiently extended to match the other operands. A barrel shifter for right shifts with sign extension is shown in Figure 3.5.8 [44]. For simplicity, the word width is limited in the figure to four bits and the maximal shift is limited to 3. When the numerical value is reduced, the sign bit a_3 is repeated accordingly. To reduce its size, the array comprises only n-channel transistors. The additional p-channel transistor at the output of the first driver stage serves to completely reconstruct the high level. This prevents a short-circuit current from flowing in the first driver stage in the case of the normally reduced high level.

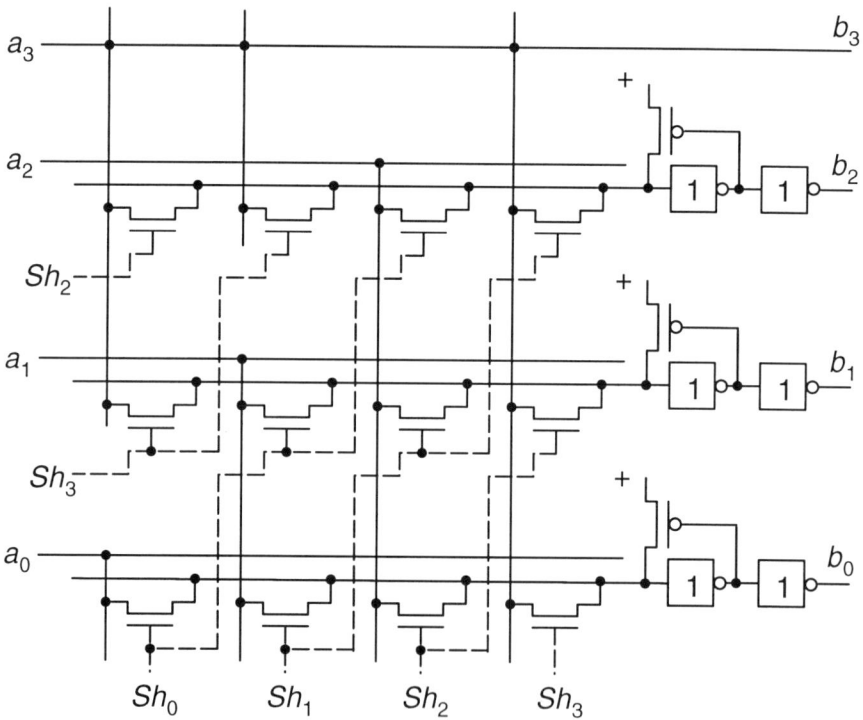

Figure 3.5.8: Barrel shifter for right shifts with sign extension

Due to its large number of transistors, a barrel shifter also has a large size. Logarithmic shifters are more efficient in terms of their costs. In this case, a shift by several bits is implemented by cascading several single shifts. Each unit can carry out either no shift or a shift by a power of 2 (Figure 3.5.9).

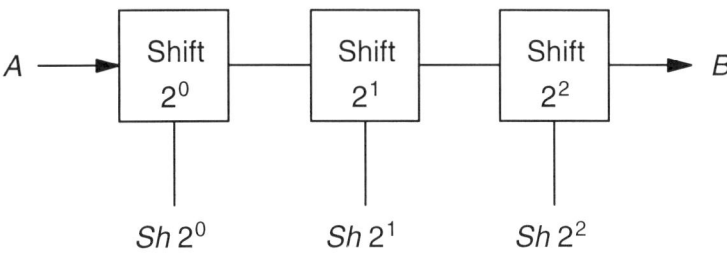

Figure 3.5.9: Logarithmic shifter

Since the number of any shift can be represented as a corresponding sum of powers of 2 (binary numbers), $\log n$ shift units must be cascaded, where n is the maximal shift.

As opposed to a barrel shifter, $\log n$ pass transistors, i.e. transmission gates, must be passed. To improve the delay characteristics, each shift unit requires a buffer at its output. A sign extension must also be implemented here at each stage. A segment of a shifter implemented using transmission gates is shown in Figure 3.5.10.

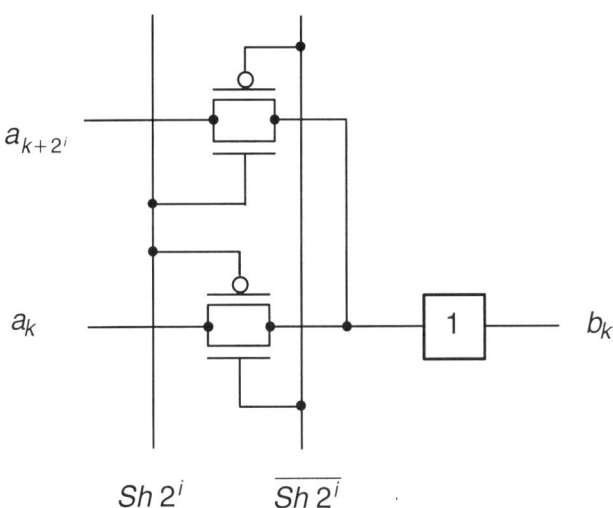

Figure 3.5.10: One-bit shift element of a logarithmic shifter

The attainable throughput is essentially dictated by the delay of the CORDIC element. For recursive CORDIC implementations, the data path delay for each iteration is given by

$$T_{D,\,CE} = T_{D,\,MUX} + T_{D,\,REG} + T_{D,\,SHIFT} + T_{D,\,ADD/SUB} \quad (3.5.45)$$

The circuit delay of a register is the set-up and hold time of a D-FF. The circuit delay for the adder can be taken from section 3.2.

The barrel shifter contributes the substantial part of the delay. Due to the sign extension, the sign is particularly problematic. In the worst case, the sign must be switched in parallel through s_m transistors, where s_m is the maximal shift. For the computation of the delay, the capacitance of s_m active transistors and $s_m^2/2$ inactive transistors must be taken into consideration at the input. At the output of the pass transistor, s_m parallel transistors (only one of which is active) contribute to the capacitance. Thus, the delay of a CORDIC element is approximately

$$T_{D,\,CE} \approx (0.65s_m^2 + 1.6s_m + 13L_z)\,\tau_L \quad (3.5.46)$$

The influence of the barrel shifter is clearly visible when large shifts s_m are to be taken into consideration. In (3.5.46), a fast-carry-chain implementation using bit-slice architecture was assumed for the adder of word width L_Z. The delay time can be reduced using logarithmic shift units. It is then approximately

$$T_{D,CE} \approx (4.4s_m + 5.4\log s_m + 13L_Z)\,\tau_L \quad (3.5.47)$$

In spite of a logarithmic number of stages, a linear component results since the capacitive loads of the sign increase exponentially for large shifts.

If CORDIC is compared with polynomial approximative implementations of functions (see section 3.5.1), it must be noted that an array multiplier including a subsequent accumulator has almost four times the delay of an adder. This neutralizes the advantage of a lower number of cycles (the number of polynomial coefficients) in many cases. The large number of attainable functions remains the particular advantage of CORDIC. For the desired accuracy, many functions in hyperbolic mode need to be approximated by rational polynomials. The division that then becomes necessary is particularly disadvantageous for the delay characteristics.

3.6 Exercises

1. An n-bit subtractor is to be designed with bit-slice structure.

   a. Determine the logic function of the subtractor cell, assuming an implementation as an addition in two's complement (see section 3.2.5).

   b. As an alternative specify the logic function of the subtractor cell, using a type 2 cell with negatively weighted carry and subtrahend bits similar to the Pezaris method.

   c. Compare the numeric representation of the n-bit difference in the implementations from a and b.

   d. Specify the generate and propagate functions for the two cases a and b.

   e. Comment on the transistor count and the circuit delay in comparison to a gate implementation like the one in Figure 3.2.3 and a transmission gate implementation such as in Figure 3.2.10.

2. The general delay characteristics of an adder are to be investigated. In Chapter 2, a general function $T_D = T_0 + FT_1$ was derived based on the circuit delay model, where T_0 is the part independent of the fan-out and T_1 is the part dependent on the fan-out.

   a. Calculate the delay components T_0 and T_1 for a gate implementation of a full adder (Figure 3.2.3). Make general statements as to what the delay components T_0 and T_1 represent and what they depend on in an arbitrary function.

   b. T_1 is initially assumed to be negligible in comparison to T_0. Specify the general delay function for a bit-slice adder as a function of the word width n. Determine the corresponding function for the carry block of a binary lookahead carry adder (see Figure 3.2.14).

   c. Now T_1 only is to be considered and T_0 suppressed. Respecify the two functions from part b.

   d. Compare the two results from parts b and c. Make some general observations about modular structures in which each module has an identical load capacitance. What happens when the load capacitances are different?

   e. Can the two partial results from b and c simply be added to calculate the total delay as a function of both T_0 and T_1 ?

3. The carry block of a lookahead carry adder is to be constructed using quarternary trees. This means that the generate and propagate signals (g_i, p_i) are to be split into 4-bit groups. Two types of functional units are to be used. The first (type α) computes the group signal (GG_j, GP_j) for generate and propagate from four sequential signals (g_i, p_i). The second (type β) takes a given block-carry function (G_k, P_k) with given (g_i, p_i) and computes the block-

carry function for three following bit levels. The hierarchical configuration of several F_α function blocks is used to evaluate the block-carry function at the positions $4^k - 1$, $k = 1, 2, ...$ In the next step, by the use of function block F_β with the available group signals (GG_j, GP_j), block-carry functions are determined at the intermediate positions $4m - 1$. In a further step all remaining block-carry functions are calculated.

Example:

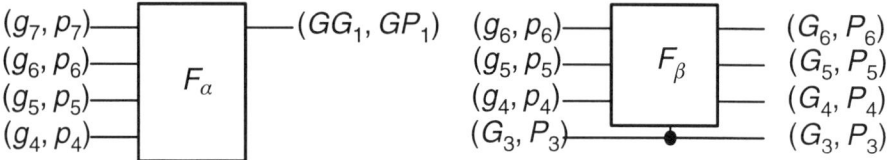

a.   Determine the logic functions F_α and F_β.

b.   Using these elements, construct a 16-bit and a 64-bit adder with tree structure analogous to the one in Figure 3.2.14.

c.   Compare the number of levels with those of a binary lookahead carry adder.

d.   Assuming a two-stage NAND–NAND implementation of the functional units, compare the transistor count and delay characteristics with a binary lookahead carry adder.

4.   A carry select adder for 12-bit operands is to be designed using adders with a word width of three bits and 2:1 multiplexers.

a.   Sketch the carry select adder.

b.   Give the logic functions of the carry select blocks.

5.   A conditional sum adder for processing 8-bit input values is to be designed.

a.   Sketch the structure of the adder and control its operation using the numbers $a = 11001110$ and $b = 11011001$.

b.   How can the adder be simplified if a set input carry is assumed?

c.   What must be observed if multiplexers with switch logic are used?

6.   The delay characteristics of ripple carry and carry save array multipliers are to be investigated. For simplicity FAs and HAs are considered identical. Furthermore, the particular influence of the load capacitances is assumed to be negligible, and

$$T_{D,FA,c'c} = T_{D,FA,c's} = T_{D,AND} = T_0$$

$$T_{D,FA,s'c} = T_{D,FA,s's} = 2T_0$$

    a.   Sketch the changes over time within the arrays of a 6×6-bit ripple carry array multiplier and use this information to generally specify the time-critical path. Show that, in general, the delay is given by $T_{D,MUL} \approx 4nT_0$. Specify operands that apply for worst-case delay.

    b.   Carry out exercise a for a 6×6-bit carry save array multiplier. In this case, the delay is given by $T_{D,MUL} \approx 3nT_0$.

7.   Design a configuration for the operation $a \cdot b + c$ in which a and b are n bits long and c is $2n$ bits long.

    a.   Extend a ripple carry array multiplier by one adder and sketch the configuration for $n = 4$.

    b.   Extend a carry save array multiplier by one adder for the same operation in such a way that only one carry propagate adder level is required. Sketch this configuration for $n = 4$.

    c.   Estimate the circuit delay of both configurations when only full adders are used. The delay of a full adder along the addition path is given by T_s. The delay from the carry input is half as large. The delay of the AND gates is negligible.

8.   Hierarchical operand reduction according to the Dadda method is to be carried out on extended counter modules. Counter modules are assumed that convert 2, 3, 4, ... 7 operand bits into a binary number.

    a.   Specify the logic function of the counter module for 4 and 5 bits of input.

    b.   Sketch the hierarchical operand reduction for the 12×12-bit multiplication as in Figure 3.3.9.

    c.   Evaluate the required number of levels of hierarchy for n operands as in Equation (3.3.14).

9.   The operation of a Booth array multiplier (see Figure 3.3.21) is to be investigated.

    a.   An 8×8-bit multiplier is assumed. Specify the values of all inputs and outputs of the sub-modules for the numbers $A = 01001100$ and $B = 10010011$.

    b.   The array multiplier shown in Figure 3.3.21 is based on a modified Booth algorithm that evaluates 3-bit groups with one bit of overlap. Design a modified Booth algorithm that evaluates 4-bit groups with one bit of overlap. Sketch the corresponding array multiplier for a 9×9-bit multiplication.

10.   Alternative structures to Pezaris array multipliers are to be examined.

    a.   Evaluate the input and output signals of the cells of a 5×5-bit Pezaris array multiplier with four types of full adders (see Figure 3.3.17) for the numbers $A = 11011 = (-5)_{10}$ and $B = 11001 = (-7)_{10}$.

b.   Through transformation of the given cells, derive a modified Pezaris array multiplier that consists solely of full adders of types 0, 1 and 2.
Hint: The full adders that take $a_n b_j$ and $a_{n-1} b_j$, $j = 0, 1, \dots n - 1$ as input can be implemented as type 1 full adders.

c.   Through modification of the structure, a modified Pezaris array multiplier is to be derived consisting solely of full adders of type 0 and 2.
Hint: Separate positive and negative summands as in Figure 3.3.18.

11. The processes within an array divider are to be examined.

a.   Carry out the division of the two positive numbers $A = 1001001$ and $D = 111$ by hand. Specify the quotient and the remainder.

b.   What are the input and output signals of the cells of a restoring array divider (Figure 3.4.2) for the given numbers?

c.   Carry out the same exercise for a non-restoring array divider (Figure 3.4.4).

12. A CORDIC unit is to be used for the evaluation of elementary functions to an accuracy of 2^{-5}.

a.   Calculate the correction factors for positive integer shift series.

b.   Specify the individual steps in the computation of $\cos z_0$ and $\sin z_0$ for $z_0 = .10110$.

c.   Work out a CORDIC unit with the given accuracy for the correction of the scaling factor of positive integer shift series.

d.   Describe its operation during the calculation of e^{z_0} and $\sqrt{w}$.

4 Measures for Increasing Performance

To achieve reliable and fully functional circuits, digital systems are synchronously clocked. This enables the correct interaction of the data path and the control circuitry. Each data transfer of results from sub-circuits into storage registers occurs at a well defined point in time. All operands of a function block are simultaneously made available. In word oriented processing, this also holds for the individual bits. The advantages of the synchronous operation are its functional reliability and the simplification of design. Two disadvantages will also be noted. In large systems, care must be taken that the clock signal is simultaneously available for all sub-circuits. The clock skew must be compensated for using special measures. A further disadvantage is that the clock rate must be fitted to the delay characteristics of the slowest sub-circuit. Propagation delay compensation between slower and faster sub-circuits is not possible. The slowest sub-circuit dictates the data throughput rate.

In the following section, architectural measures for increasing the throughput are presented, whereby pipelining is of particular importance. In the second section, criteria for evaluating these architectural measures are discussed.

4.1 Parallel Processing and Pipelining

The delay time of the slowest sub-circuit is decisive in specifying the throughput of a system. The storage register also contributes to the delay. Figure 4.1.1 will be used for explaining this. It shows a segment out of a larger synchronous circuit. The combinational logic unit F_i is connected to two dynamic D-FFs. The input D-FF supplies sequential, clock-synchronous data to the sub-circuit. The results of the sub-circuit are accepted by the output D-FF sequentially and clock-synchronously, as with the input data, and are then passed on. A dynamic D-FF is a simplification of the quasi-static D-FFs of Figure 2.2.14. This simplification is an appropriate solution for continuously clocked MOS circuits with a clock frequency in the MHz range. A non-overlapping, two-phase clock is assumed. Such clocking is particularly safe and is often used within integrated circuits [14], [15], [16], [17]. The clock period must fulfil the following requirement:

$$T_{CLK} \geq T_{D,\Phi_2} + T_{D,F_i} + T_{D,\Phi_1} + T_{\Phi_1\Phi_2} \qquad (4.1.1)$$

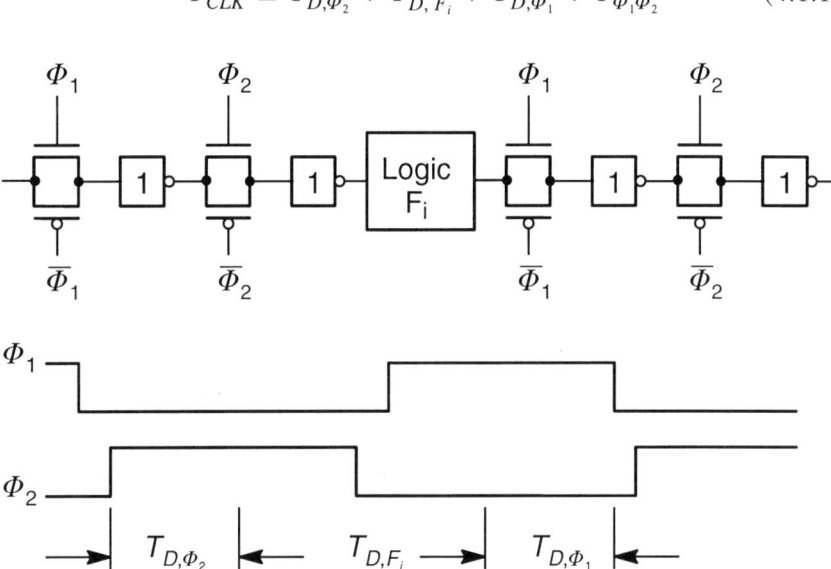

Figure 4.1.1:   Control and data transfer for a logic block F_i using dynamic D-FFs. Block diagram and timing

Here, T_{D,Φ_2} is the delay time of a piece of data between the transmission gate clocked by Φ_2 and the input of the sub-circuit F_i. T_{D,F_i} is the delay time of the sub-circuit F_i. T_{D,Φ_1} is the delay time between the output of F_i and the input of the inverter behind the transmission gate clocked by Φ_1. $T_{\Phi_1\Phi_2}$ is the clock pause between Φ_1 and Φ_2. It is to be noted that the delay times T_{D,Φ_2} and T_{D,Φ_1} depend on the characteristics of the sub-circuit. The input capacitance of F_i influences T_{D,Φ_2} and the output resistance of F_i influences T_{D,Φ_1}.

The total delay resulting from the D-FFs is given by

$$T_{D,FF} = T_{D,\Phi_2} + T_{D,\Phi_1} + T_{\Phi_1\Phi_2} \qquad (4.1.2)$$

A corresponding delay can also be given for other clock systems. In single-phase clock systems with edge triggered FFs, for example, the sum of the hold and set-up time must be substituted into the equation.

In the following, a simplified representation of synchronously clocked functional units as in Figure 4.1.2 will be used. Here, the D-FFs are symbolized by a simple dot in the wiring. The delay between the input and the output of the D-FF can be described by a delay operator with delay D. In case of word oriented processing the delay operator represents a register of D-FFs.

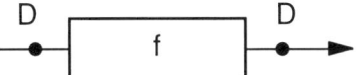

Figure 4.1.2: Simplified representation of the D-FFs (delay operator D) at the input and output of a functional unit f

The achievable throughput R_T of a system in bits per unit time is proportional to the clock rate, i.e.

$$R_T \sim \frac{1}{T_{CLK}} \tag{4.1.3}$$

On the other hand, the clock period that determines the maximal throughput is specified by the least favourable sub-circuit.

$$T_{CLK} = \max_i \left(T_{D, F_i} + T_{D, FF_i} \right) \tag{4.1.4}$$

For high throughput, small delays are essential. Modest delays can be achieved through technological measures. By shrinking the geometric structures (scaling), the capacitances can be reduced, thus achieving a reduction in the delay. The effects of such scaling are discussed in the literature [15], [16].

Besides technological measures, circuit techniques for increasing the throughput are also possible. Various circuit structures for the implementation of elementary operations were presented in Chapter 3. The alternatives shown demonstrate diverse delay characteristics. According to (4.1.4), the maximal delay of the sub-circuits is to be minimized. This means that only the slowest module must be improved. For example, in a signal processing task using multiple additions and multiplications, only the multiplier would have to be optimized in its propagation delay. This would be pointless for the adder. Thus, architectural measures for increasing the throughput are sought with which the dominance of the slowest module can be defeated.

It seems obvious that if a module cannot deliver the required through-put, several modules must carry out the processing simultaneously. Parallel processing as in Figure 4.1.3 solves this problem. The block DMUX distributes the data to be processed onto identical, parallel modules. The block MUX merges the parallel results back to one data stream. Either the data in the parallel modules is processed with a temporal offset of T_{CLK} / N or the units DMUX and MUX must be equipped with data delay units so that simultaneous processing is possible. In addition, the structure shown requires that the operations of the unit f do not depend on the results of previous operations. The throughput of a system with N paths is N times that of a simple system.

$$R_{T,N} = N \cdot R_{T,1} \qquad (4.1.5)$$

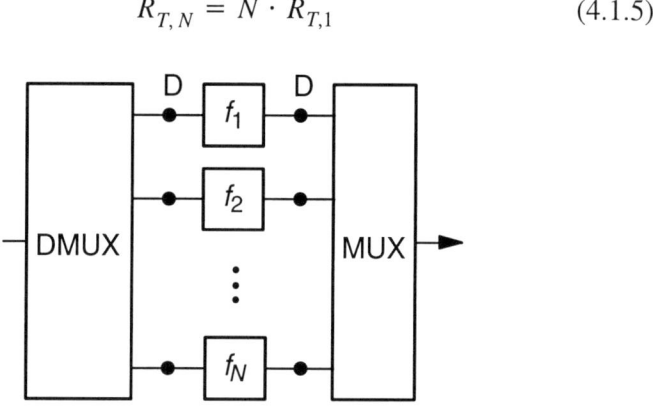

Figure 4.1.3:   Parallel implementation of a function f through N identical functional units

Parallel processing is possible not only for N identical, but also for varying sub-circuits. The data dependency of the sub-function and the evaluation of the result must be observed for each individual case. The equation

$$y = F(x) = \sum_{i=1}^{N} f_i(x) \qquad (4.1.6)$$

will be taken as a simple example. The data dependency is very simple. All sub-functions receive the same argument x, i.e. the input value x is sent to all sub-modules in parallel. The sub-modules are implemented in parallel and the total result created through addition of the partial results. Although the sub-modules work in parallel, the final, multi-operand adder is decisive for the throughput. A configuration for implementation of (4.1.6) is shown in

Figure 4.1.4. The throughput is determined by the slowest sub-module f_i and the multi-operand adder.

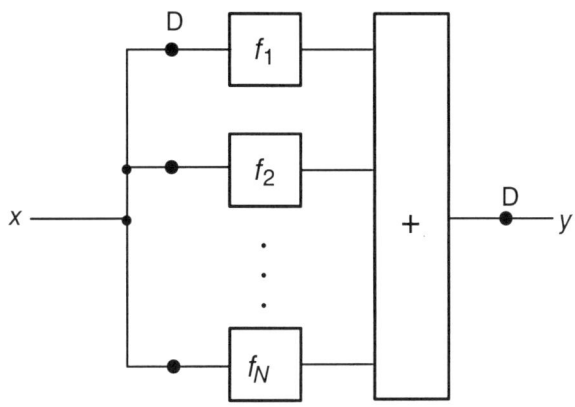

Figure 4.1.4: Parallel implementation of the function $y = \displaystyle\sum_{i=1}^{N} f_i(x)$

$$R_{T,N} \sim \frac{1}{\max\limits_{i} T_{D,f_i} + T_{D,ADD} + T_{D,FF}} \tag{4.1.7}$$

As an alternative to parallelization, in many cases it is possible to describe a function F as a series circuit of sub-functions f_i. The result of one sub-function is used as the input signal of the following sub-function.

$$F = f_N (f_{N-1} \ldots f_2 (f_1 (\cdot)) \ldots) \tag{4.1.8}$$

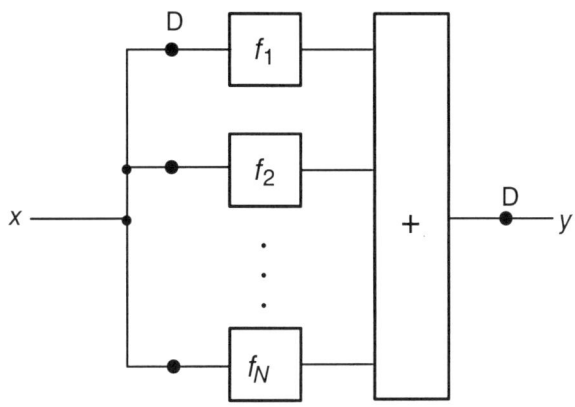

Figure 4.1.5: Implementation of a function through a pipelined cascade of sub-functions f_i

Through the insertion of synchronously clocked, intermediate storage registers, an architecture analogous to an assembly line (Figure 4.1.5) is created. This is called pipelining. While one sub-module is accepting and processing the intermediate result of its predecessor, its predecessor is simultaneously receiving new data to be processed. The slowest sub-module of the chain determines the throughput. The throughput is given by

$$R_{T,N} \sim \frac{1}{\max\limits_{i} T_{D,f_i} + T_{D,FF}} \qquad (4.1.9)$$

If all sub-modules f_i are identical,

$$R_{T,N} = R_{T,1} \cdot N \frac{T_{D,f} + T_{D,FF}}{T_{D,f} + N T_{D,FF}} \qquad (4.1.10)$$

This means that, in particular for large N, due to delay in the registers, the throughput no longer increases in proportion to N. Methods for extracting pipeline architectures will now be further considered. For the following observations, the temporal behaviour of synchronously clocked logic blocks will be abstractly formulated. If the clock rate is chosen correctly, the delay of the logic blocks will always be smaller than one clock period minus the transfer times of the flip-flops (4.1.1). Seen as a unit, this configuration seems to take the input signal of the logic block and convert it to the result at the output of the data transfer flip-flop within one clock period. A similar behaviour is described if one regards the logic block's computation as being instantaneous and assigns the total delay of one clock period to the data transfer flip-flop. It follows that the function of a D-FF can be treated abstractly as a delay operator.

Given that $D[\cdot]$ is a delay operator with a delay of one clock period T,

$$D\,[f(t)] = f(t - T) \qquad (4.1.11)$$

Repeated application of the delay operator is described by

$$D^n\,[f(t)] = f(t - nT) \qquad (4.1.12)$$

Note that this delay operator is defined in the time domain. In the signal processing literature, such delay operators are usually described using the Z-transform.

$$f(t) \quad \bullet\!\!-\!\!\circ \quad F(z)$$
$$f(t - n) \quad \bullet\!\!-\!\!\circ \quad z^{-n}\,F(z) \qquad (4.1.13)$$

Then the clock period is normalized to 1, and the delay operation corresponds to multiplication by z^{-1}. Both ways of describing the delay operator in clock systems will be used here.

Using the defined delay operator, a logic block with two input variables x_1, x_2 is given by

$$f(t) = f(x_1(t), x_2(t))$$
$$y(t) = D[f(t)] = f(t - T) \tag{4.1.14}$$

where y is the resulting signal at the output of the transfer D-FF (Figure 4.1.6). It is obvious that the distributivity of the function operations and the delay operation holds, i.e.

$$D[f(x_1(t), x_2(t))] = f(D[x_1(t)], D[x_2(t)])$$
$$= f(x_1(t - T), x_2(t - T)) \tag{4.1.15}$$

This means that in a clocked system, the delay of the results through a D-FF has the same effect as the delay of the arguments. As shown in Figure 4.1.6, this can be described as a delay transfer from output to input and vice versa. Formally, a delay transfer can also be treated as the addition and subtraction of delay times T [62]. If, in the left portion of Figure 4.1.6, T is subtracted at the output and T is added at the input, one arrives at the right-hand portion of Figure 4.1.6.

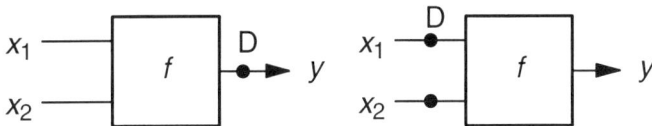

Figure 4.1.6: Delay transfer between input and output

In addition to delay transfer, time scaling is an important process in extracting pipeline architectures [63]. Through oversampling by an integer factor of n, a delay interval T is converted into several delays in a system scaled by a factor of n.

$$T = n T' \tag{4.1.16}$$

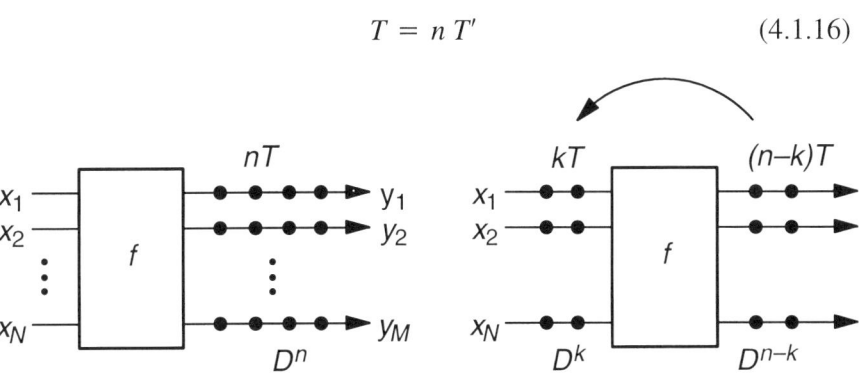

Figure 4.1.7: Generalized delay transfer, where D^n is the n-fold delay operator, and nT is the corresponding delay

However, this also means that a D-FF in the original system can be converted into n D-FFs in a system with n-fold oversampling. As shown in Figure 4.1.7, both delay transfer and scaling can be combined.

In Figures 4.1.6 and 4.1.7, the functional module was characterized by a uniform block. This block, as implied in Figure 4.1.8, can comprise a network of several sub-function blocks (processing elements).

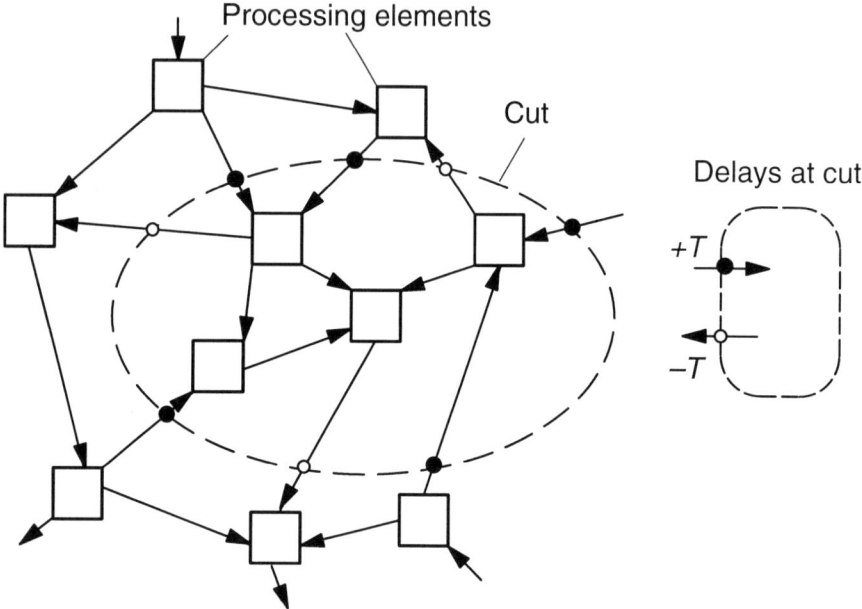

Figure 4.1.8:   Representation of the cut–set method. Neutral insertion of delays along the boundary of a cut–set of processing elements

A cut separates a set of processing elements. Delay transfers can now be applied such that all paths leading to the set receive an additional positive delay ($+T$) and all paths leading from the set are given an additional negative delay ($-T$). This is called the cut–set method. In spite of the delay transfers, the cut–set method leads to a system that exhibits the same external behaviour [63]. Negative delays do not exist. Thus, as a rule, the negative delays must be compensated for by delays within the processor elements. Negative delays at the inputs and outputs of a system can be treated specially. A delay ($-nT$) at an input can be implemented through data input that has been advanced by n clock cycles. A delay ($-nT$) at an output can be ignored. This leads to a delay of n clock cycles. In the end, ignoring negative delays at the input and output leads to an increase in the latent period of the system. This correlation is shown in Figure 4.1.9.

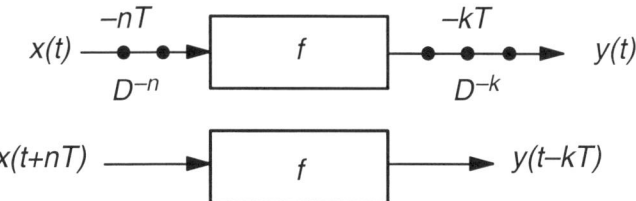

Figure 4.1.9: Elimination of negative delays at the inputs and outputs of logic blocks

Often, the delay transfer is combined with time scaling in the cut–set method [63]. In this case, fractions of T, for instance kT' where $0 < kT' < T$, are added or subtracted instead of $+T$ or $-T$. Such systems must be operated in an over-sampling mode. The cut–set method described above will now be applied to two examples.

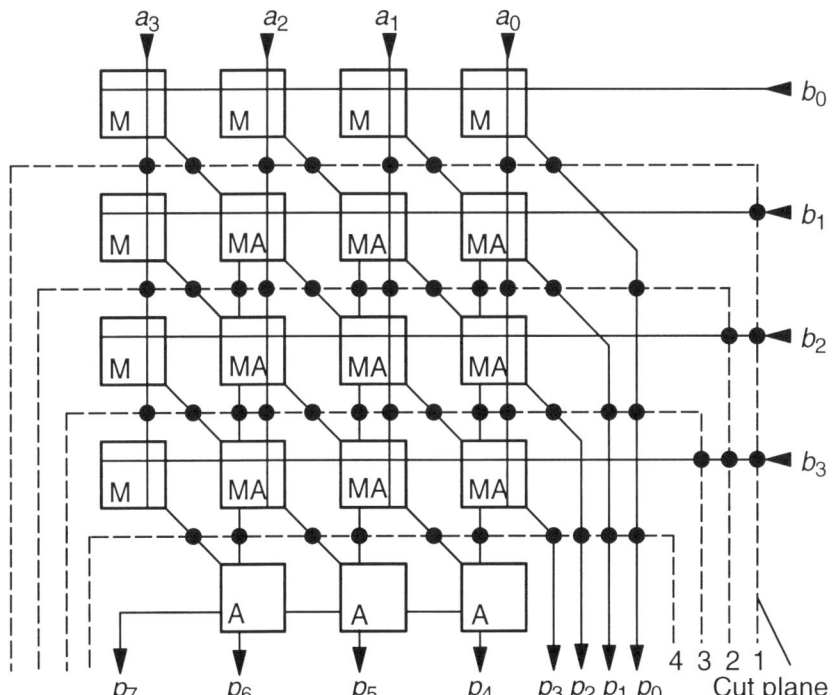

Figure 4.1.10: Creation of pipeline stages (for a 4×4-bit carry save array multiplier in this example) using the cut–set method

The cut–set method is shown in Figure 4.1.10 using a 4×4-bit carry save array multiplier as an example. As in Figure 3.3.6, the array is rectangularly structured. The processor cells of type M are 1×1-bit multipliers and those of type A are adder cells (full and half adders). The processor cells of type MA are a combination of both. The cut paths are chosen such that the array is cut horizontally and that the paths close along the output. All signals that cross the cut path within the array have the same direction. The D-FFs are inserted at these crossings. The product signals at the output are delayed in accordance with the number of cut paths. It should be noted that the operand bits are also delayed by D-FFs.

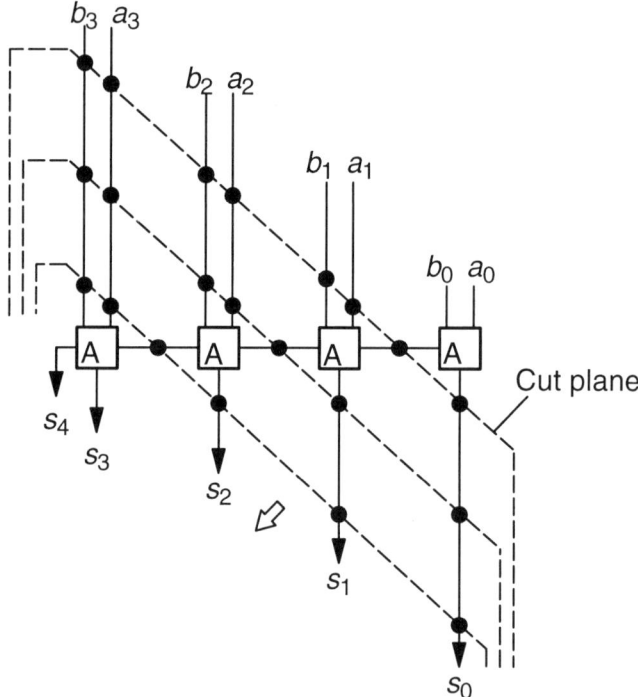

Figure 4.1.11:   Creation of pipeline stages (for a 4-bit carry ripple adder in this example) using the cut–set method

Figure 4.1.11 shows how D-FFs for pipelining are inserted into a carry ripple adder. The time-critical path runs along the carry path. Thus, the FFs must be inserted here. However, the cut–set method also shows that the input operands and the sum bits must be delayed. In order for the carry bits and the operand bits to arrive at an adder cell simultaneously, appropriate delaying of the operand bits is necessary (pre-skewing). In order for all sum bits to be simultaneously transferred to the next unit, early computed sum bits must be delayed (de-skewing). The cut–set method distinguishes itself by elegantly and automatically evaluating the delays necessary for the operand and sum bits.

One particular variant of pipelining is interleaving. It can always be applied when several parallel data streams exist that are processed identically. Typical examples are the data of parallel, identical sensors such as video sensors or even microphones. Video signal components such as (R,G,B) (red, green, blue) or (Y, C_R, C_B) (luminance, chrominance) as well as stereo signal components for video and audio signals also belong to this group.

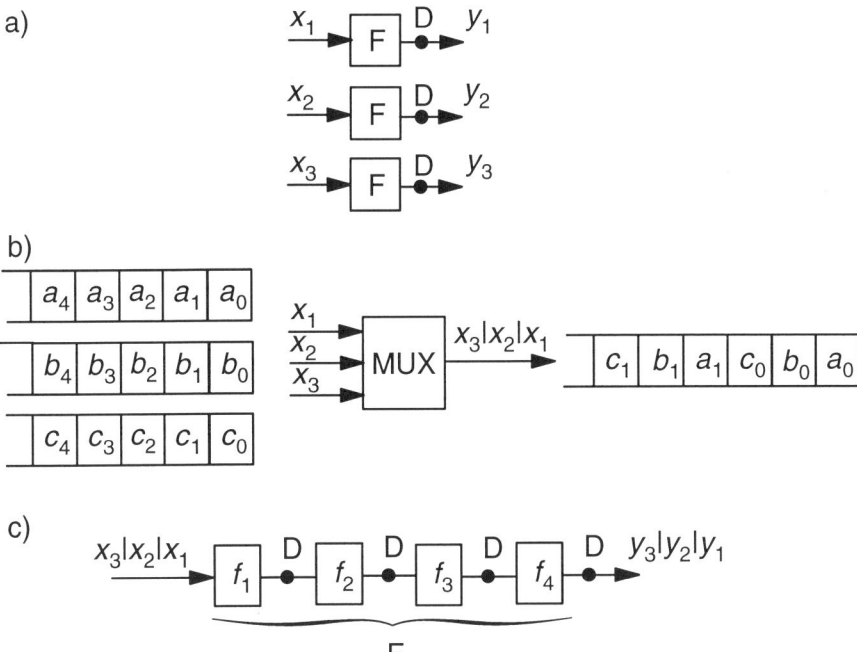

Figure 4.1.12: Interleaving: (a) parallel processing of the data streams; (b) multiplexing; (c) interleaved processing

During interleaving, the parallel data are mixed to one data stream (multiplexed), and this new data stream is fed into a processing unit. The multiplexing can be carried out based on the samples (sample interleaving) or based on the data blocks (block interleaving). The interleaving of three data streams is shown in Figure 4.1.12, whereby pipelining is simultaneously carried out in the processing unit. The number of pipeline stages can be chosen independently of the number of data streams. In the case of interleaving, the throughput is

$$R_{T,\,(K,N)} = R_{T,1} \cdot \frac{N}{K} \qquad (4.1.17)$$

where K is the number of parallel data streams and N is the number of pipeline stages. For simplicity, the influence of the pipeline registers has been ignored in this equation. Otherwise, this equation would have to be extended in a manner similar to (4.1.10).

4.2   Assessment of the Architecture Alternatives

Various measures can be chosen for the assessment of architecture alternatives. Throughput, computational performance, the computable classes of algorithms and flexibility can all be used as a measure of performance. It makes sense that a component that supports a larger number of algorithms is to be more highly valued than one that is only implemented for a specific algorithm. The spectrum of filter blocks is a good example of this. A dedicated filter block for set filter coefficients and a set number of filter taps offers absolutely no flexibility. A block with loadable filter coefficients offers higher flexibility. The flexibility is also increased via a variable number of filter taps.

Besides the measures of performance, measures of cost such as manufacturing costs, development costs and operation costs are essential for assessment. Material costs and processing costs belong to the manufacturing costs. In the large scale production of semiconductor components, the number of chips, the area of silicon and the packaging of the individual chips contribute significantly to the manufacturing costs. The production costs are also influenced by the yield of the semiconductor process and the test costs. Not only the work hours of the engineers and the computation time of the workstations, but also the expenditures for CAD tools must be accounted for in the development costs. To reduce the development costs of semiconductor components, special semi-custom development processes have been introduced, such as standard-cell designs and gate arrays with predeveloped cells or prefabricated structures. Due to the particular design stipulations, automatic design tools can be used to a large degree in semi-custom design processes. The development and production costs are generally not independent of one another. For large numbers of chips, the manufacturing cost of semi-custom components is higher than that of full custom components. The development costs are also influenced through the regularity and modularity of an architecture. It makes sense that the repeated use of identical modules and cells simplifies a design. The operating costs are influenced to a large degree by the power dissipation.

In addition to the measures of performance and cost, there are other measures of quality that are not covered by the above. Reliability, lifespan and sensitivity to external influences (electromagnetic, thermal and chemical influences) belong to this category.

Of particular interest are signal processing implementations with high efficiency, i.e. in which one achieves high performance at low cost. The quotient of performance to cost can be used in this sense for formulating the efficiency.

$$\eta = \frac{\text{Performance}}{\text{Cost}} \qquad (4.2.1)$$

Note that the measure of efficiency chosen here differs from that used in the literature for multiprocessors. There, the quotient of the processing time for a single processor to the n-fold processing time of a multiprocessor with n processors is created.

However, the measure of efficiency chosen here makes it necessary to summarize the large number of influencing parameters into singular measures for performance and cost. This is extremely difficult due to the high complexity of the relationships. The assessment of alternative architectures often takes place in the early stages of development. Many parameters for the cost function are not yet known at this stage of development. Thus to simplify the process, the number of parameters to be considered is limited to a few significant parameters. The architecture efficiency is often computed from the quotient of throughput to silicon area. In this case, the throughput is the measure of performance and the chip size is the measure of cost. This measure of architecture efficiency is discussed further in the following section. Due to the particular importance of power dissipation, ways of influencing the power dissipation such as parallel processing, pipelining and voltage scaling are explored in an additional section. The quotient of throughput to power dissipation is used there as an assessment criterion. General approaches for assessment with more than two criteria (multicriteria) are presented in a final section.

4.2.1 Architecture efficiency

Increasing the performance is the goal of architecture optimization. Throughput is used as a measure for the performance. Measuring the throughput is often problematic. In the course of the signal processing, for instance, the amount of parallel data and the word widths increase such that a special interface must be used as a common reference point for the comparison of architectures. The data input rate or the result rate are obvious choices. In many cases of signal processing, the computational power R_C in operations per unit of time is used as a measure of the performance instead of the throughput R_T. However, it is to be noted that the underlying, fixed word width must be regarded when computing the computational power. It is obvious, for example, that 8-bit and 32-bit operations cannot be considered equal in terms of their performance.

In synchronously clocked systems, the clock period T_{CLK} is a reference for evaluating both the performance and the throughput. The performance in terms of the computational rate is given by

$$R_C = \frac{n_{OP}}{T_{CLK}} \qquad (4.2.2)$$

where n_{OP} is the number of operations carried out in the time interval T_{CLK}. The throughput is given by

$$R_T = \frac{n_S}{T_{CLK}} \qquad (4.2.3)$$

where n_S is the number of samples input or output in parallel in the time interval T_{CLK}. The throughput is then given in samples per second. For specification in bits per second, this value must be multiplied by the number of bits per sample. Due to the common reference value T_{CLK}, computational rate and throughput rate are often proportional.

$$R_C = \frac{n_{OP}}{n_S} R_T \qquad (4.2.4)$$

The factor of proportionality between the computational rate R_C and the throughput R_T is the number of operations per sample. The relationship between (throughput, performance) and chip size is used to measure the efficiency of an architecture [64], [14]. In integrated circuits, the chip size represents a measure of costs. According to (2.3.1), on the other hand, the chip size is roughly proportional to the transistor count. From the previous statement, the efficiency of an architecture as given by (4.2.1) is defined as follows:

$$\eta_T = \frac{R_T}{A_{si}}$$

$$\eta_C = \frac{R_C}{A_{si}} \qquad (4.2.5)$$

As long as proportionality according to (4.2.4) holds, optimization of the efficiencies η_T and η_C leads to the same solution. In these cases, differentiation is unnecessary. From (4.2.2), (4.2.3) and (4.2.5), it follows that

$$\eta \sim \frac{1}{A_{si} \, T_{CLK}} \qquad (4.2.6)$$

This leads to the commonly used AT product [64], [14]. The smaller the AT product, the larger the efficiency. Comparisons of the efficiency of some basic structures will now be carried out. To simplify the representation, the index CLK for the clock period will be left out whenever it is unambiguous.

Firstly, the parallel implementation of identical logic modules as in Figure 4.1.3 will be compared for various parallelisms N. In the following, the indexing is so chosen that index 1 applies to the original system with parallelism 1 and the index N specifies the N-fold parallelism. In accordance with (4.1.5), the throughput increases in proportion to N. The additional area A_Z for data distribution and data merging must be taken into account for the chip size. Under the assumption that the additional area is proportional to the degree of parallelism exceeding 1,

$$A_N = NA_1 + (N - 1)\, A_Z$$

$$= A_1 \left[N + (N - 1)\frac{A_Z}{A_1} \right] \qquad (4.2.7)$$

Through combination of the two equations (4.1.5) and (4.2.7), it follows that the efficiency in terms of the parallelism N is

$$\eta_N = \eta_1 \frac{1}{1 + \left(1 - \frac{1}{N}\right)\frac{A_Z}{A_1}} \qquad (4.2.8)$$

Thus, in the case where additional hardware is necessary for data distribution and merging, the efficiency is not improved through parallel processing, but actually worsened.

In a pipeline structure as in Figure 4.1.5, the influence of the pipeline registers on both the delay time and the chip size must be taken into consideration. If the logic is split into sub-functions such that each has the same delay, the essential time for calculating the throughput is

$$
\begin{aligned}
T_N &= \frac{T_1 - T_{D,REG}}{N} + T_{D,REG} \\
&= \frac{T_1}{N}\left[1 + (N-1)\frac{T_{D,REG}}{T_1}\right]
\end{aligned}
\tag{4.2.9}
$$

In this case, index 1 symbolizes an implementation with one final register; the index N is for an implementation with N pipeline stages. The chip size for N pipeline stages is increased through the additional pipeline registers that are necessary. It follows that

$$
\begin{aligned}
A_N &= A_1 + (N-1)\,A_{REG} \\
&= A_1\left[1 + (N-1)\frac{A_{REG}}{A_1}\right]
\end{aligned}
\tag{4.2.10}
$$

The following equation for the efficiency is derived from the two equations (4.2.9) and (4.2.10).

$$
\eta_N = \eta_1 \frac{N}{\left[1 + (N-1)\frac{A_{REG}}{A_1}\right]\left[1 + (N-1)\frac{T_{D,REG}}{T_1}\right]}
\tag{4.2.11}
$$

The result shows that the efficiency increases by a factor of N as long as N is not too large and as long as the contributions of the pipeline registers to the size and the delay are insignificant. For very large N, the efficiency sinks nearly to zero. Taking the derivative with N yields

$$
N_{opt} = \sqrt{\frac{\left(1 - \frac{A_{REG}}{A_1}\right)\left(1 - \frac{T_{D,REG}}{T_1}\right)}{\frac{A_{REG}\,T_{D,REG}}{A_1\,T_1}}}
\tag{4.2.12}
$$

as the optimal value. In general, the contributions of a pipeline register to the size and the delay are significantly smaller than the contributions of the logic block. In this case, N_{opt} is clearly larger than 1 as is η_{Nopt}.

It cannot be inferred from the previous statements alone whether parallel processing or pipelining is to be preferred. To compare the two approaches, the throughput is represented in a diagram as a function of the chip size. Equations (4.1.5) and (4.2.7) offer the desired dependency, yet in a parametric representation with the parallelism N as the parameter. The same is true for equations (4.2.9) and (4.2.10). Elimination of the parameter N yields the desired function. This is depicted in Figure 4.2.1 for particular parameter values. The efficiency can be expressed in terms of the chip size in a similar manner. In this case, it is necessary to combine equations (4.2.7) and (4.2.8) or (4.2.10) and (4.2.11) and eliminate the parameter N. This dependency is illustrated in Figure 4.2.2 for particular parameters.

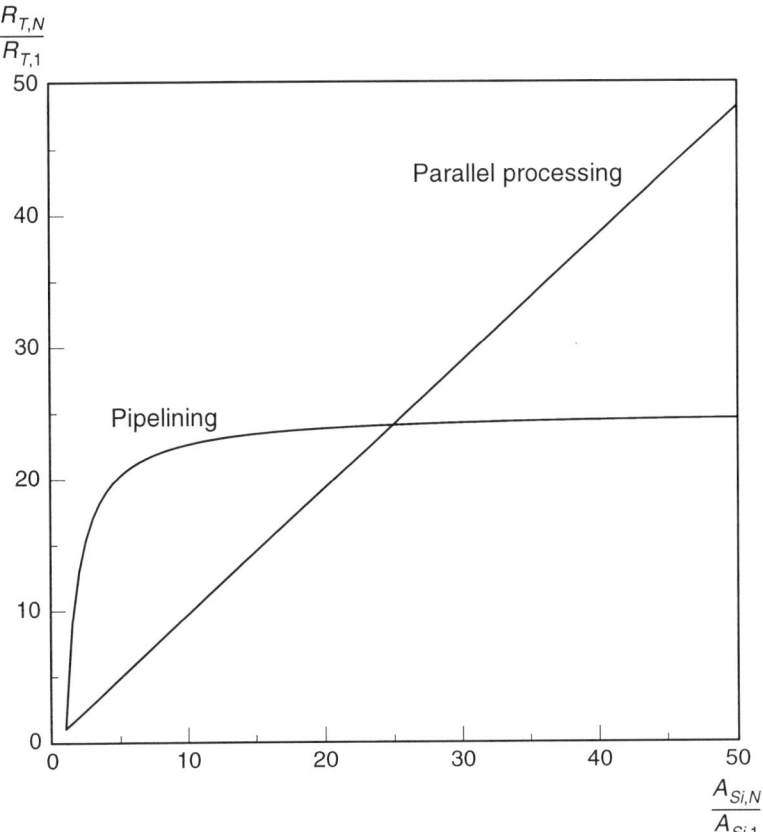

Figure 4.2.1: Throughput as a function of the chip size for parallel processing and pipelining. Parameters: $A_z/A_1 = T_{REG}/T_1 = A_{REG}/A_1 = 0.04$

From Figure 4.2.1 it can be inferred that, in parallel processing, a practically linear correspondence exists between the throughput and the chip size.

This can be explained by the fact that the throughput is increased with each additional unit accordingly. With pipelining, the throughput increases significantly at the beginning, but then shows saturation. Low additional costs for pipeline registers lead to significant increases in the throughput. For large numbers of pipeline stages, the delay of the pipeline registers dominates the throughput rate, i.e. the throughput is limited.

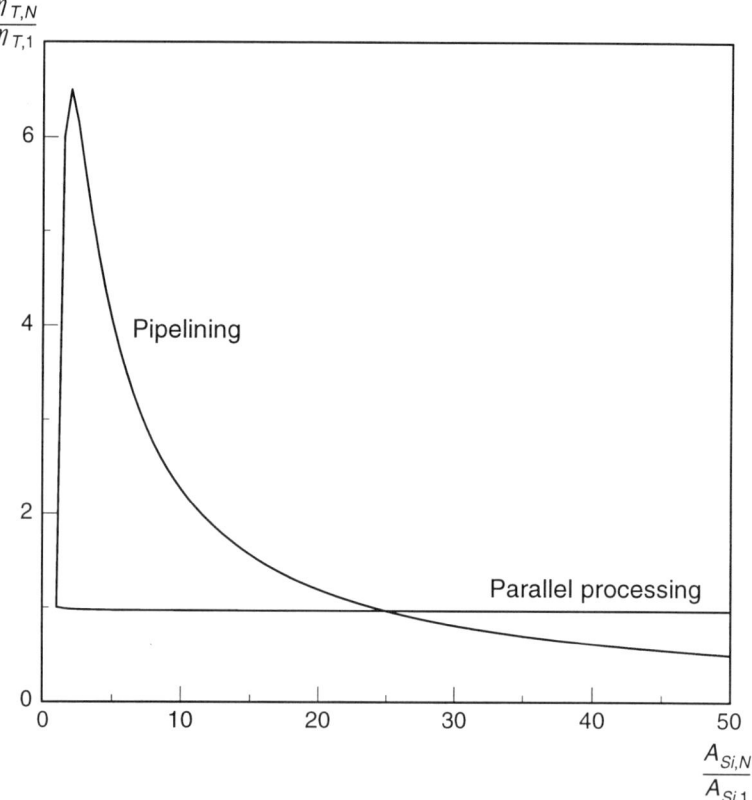

Figure 4.2.2: Efficiency as a function of the chip size for parallel processing and pipelining. Parameters: $A_z/A_1 = T_{REG}/T_1 = A_{REG}/A_1 = 0.04$

The efficiency can be inferred from the behaviour of the throughput rate as a function of the chip size. Through application of the exact differential on (4.2.5), it can be derived that

$$\frac{d\eta}{\eta} = \frac{dR_T}{R_T} - \frac{dA_{Si}}{A_{Si}} \qquad (4.2.13)$$

holds. This means that the efficiency increases as long as the relative change in the throughput is larger than the relative change in the chip size. A diagram of the efficiency for the example of Figure 4.2.1 is shown in Figure 4.2.2. Pipelining demonstrates clear increases in efficiency when the increase in chip size is small (low number of pipeline stages). For large increases in area (large number of pipeline stages), a decrease in efficiency occurs due to the saturated behaviour in the throughput. From the figure, one can also see that parallel processing cannot lead to an increase in efficiency since each increase in throughput is coupled to an equivalent increase in chip size. From this model, one can infer as a result that pipelining is to be preferred for moderate increases in throughput, whereas parallel processing is better for extreme increases in throughput. It makes sense that the best solution for very high throughput rates is obtained by combining pipelining and parallel processing.

The previous statements make the assumption that the total delay is split up equally through the insertion of pipeline stages. This is not always possible in real systems. Furthermore, additional aspects such as controller complexity and latency among others must be considered in the implementation of systems. This often changes the situation. Many references to the efficiency of parallel processing and pipelining in digital systems exist in the literature. Two representative references are [65] and [66]. The definition of efficiency used here will now be applied to several example architectures from Chapter 3.

First, a bit-serial addition (Figure 3.2.5) and a bit-parallel addition (Figure 3.2.1) will be compared. For the implementation of a full adder as in Figure 3.2.10, 24 transistors are required. If dynamic D-FFs (eight transistors) are used, a bit-serial addition requires 52 transistors and a bit-parallel addition requires $32n$ transistors. Thus, the chip size of the bit-serial addition is smaller by a factor of roughly $1.6/n$. The number 1.6 results from the additional circuitry for the bit-serial adder cell comprising 2 D-FFs and a NOR2 gate. Furthermore, it should be noted that in the case of the bit-serial addition, the slower sum path, and in the case of the bit-parallel, the most favourable carry path is decisive for the delay. Including the delay of the D-FFs, the delay of the bit-serial addition is $44.6\tau_L$ per bit. For the bit-parallel addition, a delay of $(13n + 35.6)\tau_L$ results including (3.2.19). Thus, the relationship of the efficiency between the two adder structures is

$$\eta_{ADD,ser} \approx (0.18n + 0.5)\, \eta_{ADD,par} \qquad (4.2.14)$$

Therefore, due to the more favourable efficiency, additions ought to be implemented bit-serial and not bit-parallel for $n > 3$. In practice, deviating from the conclusions about efficiency, bit-parallel implementations are preferred. One reason for this is the more complex control circuitry of bit-serial signal processing when several signal processing modules are wired together.

Whether or not hierarchical adder structures are also more efficient is of interest, too. This would mean that the additional costs of hierarchical structures are less significant than the yield from decreasing the delay. From the AT product of the results in section 3.2.4 it can be deduced that for word widths larger than 16 bits, hierarchical structures are more efficient than bit-slice structures (TG adders).

In addition, a sequential ADD/shift multiplier implementation (Figure 3.3.11) is to be compared with a Braun multiplier in terms of its efficiency. Greatly simplified, the array multiplier is n times as hardware intensive with a delay reduction of $3/n$. Thus, the sequential method ought to be more efficient. An exact analysis of the AT product shows that the sequential method is more favourable only for $n > 8$.

The result of the last observations shows that, in most cases, the fast architectures derived in sections 3.2 and 3.3 do not represent very efficient solutions in their structure as shown. They have only been optimized in terms of their delay. However, due to the general observations about pipelining (4.2.11), an improvement in efficiency can often be achieved through pipelining. Carry-ripple adders (Figure 4.1.11) are to be examined first.

For an adder without pipelining,

$$
\begin{aligned}
T_1 &= n\, T_{D,FA} + \Delta T_{D,FA} + T_{D,FF} \\
A_1 &= n\, (A_{FA} + A_{FF})
\end{aligned}
\qquad (4.2.15)
$$

In the first equation, $T_{D,FA}$ is the delay of a full adder along its carry path, $\Delta T_{D,FA}$ the portion of the delay independent of n of an n-bit adder (see (3.2.19)) and $T_{D,FF}$ the delay of a D-FF. The delay $\Delta T_{D,FA}$ results from the difference $T_{D,FA,ac} - T_{D,FA,cc}$. The variables A_{FA} and A_{FF} represent the corresponding silicon areas.

It is assumed that n is a power of 2 and that pipeline stages are introduced through continuous halving of the carry path as in Figure 4.1.11. The splitting of the adder into k segments yields

$$T_k = \frac{1}{k} T_1 + \left(1 - \frac{1}{k}\right)\left(\Delta T_{D,FA} + T_{D,FF}\right)$$

$$A_k = A_1 + \frac{3}{2} n (k - 1) A_{FF} \qquad (4.2.16)$$

The second term represents the additional contributions of the pipeline registers. The factor of 3/2 results from the doubled number of inputs. Optimization of the AT product leads to the following best value for k.

$$k_{opt}^2 = n \frac{T_{D,FA}}{\Delta T_{D,FA} + T_{D,FF}} \cdot \frac{2 A_{FA} - A_{FF}}{3 A_{FF}} \qquad (4.2.17)$$

Substitution of the transistor count and use of the delays evaluated from the delay model yields the approximation

$$k_{opt} \approx 0.8 \sqrt{n} \qquad (4.2.18)$$

This means that highest efficiency lies clearly below complete pipelining due to the contributions from the pipeline registers. It follows that complete pipelining is only desirable when the maximally achievable throughput is the goal.

The carry-save array multiplier (Figure 4.1.10) will now be examined as a more complex example. According to (4.1.4), the throughput is determined by the maximal delay between two D-FFs. In the case of the example in Figure 4.1.10, the final carry-ripple adder dominates the delay and thus the throughput. Since the carry-ripple adder alone causes nearly a third of the delay of a carry-save array multiplier, the array should be split by two cut-paths into three areas with roughly equal delays. With two pipeline stages, the contributions of the pipeline registers are secondary, and the efficiency is nearly $3\eta_1$. Further improvement is only possible if pipeline registers are introduced into both the carry-save adder array and the final carry-ripple adder. Thus, as shown in Figure 4.1.11, D-FFs must be included in the carry-ripple adder for pre-skewing and de-skewing. The efficiency is in fact dampened by the additional register costs, yet, in the case of this multiplier, an additional boost in efficiency can be achieved through insertion of further pipeline stages. When, in the end, a pipeline register has been introduced into each horizontal plane of the carry-save adder array (see Figure 4.1.10), the entire multiplier contains $1.5n$ pipeline stages – n stages in the carry save array and $0.5n$ stages in the final carry-ripple adder. The costs in terms of area for all the D-FFs are then of the same order as the costs for the multiplier and adder cells, i.e.

$$A_k \approx 2A_1 \qquad k = 1.5n \qquad (4.2.19)$$

The delay of a dynamic D-FF is roughly half as large as the delay between the input and the sum output of a full adder. The delay of the whole multiplier (T_1) is approximately $3n$ times the delay $T_{D,FA,cc}$ of a full adder along its carry path. The delay $T_{D,FA,as}$ of a full adder is three times the delay of $T_{D,FA,cc}$ plus the register delay $T_{D,FF}$. Thus

$$T_k \approx \frac{3}{3n}T_1 = \frac{T_1}{n} \qquad k = 1.5n \qquad (4.2.20)$$

It follows that the efficiency of a multiplier with $k = 1.5n$ pipeline stages is

$$\eta_k = \frac{n}{2}\eta_1 \qquad k = 1.5n \qquad (4.2.21)$$

In comparison to the linear increase, the efficiency has already been dampened by a factor of 1/3. The example shows that the entire configuration must be examined for the efficiency. On its own, a carry-ripple adder already exceeds the maximum efficiency when a few pipeline stages have been introduced. In connection with the multiplier, the maximal efficiency is not reached until complete pipelining is done.

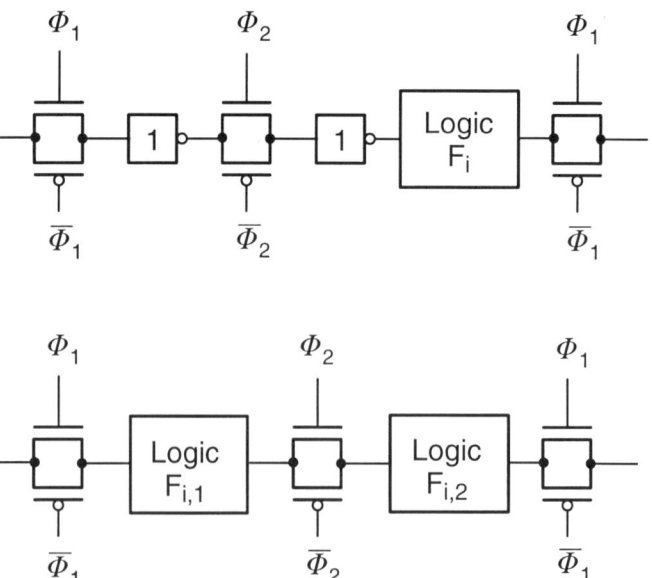

Figure 4.2.3:   Reduction of the clocking costs by splitting up of a logic block

The smaller the contributions of the pipeline registers are to the chip size and the delay, the larger the gains in efficiency through pipelining. It follows that by changing from quasi-static D-FFs (Figure 2.2.14) with 16 transistors to dynamic D-FFs (Figure 4.1.1) with eight transistors, the pipelining becomes more efficient. Through the splitting up of combinatoric logic (Figure 4.2.3), it can be shown that, in the end, the D-FF can be reduced to two transmission gates [15]. Pipeline stages reduced in this manner have clear advantages in terms of area and delay.

One particular aspect of extracting cut paths still needs to be discussed. The maximal delay between two cut paths determines the throughput. This delay is lessened when the cut paths are also equitemporal planes of the array. In the case of the carry-save array multiplier, the horizontal cut path was also an equitemporal plane of the delay and thus favourable. This does not hold for the final carry-ripple adder, for which reason the cut paths must be re-routed. A horizontal cut path is inappropriate for a carry-ripple array multiplier since the planes of constant delay run diagonally through the array. Figure 4.2.4 exemplifies this. The maximal delay here between two planes is larger than the maximal delay divided by the number of cut paths. In the picture, the maximal delay without pipelining is 5τ. After introduction of the two pipeline stages, the maximal delay is 3τ. This means that the pipelining does not yield the expected efficiency in this case. If, in a carry-ripple array multiplier, for example, $n-1$ horizontally aligned pipeline stages are introduced, the throughput is nonetheless only raised by a factor of approximately 4. Here, diagonal cuts corresponding to equitemporal planes are more efficient.

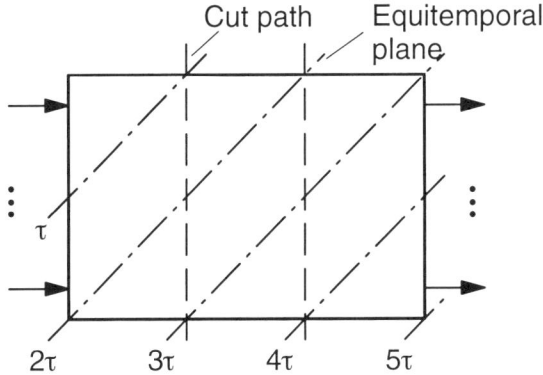

Figure 4.2.4:   An example with skewed equitemporal planes and cut paths for the introduction of pipeline registers

4.2.2   A comparison of power dissipation

Power consumption is an important aspect of densely integrated circuits. In portable, battery driven equipment, long operating times are achieved through low power consumption. But even for stationary, plug-in equipment, the power required by the system should be as low as possible so as to hold down the size and price of the power supply. This also allows the dimensions of the on-chip supply lines to be reduced. An additional problem is the dissipation of the heat produced by the IC. Heat sinks and fans are employed to improve heat dissipation. Such measures increase the production costs of such systems. Furthermore, higher temperatures promote many sources of errors within ICs.

The results of section 2.5 show that the power dissipation depends on the supply voltage V_{DD}, the capacitances C_i, the clock frequency f_{CLK} and the transition activity factor α_i. Since these variables also influence the performance characteristics of an architecture, the quotient of throughput to power dissipation will be used to judge architectural measures.

$$\zeta = \frac{R_T}{P} \qquad (4.2.22)$$

In the sense of (4.2.1), the power dissipation will be interpreted here as a cost factor. To start with, the influence of parallel processing and pipelining on the proportion ζ will be surveyed. It will be first assumed that the supply voltage is unchanged.

As in Figure 4.1.3, a parallel implementation of identical functional units is assumed. In accordance with (4.1.5), the throughput increases proportional to the number of parallel units. An equation holds for the power dissipation that is similar to the one for the chip size (4.2.7).

$$P_N = NP_1 + (N - 1)P_Z \qquad (4.2.23)$$

The first term directly results from the assumption of N identical and parallel units, driven at the original clock rate. Since the capacitances are roughly proportional to the area, additional power dissipation for the multiplex and demultiplex circuitry results that is proportional to the degree of parallelism exceeding 1. The following holds for the quotient ζ accordingly:

$$\zeta_N = \zeta_1 \frac{1}{1 + (1 - \frac{1}{N})\frac{P_z}{P_1}} \qquad (4.2.24)$$

This means that the power dissipation increases somewhat faster than the throughput and that the quotient ζ decreases only slightly. A change in the clock rate influences the throughput and the power dissipation to the same degree, so that would not alter the quotient ζ.

In the case of ideal pipelining with division into identical subfunctions, equations (4.2.9) and (4.2.10) form the basis for further analysis. The critical time T_N determines the clock rate and influences the throughput and the power dissipation to the same degree. A further portion of the increase in power dissipation is brought about by the capacitances in the additional pipeline registers. An equation similar to (4.2.24) follows, namely

$$\zeta_N = \zeta_1 \frac{1}{1 + (1 - \frac{1}{N})\frac{P_{REG}}{P_1}} \qquad (4.2.25)$$

The above equation ignores the fact that the logical depth between two clocked registers is changed through the pipelining and that the probability of glitching is reduced accordingly. The word glitching describes short term charge/discharge effects that result from non-uniform propagation times (hazards) in networked combinatoric logic. The occurrence of such a hazard is illustrated in Figure 4.2.5.

The number of transitions of complex circuits can be split into those logically needed because of change of arguments and those caused by delay hazards within the composite circuit. In case of a carry-ripple adder the power dissipation caused by glitches is in the order of 22% [67]. For complex circuits like multipliers the possibility of glitches increases with the depth of the circuit. By insertion of pipeline registers the logical depth is reduced and thus the probability of glitches. The evaluation of the power dissipation resulting from glitching requires complex simulations [68].

In accordance with equation (2.5.10), the power dissipation is proportional to the square of the supply voltage. It thus seems sensible to reduce the power dissipation through a reduction of the supply voltage. However, the delay along the time-critical path that determines the maximal clock rate is also influenced by a voltage reduction. A reduction in the clock rate also reduces the throughput.

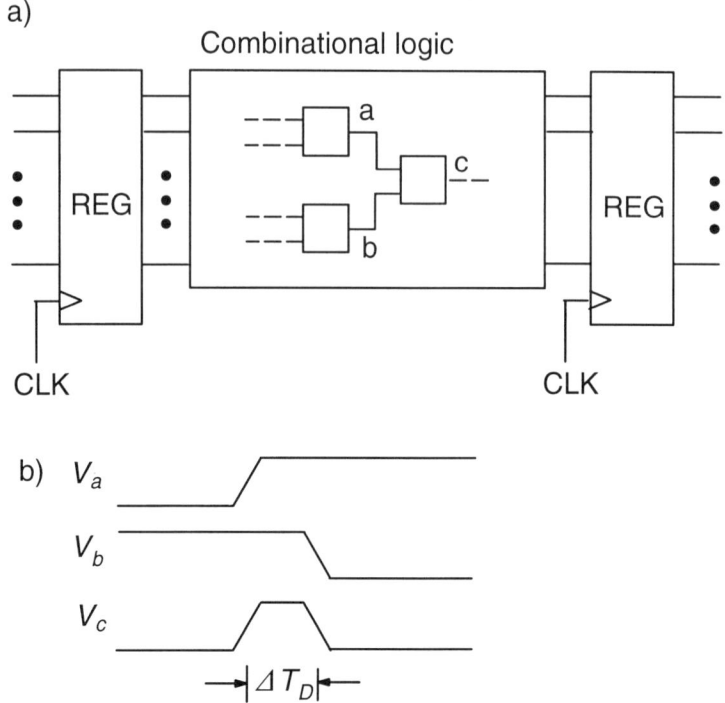

Figure 4.2.5: Hazard generation due to non-uniform propagation in combinatoric logic: (a) circuit diagram; (b) voltage waveforms with delay difference ΔT_D

The degree to which a voltage reduction alters the delay is now to be roughly evaluated. It is assumed that the effective channel resistance for evaluating the delay τ is proportional to the resistance at the point of saturation, i.e.

$$\tau \sim \frac{V_{DD}}{I_{DS,sat}} \quad \text{for} \quad V_{GS} = V_{DD} \tag{4.2.26}$$

If the voltage is now linearly reduced by a factor of γ, it follows that

$$V'_{DD} = \gamma V_{DD} \tag{4.2.27}$$

$$\frac{\tau'}{\tau} = \gamma \frac{(V_{DD} - V_T)^2}{(\gamma V_{DD} \div V_T)^2} \tag{4.2.28}$$

The latter equation is derived using (2.1.3) and (4.2.26).

The change in the delay leads to an alteration in the clock frequency that reduces both the throughput as well as the power dissipation. The power dissipation is also decreased through the voltage reduction so that only the voltage factor remains in the ratio of the quotient ζ.

$$\frac{\zeta'}{\zeta} = \frac{1}{\gamma^2} \tag{4.2.29}$$

Example 4.2.1 Possible ratios are to be approximated for specific numeric values. The supply voltage is to be reduced by a factor of 5, i.e.

$$\gamma = 0.2$$

The threshold voltage is not to sink below half the supply voltage in order to limit the increase in the leakage current.

$$V_T = 0.1 V_{DD}$$

According to (4.2.28), the ratio of the delays is given by

$$\frac{\tau'}{\tau} \approx \frac{R_T}{R_T'} \approx 16$$

In addition, the altered supply voltage must be considered in the power dissipation. Thus

$$\frac{P'}{P} = \frac{1}{400}$$

A significant reduction in power dissipation is possible with only moderate change in the throughput.

According to previous considerations, the reduction in throughput can be compensated for by pipelining. This changes the ratio of the two ζ's only insignificantly. In this manner, a reduction in power dissipation can be achieved for the same throughput.

4.2.3 Multicriteria assessment

In sections 4.2.1 and 4.2.2, for the sake of simplifying the context, the assessment was always reduced to two criteria – the ratio of throughput to area in the first section and the ratio of throughput to power dissipation in the second section. As explained in the introductory paragraphs of section 4.2, a variety of criteria exist that can be applied to the assessment of architectures. Thus, assessment functions with several criteria are sought that allow variable weighting of the individual criteria. Multiplicative and additive functions are often used [69].

A multiplicative assessment function is given in general by

$$\zeta = \prod_{i=1}^{n} \xi_i^{w_i} \qquad (4.2.30)$$

where the ξ_i's are the individual criteria and the w_i's are the corresponding weight coefficients. The criteria are characteristics with diverse physical units. To eliminate the influence of the particular units, the criteria are formulated as normalized, scalar variables. For example, the criterion for the chip size could be formulated as the quotient of the true area relative to a reference size.

$$\xi_A = \frac{A_{Si}}{A_{Si,0}} \qquad (4.2.31)$$

The values from a particular architecture, upper and lower limits or target values are reasonable reference values.

For each individual, relative criterion, the sign of the exponent should designate if its enlargement contributes to the improvement or the deterioration of the total weighting. Furthermore, one must consider if the optimum is to be extracted through minimization or maximization of the total assessment.

The exponent in (4.2.30) supports a prioritization of the criteria. The larger the exponent, the larger its contribution to the total weighting. An assessment in accordance with (4.2.5) and (4.2.22) corresponds to the approach of (4.2.30). It represents the case of two criteria with exponents 1 and -1. In the examples discussed in the previous sections, the reference values were not introduced with the individual criteria, but rather in the assessment function.

As an example, a criterion is presented based on throughput, chip size and power dissipation and whose maximization leads to the best solution.

$$\zeta = \xi_R^{w_R} \xi_A^{-w_A} \xi_P^{-w_P} \qquad (4.2.32)$$

The function demonstrates expected behaviour since increases in the throughput lead to an improvement and increases in the chip size and power dissipation lead to a depreciation of the total weighting. Furthermore, the function has the desired compensating behaviour since, in general, many criteria change simultaneously in a modified architecture.

In addition to the multiplicative approach, an additive approach is sometimes used in the literature. In general, the additive approach is given by

$$\zeta = \sum_{i=1}^{n} w_i \xi_i \qquad (4.2.33)$$

The role of the weight coefficients in this equation is analogous to that of the multiplicative approach discussed above. A particular disadvantage of the additive approach is that increases and decreases are evaluated differently for large changes. If a relative criterion is enlarged by a factor of 4, this has a stronger influence than a decrease by a factor of 1/4. If such changes are to have the same influence in the positive and negative direction, a multiplicative criterion must be used.

A disadvantage of the previous formulation of a total assessment in accordance with (4.2.30) and (4.2.33) is that limiting values and the fulfilment of desired values cannot be respected. Given a maximal throughput, it is not meaningful to weight any increase beyond this desired value as an improvement of the criterion. Also, a surplus beyond a limiting value may not be compensated for in the total weighting. This is true, for instance, for chip size when the maximal size is limited due to the particular choice of housing.

One possibility for taking limiting values and the degree of fulfilment of desired value into consideration consists of the introduction of non-linear assessment functions. These non-linear functions are designated here by q_i. Using such functions, the total assessment is given by

$$\zeta = \prod_{i=1}^{n} q_i(\xi_i) \qquad (4.2.34)$$

The general form of such assessment functions is illustrated in Figure 4.2.6. With the first criterion, improvements are achieved up to a threshold of $\xi_{U,1}$. This function leads to a maximization of the criterion up to its fulfilment above the given threshold. In the second criterion, the threshold $\xi_{U,0}$ represents an upper limit. If the limit is exceeded, the weighting drops to zero. Through the given function, a minimization of the criterion is achieved. For the third criterion, an interval $[\xi_{L,1}, \xi_{U,1}]$ for fulfilment of the criterion is given. Including the threshold regions, an interval $[\xi_{L,0}, \xi_{U,0}]$ exists with positive weighting. During maximization of the assessment function, the criterion is forced into a zone with positive weighting. In the illustration, the transitions are given as straight lines. In principle, other transition functions are also possible here. The assessment method presented here can be treated mathematically with fuzzy multicriteria analysis methods [70]. The formulation of useful criteria and their derivation from the architecture are one of the main problems.

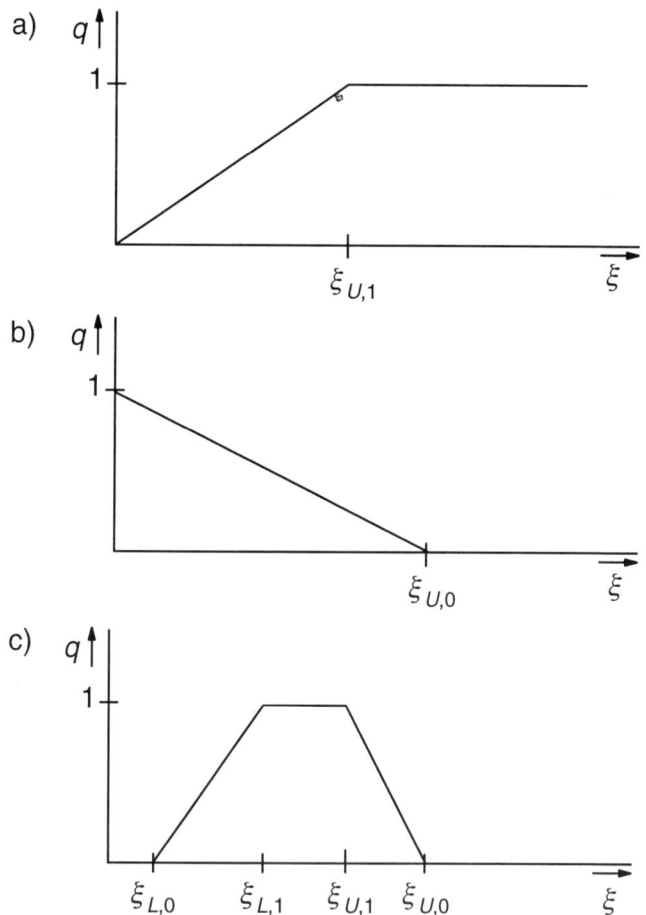

Figure 4.2.6:   Examples of non-linear assessment functions: (a) maximization with a desired value $\xi_{U,1}$; (b) minimization with a limiting value $\xi_{U,0}$; (c) tolerance range

Example 4.2.2 Using non-linear assessment functions, the pipelining of a logic unit is to be optimized. Through insertion of pipeline stages, the throughput is to be maximized and the power dissipation is to be simultaneously minimized through reduction of the supply voltage. The increase in throughput is assumed to be met when a factor of 3 has been exceeded. The power dissipation due to pipelining should be smaller than 1.5 times that of the unit without pipelining. To simplify the relationships, it is assumed that the contributions of the pipeline registers to the delay and the area are negligible in the range of interest. The decrease in the clock rate due to the voltage reduction is to be modelled by the function $1.4\gamma - 0.4$ in the interval $0.5 \leq \gamma \leq 1$. The number of pipeline stages is given by N. If the logic unit without pipelining is used as a reference, the following criteria hold for the throughput, the area and the power dissipation:

$$\xi_R = N \cdot (1.4\gamma - 0.4)$$

$$\xi_A = 1$$

$$\xi_P = N \cdot \gamma^2(1.4\gamma - 0.4)$$

Since the criterion for the area does not depend on the sought parameters N and γ, it can be dropped from further observations. From the given information, assessment functions for the throughput and power dissipation can be determined similar to the functions in Figure 4.2.6.

$$q_R = \begin{cases} \frac{1}{3}\xi_R & 0 < \xi_R \leq 3 \\ 1 & 3 < \xi_R \end{cases}$$

$$q_P = \begin{cases} 1 - \frac{2}{3}\xi_P & 0 < \xi_P \leq \frac{3}{2} \\ 0 & \frac{3}{2} < \xi_P \end{cases}$$

The target function $q_R \cdot q_P$ can be expressed as a function of the parameters N and γ, in accordance with the aforementioned definition. The relative maximum in the interval of definition can be determined via roots of the derivative function. No common roots in the derivative function are possible in the interval of definition. The root of the derivative with respect to N yields $\xi_P = 3/4$. The root of the derivative with respect to γ yields approximately $\xi_P \approx 0.5$. Thus, the maximum occurs along the edge of the interval of definition. The optimum of the total cost function occurs for the following values:

$$
\begin{aligned}
\xi_R &= 3 & \xi_P &= \frac{3}{4} \\
\gamma &= \frac{1}{2} & N &= 10
\end{aligned}
\tag{4.2.35}
$$

The result shows that halving the supply voltage and the introduction of 10 pipeline stages increases the throughput by a factor of 3 and reduces the dissipation power by 1/4.

In later Chapters, the efficiency given as the quotient of throughput and size will be used as the cost criterion due to its simple derivation from the architecture parameters. The discussion in this section is only to point out the general problems in evaluating architectures.

4.3   Exercises

1. The use of pipeline registers in hierarchical adder stages is to be evaluated.

    a.   Pipeline stages are to be introduced into a binary lookahead carry adder. The number of D-FFs required for complete pipelining is sought, i.e. D-FFs are to be inserted between all submodules shown in section 3.2.4. Even when possible, no pipeline stages will be introduced to the submodules. The delay components that determine the throughput are to be determined via a propagation estimation of the submodules.

    b.   A conditional sum adder is to be evaluated in accordance with a.

    c.   The results from a and b are to be compared with those of a carry ripple adder (Figure 4.1.11). A rough comparison of efficiency is to be carried out, and one should state whether or not extensive pipelining is useful for hierarchical adder structures.

2. A comparison of efficiency is to be carried out for the two accumulator structures as in Figure 3.2.23.

    a.   The transistor count for the accumulators is to be computed as a function of the word width m and the number N of additions A_k. The CPA adder is to be implemented as a carry ripple adder with 26 transistors per full adder.

    b.   The total computation time for the accumulators is also to be given as a function of m and N. The delay of a full adder for the operand input to the sum output or carry output is given to be $T_{D,FA}$. The delay from the carry input to the outputs is presumed to be $T_{D,FA}/2$; this also holds for a D-FF.

c. The values of efficiency are to be computed from the results of a and b. Determine above which word width m and number N of CSA accumulators the more efficient implementation occurs.

3. The throughput of array multipliers is to be raised through the use of pipelining.

a. Firstly, a Booth multiplier is to be evaluated. The cut-lines for raising the throughput by factors of 2, 3 and 4 are to be determined under the simplifying assumption of a constant delay time $T_{D,FA}$ for full adders and $T_{D,FF}$ for registers.

b. The same evaluation is to be carried out for a Pezaris array multiplier.

c. A comparison of efficiency is to be made for the two array structures from a and b. Which of the two structures should be used for an array multiplier?

4. An optimization of the parallel processing is to be carried out similar to Example 4.2.2. The number of parallel units and the voltage reduction factor N are sought for a maximization of the throughput while simultaneously minimizing size and dissipation power. The hardware for multiplexing and demultiplexing the data streams is to be neglected. In accordance with Figure 4.2.6, cost functions with linear transitions are to be designed, whereby the target value for the throughput is the threefold value and the limiting value for the size and the dissipation power are sixfold and one and a half fold, respectively.

5   Array Processor Architectures

Regular and recursive algorithms are often applied in signal processing. These are algorithms that repetitively apply identical series of operations. Furthermore, the operations to be carried out are predetermined, independent of the input data and the results. Such algorithms are classified as low-level algorithms and are often used in preprocessing. The number of operations to be carried out per unit of time is directly proportional to the sampling rate. Regular architectures based on identical processing elements, so-called array processors, are a natural choice for application-specific implementations of such processes.

In the previous chapter, it was shown how the throughput in signal processors can be raised through the use of pipelining in given structures. Furthermore, a measure of efficiency was defined that represents the quotient between performance and costs. By optimizing the efficiency, the use of pipelining is limited to achieve high throughput for a given circuit size. However, if high throughput is sought, regardless of the costs (circuit size), then hardware architectures with massive parallel processing and intensive pipelining are the optimal solution.

Parallel processing and pipelining can be implemented with programmable and application-specific processors. The terminology is such that systems with programmable processors are called multiprocessor systems and systems with application-specific processors are called array processors. Two particular forms of implementing array processors are systolic arrays and wavefront arrays [63]. The characteristics of these two implementations are presented in the following section. Array processor architectures are application-specific, i.e. the hardware structures conform to the particular algorithm. Methods for directly mapping the algorithm into an architecture have been developed for this case [63], [71]. The starting point for such mappings is a description of the algorithm. For this reason, alternative algorithm representations will be presented in an ensuing section. The subsequent description of an underlying method for mapping algorithms to arrays follows.

First, regular algorithms are formulated as data dependency graphs. These data dependency graphs are mapped to signal flow graphs that can be easily converted to array processors. In the final section, extensions of these underlying methods are presented.

5.1 Implementations of Array Processors

Array processor are computational networks with distributed data storage and a distributed arrangement of processing elements. To achieve high throughput, a high number of operations must be carried out simultaneously. This can be reached through massive use of parallel processing and pipelining. Short connecting wires enhance the throughput and are thus to be preferred. In order to hold down the development costs, regular, modular structures are desirable, i.e. a basic structure is sought that can be applied repetitively. According to these considerations, array processors should have the following characteristics [63], [72], [73]:

- parallelism
- locality
- regularity and modularity

Parallelism is relevant not only for the data operations but also for the data transfers. The term locality implies that connections exist only to directly neighbouring processing elements. The regularity should encompass the general arrangement and the internal configuration as well as the communications structures. The term processing element implies a basic structure for the execution of operations that can be repetitively integrated. Thus, a processing element can be limited to the implementation of a single operation such as addition or multiplication or it can be a complex configuration of several operators and their corresponding control circuitry. In the following, processing elements will be abbreviated PE. If an array consists of only one type of PE, it is termed a homogeneous array. Array processors that are both local and regular are termed cellular arrays.

The principal structure of an array processor is shown in Figure 5.1.1, in this case a two-dimensional (2-D) array. One-dimensional arrays are frequently termed linear arrays. Local locality with geometrically short connections of equal length is only possible for 1-D and 2-D arrays in VLSI chip or printed circuit board (PCB) implementations. Through special mounting techniques such as MCM (multi-chip modules), 3-D arrays with spatial locality are conceivable in principle. In general, however, arrays with dimension higher than two cannot be implemented with spatial locality.

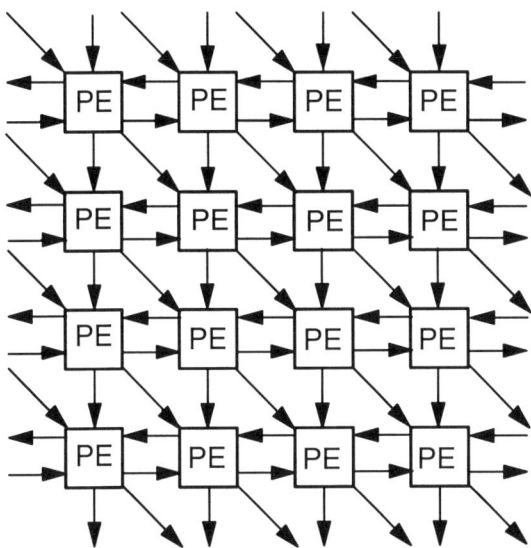

Figure 5.1.1: Principal structure of an array processor

Two particular types of array processor implementations are systolic arrays and wavefront arrays. Systolic arrays are array processors with synchronous clocking and synchronous control signals. Analogous to the circulatory system with its two phases systole and diastole, systems that are entirely pipelined and have periodic computation and data transfer cycles are termed systolic [72]. A systolic array is characterized by the following features:

- synchrony
- regularity and modularity
- spatial and temporal locality
- pipelinability

Synchrony describes processing and data transfer that is controlled by a global clock. Spatial locality refers to the juxtaposition of their connection points for the wiring in terms of their geometric configuration. Temporal locality implies that a piece of data is immediately transferred to its neighbouring PE and processed there, i.e. there is no wait state in the transfer. Pipelinability refers to the extendability with a linear increase in the throughput. In comparison to a system with one PE, a systolic array with M PEs has an increase in throughput of the order $O(M)$. Pipelinability is also called scalability. In order for pipelinability to hold, a signal transaction from one node to the next must have a delay of at least one clock cycle. This minimal delay is included in the definition of temporal locality.

A disadvantage of systolic arrays is the high proportion of registers in their implementation. For processor arrays of moderate size, so-called semi-systolic arrays are used in practice. These arrays do not fulfil the definitions of temporal locality and pipelinability. In general, they contain global communication by broadcast wiring. Broadcast wiring supplies all or a group of PEs with identical data simultaneously. Figure 5.1.2 shows a 1-D array with broadcast wiring. Degradation in the achievable throughput due to long wires and high capacitances in the broadcast line can be limited through special dimensioning of the drivers.

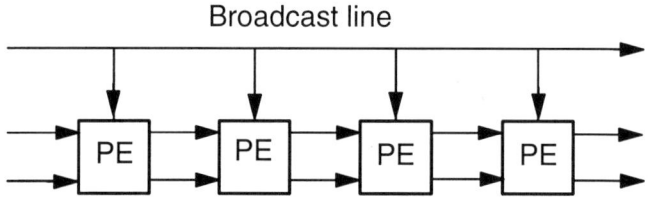

Figure 5.1.2: 1-D array processor with global communication

For very large arrays, synchronous clocking and processing via a global clock can become problematic. The clock pulses reaching the PEs are skewed due to the size of the array. The effects of clock skew must be reduced through special synchronization mechanisms. In addition to arrays with synchronous data transfer, asynchronous data transfers are also possible through the use of self-timed PEs. Special handshaking mechanisms (protocols) control the data transfer. Besides the data wiring for the operands, there are wires that signal the need for new data (request) or the availability of data (acknowledge). Since the data cannot always be immediately processed at the next stage on account of the various transfer times, buffers (FIFO buffers: first-in-first-out) are required. Asynchronous transfers lead to data driven implementations that follow the data flow principle. Corresponding array processors are called wavefront arrays. The characteristics of wavefront arrays are:

- self-timed, data driven computation
- regularity and modularity
- local interconnections
- pipelinability

In data flow processors, a token indicating the validity of the data is communicated along with the data. Data flow processors can be expanded in such a way that the operation to be carried out is indicated along with the token. This enables data flow controlled programmability.

It can be shown that the equitemporal data in a wavefront correspond to a large degree to the equitemporal planes in systolic array processors [63]. Wavefront arrays can also be developed with the design methods of synchronously clocked arrays [63]. The following section limits itself to the design of synchronously clocked array processors.

5.2 Algorithmic Representations

An algorithm is a computational procedure for solving a signal processing task in a finite number of steps. Algorithms for which the operations to be carried out and the operands to be processed are fixed prior to run-time are particularly suited for representation as array processors. For this reason, adaptive algorithms in which the particular operations and operands depend on the data to be processed will not be treated further. Regular, iterative algorithms are especially suitable for mapping to array processors. These consist of several iterations of identical series of instructions.

Most digital signal processing algorithms can be formulated as regular, iterative algorithms. Examples of algorithms that can be formulated in this way are matrix operations, filtering and transformations. As a rule, it is necessary to convert the mathematical, signal processing expression to a regular algorithm. In the following, the various representations of algorithms will be discussed using the multiplication of two matrices and matrix–vector multiplication as examples. The starting point will be the usual mathematical expressions. For the product of two $n \times n$ matrices

$$C = AB \qquad (5.2.1)$$

each element of the resulting matrix is given by

$$c_{ij} = \sum_{k=1}^{n} a_{ik} b_{kj} \qquad 1 \le i, j \le n \qquad (5.2.2)$$

The matrix–vector product

$$c = Ab \qquad (5.2.3)$$

is given by

$$c_i = \sum_{k=1}^{n} a_{ik} b_k \qquad\qquad 1 \leq i \leq n \qquad\qquad (5.2.4)$$

accordingly.

Since signal processing algorithms are often analysed and verified by computer, the obvious choice is to represent the algorithm in a typical programming language. In the programing language PASCAL, the matrix product is given by the following program segment:

```
FOR i : = 1 TO n DO
    FOR j : = 1 TO n DO
        BEGIN
        c[i,j] : = 0 ;
        FOR k : = 1 TO n DO
            c[i,j] : = c[i,j] + a[i,k] * b[k,j] ;
        END;
```

In this case, the code is sequential. The accumulation is carried out via value assignment. The variable $c[i,j]$ is set to NULL . In a DO-loop, the product $a[i,k] * b[k,j]$ is added to the variable as a partial result and the sum is reassigned to the variable $c[i,j]$. This means that the variable is overwritten several times, i.e. a value is assigned to it numerous times.

Recursive algorithms are needed for mapping to array processors. In recursive algorithms, identically named variables are differentiated along the course of the procedure through a recursion index. Due to this differentiation through the recursion index, in the end each variable is assigned only one value. By nature, recursive algorithms consist of single-assignment code. Single-assignment code (SAC) is a structure in which each variable is assigned a value only once in the course of the algorithm.

In the program above, three-dimensional indexing of the element $c[\cdot]$ would provide an additional index and make single-assignment code possible. The program for single-assignment code of a matrix product is:

```
FOR i : = 1 TO n DO
    FOR j : = 1 TO n DO
        BEGIN
        c[i,j,0] : = 0 ;
        FOR k : = 1 TO n DO
            c[i,j,k] : = c[i,j,k−1] + a[i,k] * b[k,j] ;
        c_out[i,j] : = c[i,j,n] ;
        END;
```

In this program, each variable is truly assigned a value only once. The triply indexed variable $c[i,j,k]$ now has the function of an internal variable. The output variable for the result is $c\_out[i,j]$, thus the final assignment of the value of $c[i,j,n]$ to $c\_out[i,j]$.

As shown in section 5.1, a systolic array should have temporal locality, i.e. data transfers should only take place between neighbouring PEs. This means that data dependencies during processing are based only on local neighbours. Algorithms that are to be mapped to systolic arrays should demonstrate this characteristic whenever possible. In the previous program, the variable $c[\cdot]$ is defined in three-dimensional index space. The elements $a[\cdot]$ and $b[\cdot]$ are doubly indexed. Thus they can be seen as global variables for N positions in 3-D index space. A localization of the data dependencies can be achieved through triple indexing of the elements and insertion of data transfers into the algorithm. A possible program for the computation of a matrix multiplication with local data dependencies is:

```
FOR i : = 1 TO n DO
    FOR j : = 1 TO n DO
        BEGIN
        c[i,j,0] : = 0 ;
        FOR k : = 1 TO n DO
            BEGIN
            IF i=1 THEN b[1,j,k] : = b_in[k,j]
            ELSE b[i,j,k] : = b[i−1,j,k];
            IF j=1 THEN a[i,1,k] : = a_in[i,k]
            ELSE a[i,j,k] : = a[i,j−1,k];
            c[i,j,k] : = c[i,j,k−1] + a[i,j,k] * b[i,j,k] ;
            END;
        c_out[i,j] : = c[i,j,n] ;
        END;
```

In this program, the matrices to be multiplied have the elements $a\_in[i,k]$ and $b\_in[k,j]$. The resulting matrix has the elements $c\_out[i,j]$. The triply indexed elements $a[i,j,k]$, $b[i,j,k]$ and $c[i,j,k]$ are internal variables. The elements $b\_in[\cdot]$ or $a\_in[\cdot]$ are inserted for the plane with $i = 1$ or $j = 1$. For all other cases, the data transfer takes place along the i- or j-axis.

The aforementioned program has single-assignment code and only local data dependencies. The program contains instructions for initialization, the input of external elements, data transfers, computation and output of the results. All of these instructions are mixed yet carried out in a special, sequential order.

Alternative descriptions for signal processing algorithms are known from the literature [74], [75], [76]. These program languages strive to restrict themselves to elementary mathematical relations and index dependencies. One possible description form is UNITY [74]. A UNITY program for a matrix multiplication is given that resembles the program description chosen by Thiele [75].

MATMUL
in
$(\langle; i,j,k : 1 \le i,j,k \le n : : (\, a\_in[i,k]; \, b\_in[k,j] \,)\rangle)$
always
$\langle \, \| \, i,j,k : 1 \le i,j,k \le n : :$
     $b[i,j,k] = b\_in[k,j]$ if i=1
            $\sim b[i-1,j,k] \, \|$
     $a[i,j,k] = a\_in[i,k]$ if j=1
            $\sim a[i,j-1,k] \, \|$
     $c[i,j,k] = a[i,j,k] \cdot b[i,j,k]$ if k=1
          $\sim c[i,j,k-1]+a[i,j,k] \cdot b[i,j,k] \, \|$
   c_out $[i,j] = c[i,j,k]$ if k=n$\rangle$
out
$(\langle; i,j,k : 1 \le i,j \le n : : c\_out[i,j])$

The meaning of several symbols can be derived through comparison with the PASCAL program above. Nonetheless, a few general statements about UNITY will be given here. A more exacting definition and nomenclature can be found in the literature [74], [75]. A UNITY program describes "what" must be done. A specific order for the operations is not formulated exactly. As with the data flow principle, an instruction cannot be carried out until the dependent arguments are defined. "When", "where" and "how" the operations are to be carried out is not defined until they have been mapped to an architecture. UNITY programs are often used in the literature to describe regular algorithms. A UNITY module consists of a name, an input section **in**, the computational section **always** and an output section **out**. The value of y is assigned to the variable x via the equation $x=y$. Equations are separated by the symbol $\|$. A set of indexed equations can be quantified through specification of the iteration space (index space). As in all programming languages, condition-controlled selection statements are possible. A UNITY program is one possible way of representing a formal description of an algorithm.

The data dependencies in an algorithm can be most compactly represented using a data dependency graph. A graph $G=[V,E]$ is a set of nodes V and a set of edges E. The set of edges E is an ordered set of the nodes V, i.e.

$$E \subseteq V \times V \qquad (5.2.5)$$

Thus, an edge $e \in E$ is given by

$$e = (i,j) \quad \text{with } i,j \in V \qquad (5.2.6)$$

In a directed graph, i is the starting point and j is the end point. One advantage of the use of graphs is that a graphical representation of the dependencies is possible in addition to their purely set-oriented definition. In many cases, the graphical representation can illustrate the correlations quite vividly and thus ease understanding.

The term dependence graph will be abbreviated to DG in the following equations. The matrix multiplication example treated earlier shows that the nodes of the dependence graph are points in multidimensional index space. The nodes of the dependence graph are given by c. The term c is intended to imply that the nodes are the locations for the computations. The set of all nodes in the dependence graph is I_{DG}. The edges of the dependence graph are given by d; they dictate the data dependencies between the nodes. Given c_1 as the starting point and c_2 as the end point of an edge,

$$d = c_2 - c_1 \qquad (5.2.7)$$

From the matrix multiplication example:

$$I_{DG} = \left\{ [i,j,k]^T \mid 1 \leq i,j,k \leq n \right\}$$
$$c \in I_{DG} \qquad (5.2.8)$$

The edges of the dependence graph describe the data dependencies. One can read from the programming examples that data travel in the direction of i and j and that partial results are transported in the direction of k. Thus, edges in three directions exist for each node. All existing edges of the dependence graph can be described by data dependence vectors and summarized in a data dependence matrix.

$$D = [d_1 \, d_2 \, d_3] = \begin{bmatrix} 1 & 0 & 0 \\ 0 & 1 & 0 \\ 0 & 0 & 1 \end{bmatrix} \qquad (5.2.9)$$

The dependence graph for matrix multiplication is shown in Figure 5.2.1. In order to keep the illustration simple, only the edges on the visible surfaces are shown. Thus the inner nodes and edges are invisible. The input data are introduced and the output data are extracted along the borders. The matrix A is introduced along the plane $j = 1$, the matrix B is introduced along the plane $i = 1$ and a matrix of zeros is introduced along the plane $k = 1$. The resulting matrix C stands available in the plane $k = n$. The operative part corresponding to each node in the dependence graph is shown in Figure 5.2.2. In addition to two data transfers, a multiplication and an addition must be assigned to each node.

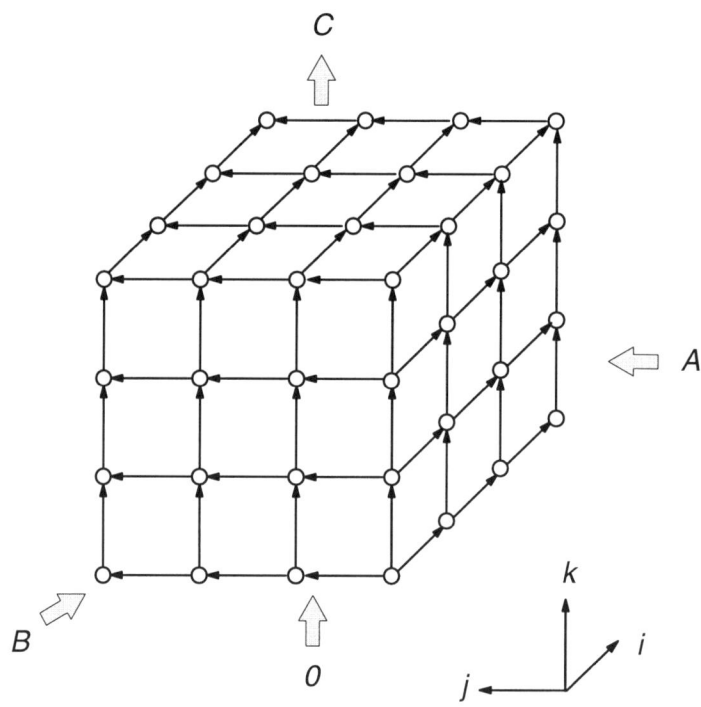

Figure 5.2.1: Dependence graph of a matrix multiplication ($n = 4$)

The example at hand shows a characteristic that many dependence graphs possess. Excepting the borders of the defined regions, the edges leading from each node are identical for all indexed points. Such a dependence graph has invariant data dependence with exception of the border nodes. Dependence graphs with this characteristic are termed homogeneous or shift-invariant.

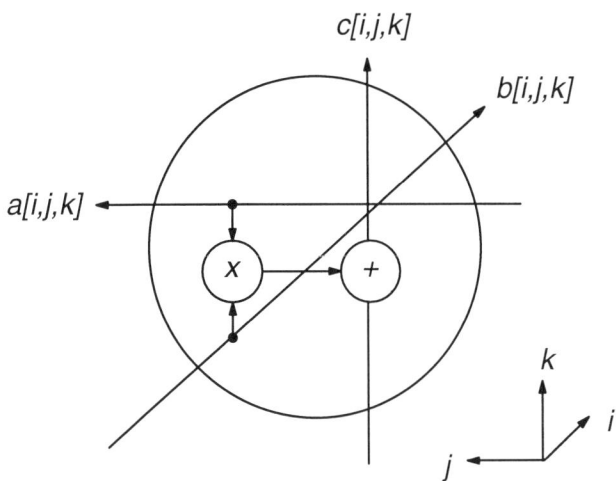

Figure 5.2.2: Node of a dependence graph for a matrix multiplication with representations of its operations

Each matrix multiplication includes a matrix–vector multiplication. According to (5.2.3), the dependence graph of a matrix–vector product is a plane with j = constant from the dependence graph in Figure 5.2.1. The resulting dependence graph for a matrix–vector multiplication is shown in Figure 5.2.3. The vector b is input along the line $i = 1$ and the result vector c is available along the line $k = n$. The corresponding matrix element $a[i, k]$ is present in each node $[i, k]^T$.

The previous representations describe algorithms for a single application of a set of data. In signal processing algorithms with high throughput, such algorithms are to be continuously carried out for new sets of data. The effective time interval until the next input of data is given by T_R. If n_S pieces of data are input in parallel, the throughput is determined by the quotient of n_S and T_R in accordance with (4.2.3). Using matrix–vector multiplication as an example, the continuous computation can be described by

$$rep_{T_R}(c) = rep_{T_R}(Ab) \qquad (5.2.10)$$

where *rep* $(\cdot)$ signifies the periodic computation (repeat) and the periodic data input or output. In the case of transformation methods, the matrix A is predetermined. After time T_R, a new vector b is injected and a new result vector c is extracted by the following unit. The continuing processing of new sets of data can be described in the algorithm through an additional index in the index space. To simplify the representation, this will be left out in many of the later discussions of algorithms. The description of the algorithm is then limited to one basic interval. A finite index space results from the avoidance of this additional index.

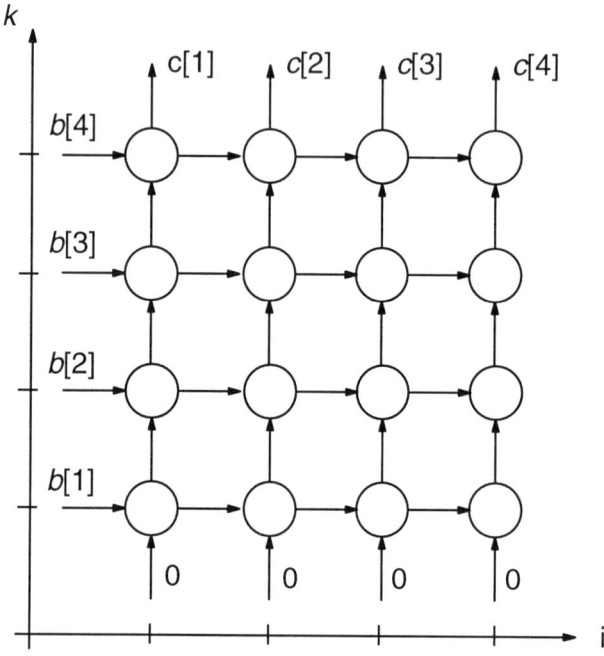

Figure 5.2.3: Dependence graph for a matrix–vector multiplication ($n=4$)

As an additional example, a sorting algorithm will be treated. The given series $\{x\_in(i)\}$, is to be resorted into a new series $\{y\_out(i)\}$. The new series is to be sorted in decreasing order, i.e. $y\_out(i) \geq y\_out(k)$ for $i < k$. One possible algorithm is as follows. A subset of the input series is already sorted. Another element of the input series is to be inserted into this sorted series. This is achieved through comparison with the given series (starting with the largest value) and determining when the new element is larger than an element of the series for the first time. The new element is inserted in this position and all following smaller elements are shifted by one place.

Given are i, the iteration index that simultaneously specifies the number of elements in the intermediate series, and j, a continuous index for the comparisons within series of length i. A corresponding algorithm as a UNITY program with single-assignment code and local data dependencies is:

SORT
in
$(\langle ; i : 1 \le i \le n : : x\_in[i] \rangle)$
always
$\langle || i,j : 1 \le i \le n \wedge 1 \le j \le i : :$
$\quad x[i,j] = x\_in[i] \qquad \text{if } j = 1 ||$
$\quad y[i,j] = - \infty \qquad\quad \text{if } i = j ||$
$\quad y[i+1,j] = max (x[i,j], y[i,j]) ||$
$\quad x[i,j+1] = min (x[i,j], y[i,j]) ||$
$\quad y\_out [j] = y[i+1] \quad \text{if } i = n \rangle$
out
$(\langle ; j : 1 \le j \le n : : y\_out [j] \rangle)$

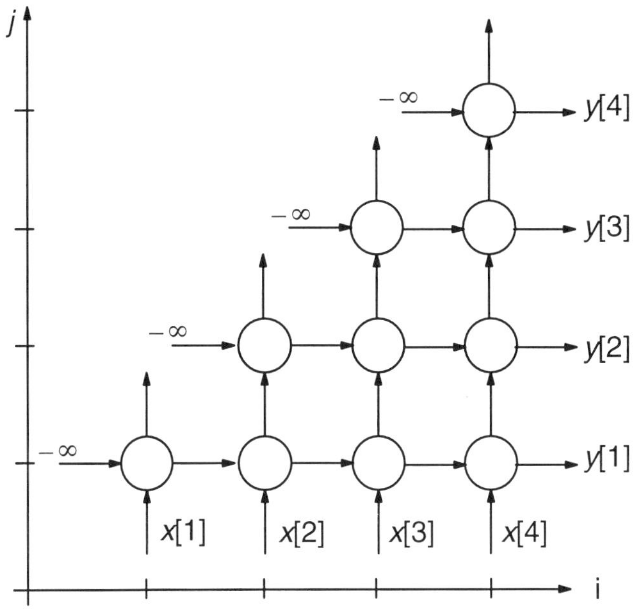

Figure 5.2.4: Dependence graph for sorting a series

The corresponding dependence graph is illustrated in Figure 5.2.4. In this case, the index space is a triangular region. The input series is introduced into the plane $j = 1$. To complete the functionality of this regular system, the most negative values are introduced along the diagonal (i,i). The resulting series is available in the plane $i = n + 1$. The dependence graph is shift-invariant since the same data dependency holds for each node. The data dependence matrix is

$$D = [d_1, d_2] = \begin{bmatrix} 1 & 0 \\ 0 & 1 \end{bmatrix} \qquad (5.2.11)$$

The implementation of a node is shown in Figure 5.2.5. The result of a comparison (CMP = compare) controls two multiplexers for extracting a maximum and a minimum.

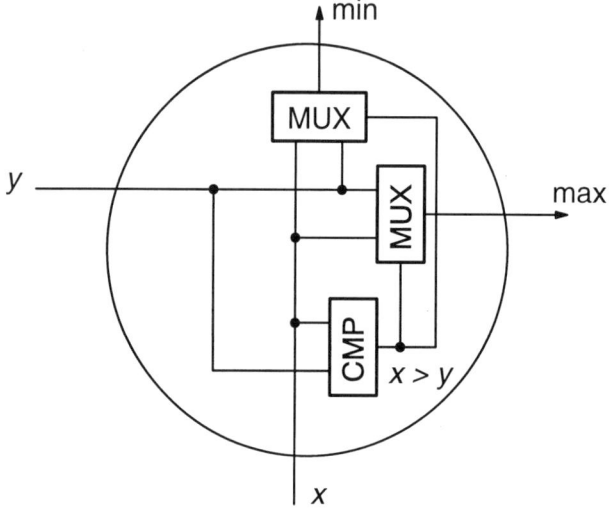

Figure 5.2.5: Node from the dependence graph of a sorter

A few short, supplementary remarks to this section now follow. Interested readers are referred to the literature [71], [77], [78]. Due to their implementation, the algorithms to be treated can be classified as linear recursive algorithms. Such algorithms are characterized by:

- linearly index variables
- limited index range (iteration space)
- single-assignment instructions

Thus, the algorithms contain instructions of the form

$$x_k(P_k i + p_k) = F_k\{..., x_l(Q_{lk} i + q_{lk}), ...\}$$
$$\forall i \in I_k \subset \mathbb{Z}^s \qquad (5.2.12)$$

It is assumed that the algorithm is based on a set of variables $x_1, ... x_k$, $... x_v$. An index vector i is assumed that lies in s-dimensional, convex index space I_k. The index dependencies are described by the matrices P_k and Q_{lk}

and the vectors p_k and q_{lk} with constant elements. F_k describes the function for computing the value of a variable x_k. In order to guarantee single-assignment characteristics and computability of these functions, each variable for a specific index point may appear only once on the left-hand side of the equation (SAC). Furthermore, an ordering of the equations exists such that a variable for a specified index point appears on the left-hand side before it appears on the right-hand side (computability).Transformations can be applied to the linear recursive algorithms that lead to both geometric and structural modifications.

For mapping algorithms to array processors with local connections, algorithms with local data dependencies are sought. Localization can be achieved by increasing the dimension of the index space by inserting new variables. The linear recursive algorithm is hereby converted to a piece-wise regular algorithm. This algorithm contains instructions of the form

$$x_k(i) = F_k \{..., x_l(i + i_{lk}), ...\}$$
$$\forall i \in I_k \subset \mathbb{Z}^s \qquad (5.2.13)$$

Each instruction can be assigned to a separate index space. A large number of alternative ways of formulating a piece-wise regular algorithm exist, in particular when the algorithm is based on commutative and associative operations. For example, the accumulation of the partial products in a matrix multiplication can be carried out in the opposite direction instead of from $k = 1$ to $k = n$.

A dependence graph consisting of regular subgraphs with regular connections can be assigned to any piece-wise regular algorithm. An index point i is assigned to each node in the dependence graph and a vector i_{lk} is assigned to each edge. In this case, within the dependence graph, the index points (nodes) are designated c and the dependencies are designated d. The two examples discussed above, matrix multiplication and sorting, possess the structure shown in (5.2.13).

5.3 Linear Mapping Methods

An application-specific array processor architecture is sought that is suitable for processing a given algorithm. The architecture should fulfil the demands regarding computational performance and throughput, and furthermore, the hardware costs should be as low as possible. The considerations in section

5.1 showed that systolic array processors make system implementations with high throughput possible. Due to their local connections and their modularity, they are well suited for VLSI implementations. To simplify designing, a systematic method is sought that exploits that interaction between the algorithm and the architecture.

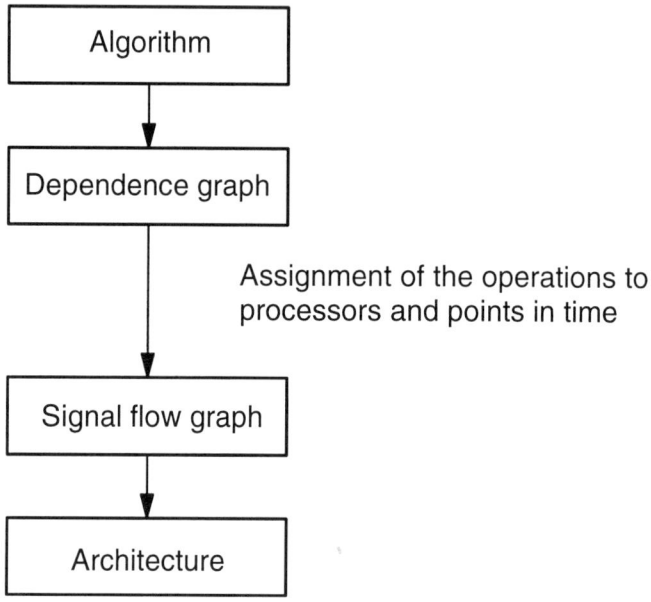

Figure 5.3.1:   Principal method for mapping algorithms to array processors

In section 5.2, it was shown that regular algorithms can be formulated through dependence graphs with local data dependencies. The target architecture possesses local connections so it is reasonable to achieve the array processor structure through transformation of the dependence graph. Figure 5.3.1 shows the principal method. A given algorithm is structurally modified in such a way that it can be represented by a loop-free dependence graph with local data dependencies. The dependence graph is then mapped to a signal flow graph. There are two basic considerations for mapping a dependence graph to a signal flow graph. The nodes of the dependence graph operations are assigned to processors represented by nodes of the signal flow graph (assignment). Further the temporal sequence of operations assigned to a processor has to be specified (scheduling). The signal flow graph is a weighted graph in which the weighting of the edges represent the delay, the nodes the operations and the edges the data transfers. This kind of signal flow graph can be easily mapped to an array processor by combining the nodes and the delays into processor elements.

The number of processing elements n_{PE} is generally smaller than the number of nodes in the dependence graph n_{DG}. For mapping, it is important to know both numbers since they specify how many nodes of the dependence graph must be mapped to a PE. Thus, the number of processing elements should be estimated before mapping.

5.3.1 Estimating the number of PEs

The processing time of a PE is given by T_{PE} including the transfer time of an FF. Here it is assumed that transfer registers are only present at the output of the PE and that within T_{PE} each PE carries out a total of $n_{OP/PE}$ operations. The computational rate of the processor array is then given by

$$R_{C,ARRAY} = n_{PE} \cdot \frac{n_{OP/PE}}{T_{PE}} \qquad (5.3.1)$$

It is assumed that the desired throughput is $R_{T,target}$ and that the number of operations per sample $n_{OP/SAMPLE}$ is known from the algorithm. From (4.2.3) together with (5.3.1), it then follows that

$$n_{PE} = \frac{n_{OP/SAMPLE}}{n_{OP/PE}} \cdot R_{T,target} \cdot T_{PE} \qquad (5.3.2)$$

The number of processing elements required can be further reduced through pipelining within the PEs. Given n_P as the number of pipeline stages along the data path of the PE, the time T_{PE} in (5.3.2) must then be replaced by T'_{PE}.

$$T'_{PE} = \frac{T_{PE} - T_{REG}}{n_P} + T_{REG} \qquad (5.3.3)$$

The number of pipeline stages can be raised as long as the smallest logic units between two registers remain meaningful. In the case of an adder this is a full adder, and in the case of a multiplier this is an AND gate in combination with a full adder (MA cell).

Example 5.3.1

The samples of a colour television signal (27 MHz effective sampling rate) are to be reformatted into sequential 8×8 blocks, and each of these blocks is to be multiplied by an 8×8 matrix (1-D transformation). The samples and the matrix coefficients are assumed to be in 8-bit representation; the result of the accumulated product is 19-bit. A PE contains 1 MUL, 1 ADD (see Figure 5.2.2) and a register at its output. The numbers of processors with and without pipelining within the PE are to be calculated. The following holds for the operations:

$$n_{OP/PE} = 2 \quad (\text{MUL, ADD})$$

$$n_{OP/SAMPLE} = 2 \cdot 8 \quad (2 \cdot 8^3 \text{ operations for } 8^2 \text{ samples})$$

$$T_{PE} = T_{MA,ARRAY} + T_{REG}$$

$$T_{MA,ARRAY} \approx (3n + 3)T_{D,FA,c'c} = 27 \cdot 16\tau_L$$

$$T_{REG} \approx T_{D,FA,c'c} = 16\tau_L$$

The processing time T_{PE} with intensive pipelining down to MA cells is given by
$$T'_{PE} = T_{MA,CELL} + T_{REG}$$

$$T_{MA,CELL} = T_{D,AND} + T_{D,FA,s's} = 42\tau_L$$

Assuming a 1 μm CMOS process with $\tau_L = 50$ ps substitution into (5.3.2) yields

$$n_{PE} = 4.8 \quad (T_{PE} \approx 22.4 \text{ ns})$$
$$n'_{PE} = 0.6 \quad (T'_{PE} \approx 2.9 \text{ ns})$$

Without pipelining in the PE, more than four PEs must be used. For the given problem it then seems reasonable to use eight PEs. Through the use of pipelining in the PE, one PE suffices. Yet due to the high clock rate, the data input and output are particularly problematic in this case.

5.3.2 Mapping without changing the number of nodes

If the approximation in accordance with (5.3.2) and based on the desired throughput yields $n_{PE} \geq n_{DG}$, then the dependence graph can be converted to a signal flow graph with the identical number of nodes. Thus

$$n_{PE} = n_{SFG} = n_{DG} \qquad (5.3.4)$$

The dependence graph yields a signal flow graph with systolic characteristics by assigning at least one delay element (D-FF) to each edge. The insertion of delay elements can be carried out using the cut–set method described in Chapter 4. Figure 5.3.2 shows the signal flow graph corresponding to the example of matrix–vector multiplication in Figure 5.2.3.

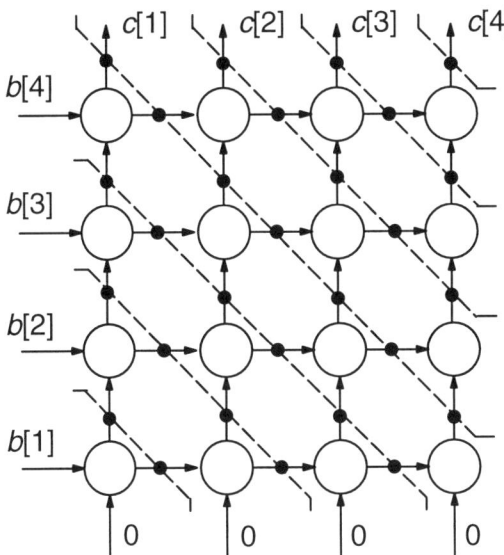

Figure 5.3.2: Signal flow graph of a matrix–vector multiplication ($n = 4$)

The dependence graph and the signal flow graph have the same index space, the same nodes and edges.

$$
\begin{aligned}
I_{DG} &= I_{SFG} \\
p &= c \qquad c \in I_{DG} , \; p \in I_{SFG} \\
E &= D
\end{aligned}
\qquad (5.3.5)
$$

The nodes of the signal flow graph are designated by p, the edges by e. The matrix E contains all the edges e of the signal flow graph. All the edges of the signal flow graph are assigned a delay:

$$\tau(e_i) = t(c_j) - t(c_k) \quad \text{with } e_i = c_j - c_k \qquad (5.3.6)$$

The delay times are integral multiples of the base clock period τ_0. The base clock period is generally normed to a unit delay with the value 1. Lines and planes of constant delay result in the signal flow graph from this delay assignment. The point in time for executing a node's operation can be determined from the scalar product with a schedule vector s.

$$t(c) = s_{\!\bullet}^T c + t_0 \qquad (5.3.7)$$

The schedule vector is always perpendicular to the equitemporal lines and planes. Since only integral delay times (number of clock cycles) are possible, the elements of the schedule vector must be integers. Using t_0, the absolute time of a specific node can be set to zero. By shifting the index space, $t_0 = 0$ can always be achieved. The relationship to execution time as in (5.3.7) holds for the first set of data. In the case of continuous computation of several sets of data (see (5.2.10)), the points in time $t(c)$ are shifted by multiples of period T_R.

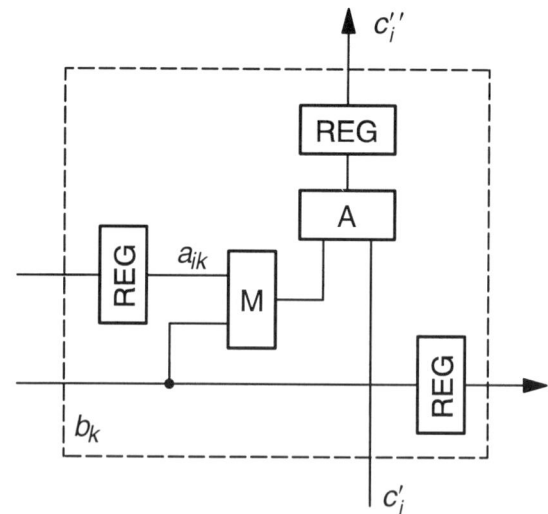

Figure 5.3.3: Processing element for matrix–vector multiplication

Assuming that all the edges in the chosen example are assigned unit delay, the schedule vector is then

$$s = \begin{bmatrix} 1 \\ 1 \end{bmatrix} \tag{5.3.8}$$

The temporal shifting of the input and result data is set simultaneously with the time assignment in accordance with (5.3.7).

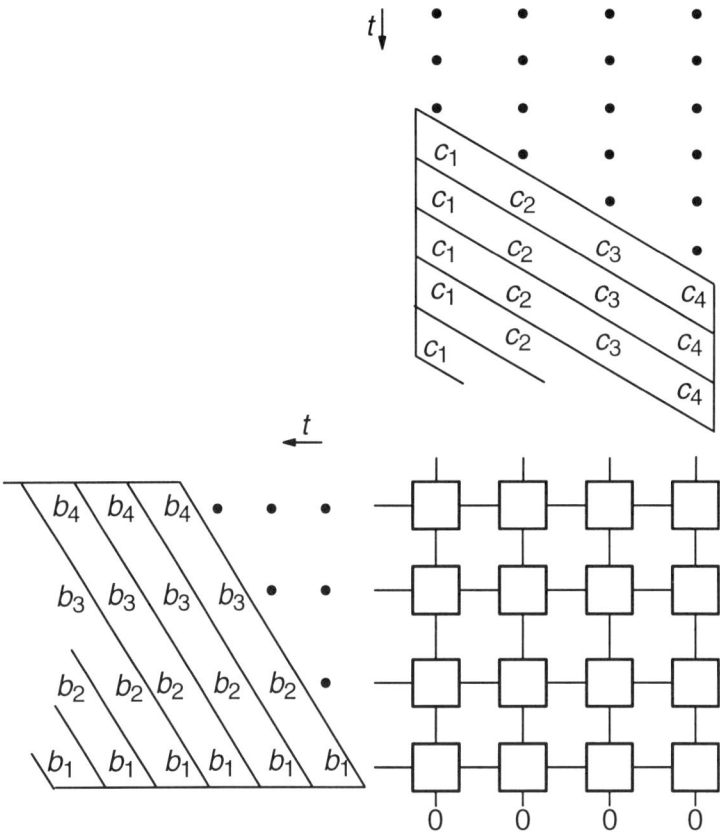

Figure 5.3.4: Systolic array for matrix–vector multiplication

The processing elements of the array processor are created from the nodes of the signal flow graph and the delay of the edges. Figure 5.3.3 shows the processing element for a matrix–vector multiplication. One matrix coefficient is stored in each PE. Figure 5.3.4 shows the complete systolic array. Vectors *b* for the computation of result vectors *c* can be sequentially inserted into the array. The figure also shows the temporal shifting of the inserted vectors. The dots indicate non-valid data.

Systolic arrays created in this manner enable maximal throughput for a given technology. If the desired throughput is higher, pipelining must be incorporated in the PEs. This can be carried out via time scaling.

Each unit delay along the edges is substituted by a k-fold delay, i.e. (5.3.6) becomes

$$\tau(e_i) = k\tau'(e_i) \qquad (5.3.9)$$

The k-fold delay can be equally distributed along the data path through delay transfer (see Chapter 4). The larger k is, the higher the achievable throughput.

5.3.3 Mapping with a reduced number of nodes

The method described above does not yield a systolic array with implementable local connections if the dimension of the dependence graph's index space is larger than two. An additional disadvantage is that, whenever only moderate requirements are made on the throughput, the hardware costs rise above what is necessary. The estimate (5.3.2) is an indicator for the desired processor count. If $n_{PE} < n_{DG}$, a mapping is sought in which one processor element takes over the processing of several nodes from the dependence graph.

One possible mapping method based on the reduction of the dimensionality of the signal flow graph using projection will be derived based on the example examined above. If the desired throughput is less than that offered by the hardware, the data input can be reduced through dummy cycles. The upper half of Figure 5.3.5 shows an example of this for the input vectors of a matrix–vector multiplication. The number of dummy cycles between the input vectors is chosen such that only one node in the vertical direction possesses valid data. If the number of nodes in the vertical direction is n, then $n - 1$ dummy cycles are required.

All nodes of the signal flow graph are identical, yet only one in n has valid data for a given point in time. It thus seems obvious to create a new signal flow graph in which each node constantly possesses valid data. This can be achieved through vertical projection. The result is shown in the lower half of Figure 5.3.5.

Since a node's output is connected with the input of the next node, projection results in a feedback loop that includes a D-FF. After projection, the node is sent a zero in the first clock cycle and the previous partial result from its output thereafter. In the original signal flow graph, each node was assigned a fixed coefficient a_{ik}. After projection, all coefficients of nodes that are projected onto one another must be supplied externally, whereby attention must be paid to the horizontal temporal shift due to the D-FFs.

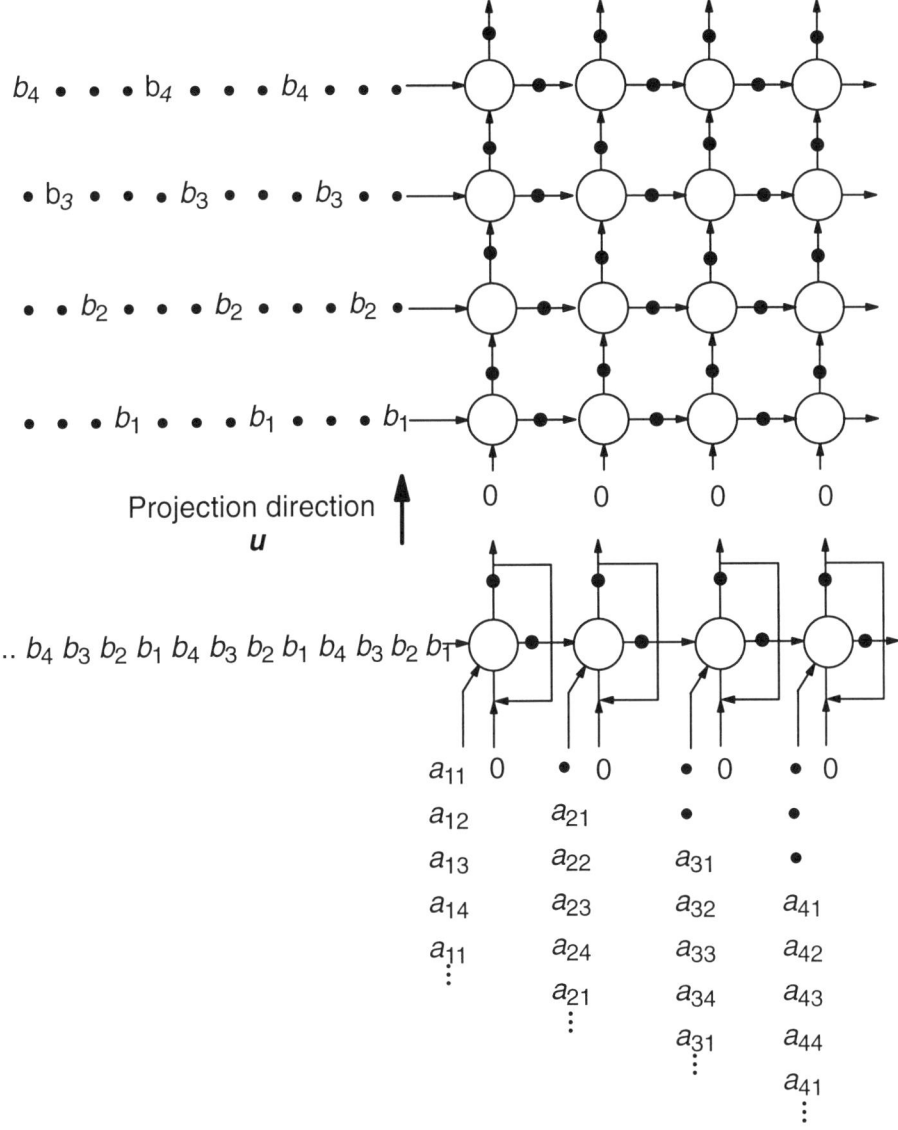

Figure 5.3.5: Projection of a 2-D array onto a 1-D array (example: matrix–vector multiplication)

The number of nodes in the original signal flow graph in the example is

$$n_{PE} = n_{SFG} = n_{DG} = n^2 \tag{5.3.10}$$

Through projection in the direction of its axis, the number of nodes is reduced by a factor of $1/n$.

$$n_{PE} = n_{SFG} = n \tag{5.3.11}$$

The throughput and the computational performance are reduced by the same factor for identical clock rates. This means that the costs and the performance are reduced to the same extent. The mapping of the nodes from the dependence graph to the signal flow graph in the example at hand is described through linear mapping by a processor base Π.

$$p = \Pi^T c + p_0 \tag{5.3.12}$$

Since integers are used for the indexing of p and c, the elements of Π must also be integers. By choosing the indices of p accordingly, $p_0 = 0$ can always be achieved. All nodes in the direction of projection are mapped to one node, i.e. shifts in the direction of projection do not change the new indexing.

$$\Pi^T(c + u) = \Pi^T c \tag{5.3.13}$$

An alternative way of stating this is that the signal flow graph has zero width in the direction of projection. From these two statements, it follows that

$$\Pi^T u = 0 \tag{5.3.14}$$

This means that the processor base Π and the projection vector u are perpendicular to one another. For the example of Figure 5.3.5:

$$u = \begin{bmatrix} 0 \\ 1 \end{bmatrix}$$

A feasible, corresponding processor base that fulfils (5.3.14) is

$$\Pi^T = [\, 1 \; 0 \,]$$

A further reduction of the number of processors can be carried out by projecting 1-D arrays to one processing element. As in the previous case, special measures must be taken so that only one in n nodes contains valid data at a given point in time. The data input must be reduced by a factor of $1/n$. Furthermore, a temporal offset and new time scale must be created between the nodes. In the example of Figure 5.3.5, n nodes must be mapped onto one another. Thus, an offset between the nodes of

$$\tau' = \tau/n \qquad\qquad (5.3.15)$$

is sufficient. The entire system undergoes time scaling with the factor n in order to yield integer delays. To achieve a minimal delay of τ' between the nodes, the schedule vector $s' = [-3]$ must be chosen. The signal flow graph that results after mapping to one PE is shown in Figure 5.3.6.

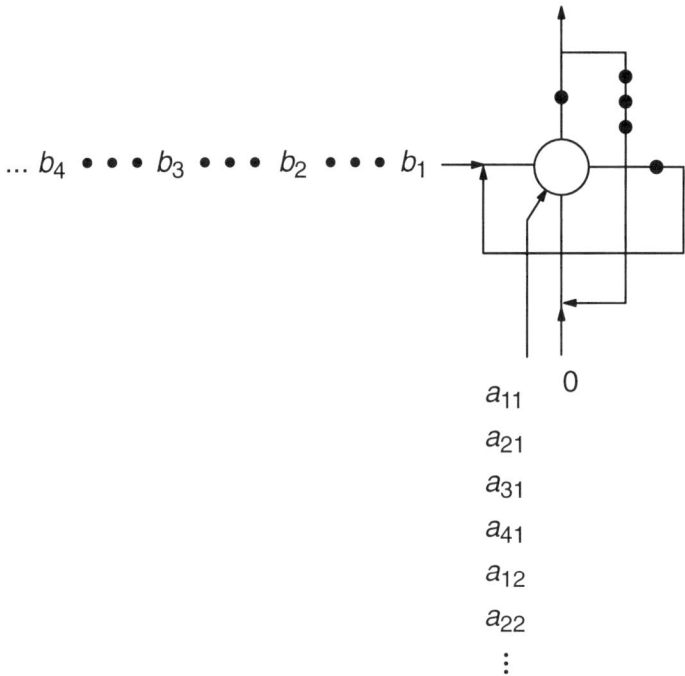

Figure 5.3.6:   Matrix–vector multiplication with one PE ($n = 4$)

After the presented second projection, the partial results are nodes strung in a row and separated by the given offset. An alternative solution is also possible in which the delay between two sequential nodes is chosen to be so large that the successor does not receive valid data until the predecessor has computed its result. In the example, this means a delay of n clock cycles. This is equivalent to time scaling the horizontal axis by a factor of n. The resulting signal flow graph with one PE is shown in Figure 5.3.7.

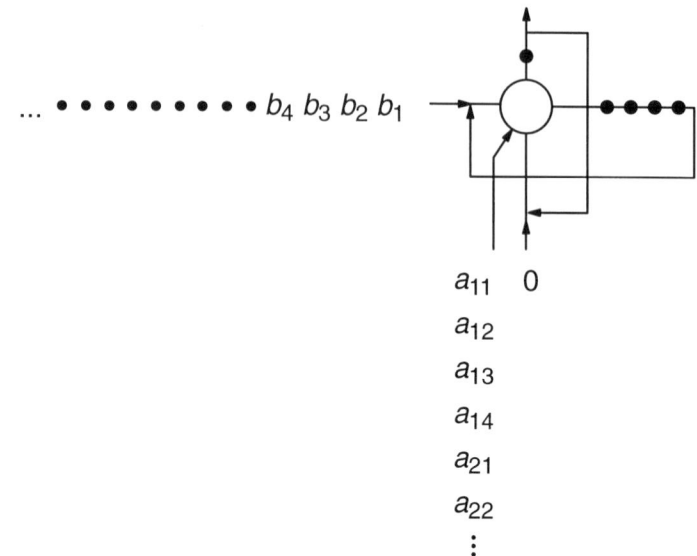

$\cdots \bullet\bullet\bullet\bullet\bullet\bullet\bullet\bullet\bullet\bullet\; b_4\; b_3\; b_2\; b_1$

a_{11} 0
a_{12}
a_{13}
a_{14}
a_{21}
a_{22}
⋮

Figure 5.3.7:   Alternative solution for a matrix–vector multiplication with one PE

The solutions shown in Figures 5.3.6 and 5.3.7 could have been derived directly without complex re-evaluation of the delay times. The key is to find a temporal distribution for the 2-D array of Figure 5.3.2 such that, given appropriate data input, only one in n^2 nodes contains valid data at a given point in time. If the schedule vector of (5.3.8) is modified to

$$ s = \begin{bmatrix} n \\ 1 \end{bmatrix} \text{ or } s = \begin{bmatrix} 1 \\ n \end{bmatrix} \tag{5.3.16} $$

the solutions of Figures 5.3.6 and 5.3.7 can be extracted via direct, two-fold projection. If the input of data is carried out accordingly, one need not feed back the b_k's of the particular PE. In this case, the b_k data must be input several times in a special order.

Solutions with n^2, n or one PE have been presented for the same problem of matrix–vector multiplication. Table 5.3.1 summarizes the essential results (throughput and transistor count) for an 8×8 matrix. The values for the delay time are taken from Example 5.3.1. Since it is possible, in principle, to carried out extended pipelining within a PE, this is also taken into account in the table. The results show that the throughput and the transistor count are nearly directly proportional to the processor count. Yet the results also show that pipelining within the PEs is very rewarding. Through complete pipelining of a PE (20 pipeline stages), the transistor costs are nearly tripled, yet the throughput is more than seven times higher. This verifies the efficiency of pipelining. From the results, it can be inferred that the throughput should first be raised using pipelining. Should this not be sufficient, the number of processors can then be increased.

Table 5.3.1:   Throughput and transistor count for a matrix–vector multiplication: 8×8 matrix with 8-bit accuracy for the samples and the coefficients

Pipelining within PE	n_{PE}	Throughput rate in Msample/s	Transistor count
No	8×8	360	180 000
	8	45	24 000
	1	6	3 000
Yes	8×8	2 700	540 000
	8	340	69 000
	1	43	8 600

5.3.4   Formal description of the projection method

The mapping of a dependence graph to a signal flow graph requires that each node c of the dependence graph be assigned to a processor $p(c)$ and to a point in time $t(c)$ for execution of its operation. These spatial and temporal assignments must fulfil special requirements. The order of computation given in the dependence graph must be retained. This leads to the **causality** criterion:

$$t(c_2) \geq t(c_1) \qquad \forall \ c_2 - c_1 \in \{d_m\} \qquad (5.3.17)$$

The d_m's are the data dependence vectors (edges) of the dependence graph. Furthermore, one node in the signal flow graph can carry out the operations of only **one** node of the dependence graph simultaneously. Thus follows the requirement of **unambiguity**.

$$\begin{bmatrix} t(c_2) \\ p(c_2) \end{bmatrix} \neq \begin{bmatrix} t(c_1) \\ p(c_1) \end{bmatrix} \qquad \text{for } c_2 \neq c_1 \qquad (5.3.18)$$

The method of linear mapping is the type most often used in the literature. It is particularly suitable for shift-invariant, homogeneous dependence graphs [63], [71]. The mapping is done via a matrix M.

$$\begin{bmatrix} t \\ p \end{bmatrix} = Mc \qquad\qquad (5.3.19)$$

Given are

$$\begin{aligned}
c &= \begin{bmatrix} i_1, \dots & i_n \end{bmatrix}^T & c \in I_{DG} \subset \mathcal{Z}^n \\
p &= \begin{bmatrix} j_1, \dots & j_k \end{bmatrix}^T & p \in I_{SFG} \subset \mathcal{Z}^k
\end{aligned} \qquad (5.3.20)$$

Even if causality (5.3.17) and unambiguity (5.3.18) are observed, a large number of different matrices M exist for carrying out the mapping. In particular, mappings are sought that respect local data dependencies and continuous data input and output. As was explained in section 5.2, the variables that are to continually receive new data are often not explicitly defined in the formulation of dependence graphs. By suppressing this dependency, the dimension of the index space is reduced and, above all, the index space remains finite. During mapping, attention must be paid to the variables that are continually replaced. These variables must be inserted along the border of the processor array.

Various strategies exist for determining transformation matrices [63], [71],[77]. One method of projection will be described that is of advantage when the dimension of the signal flow graph is one less than the dimension of the dependence graph.

$$I_{DG} \subset \mathcal{Z}^n \qquad I_{SFG} \subset \mathcal{Z}^{n-1} \qquad\qquad (5.3.21)$$

As in (5.3.7) and (5.3.12), the matrix M can be split into a schedule vector and a processor base.

$$M = \begin{bmatrix} s^T \\ \Pi^T \end{bmatrix} \qquad\qquad (5.3.22)$$

A schedule vector is allowable when the causality criteria can be upheld, i.e.

$$\tau(e_m) = s^T d_m \geq 0 \qquad \forall d_m \qquad d_m = col_m(D) \qquad (5.3.23)$$

The term $col_m(\cdot)$ stands for the mth column vector. In systolic solutions, each edge of the dependence graph is assigned a delay, in which case the equality relation above does not hold for all data dependence vectors d_m. The relation (5.3.23) specifies the delay that is assigned to the edges e_m of the signal flow graph.

Through projection in the direction of a projection vector u, the n-dimensional dependence graph is to be converted to a signal flow graph of dimension $n - 1$. The processor base must be orthogonal to the projection vector. This leads to

$$\Pi^T u = \begin{bmatrix} \pi_1^T \\ \vdots \\ \pi_{n-1}^T \end{bmatrix} u = 0 \qquad (5.3.24)$$

The processor base is derived through projection onto the plane spanning $(\pi_1, \pi_2, ...\pi_{n-1})$. Figure 5.3.8 shows this relationship for $n = 3$.

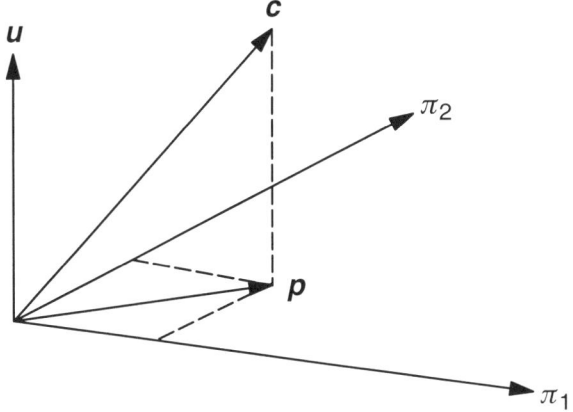

Figure 5.3.8:   Projection of the vector c onto the plane spanning (π_1, π_2)

Unambiguity as in (5.3.18) requires a matrix M of rank n. Since the processor base Π is orthogonal to the projection vector u, this cannot hold for the schedule vector. It thus follows that

$$s^T u \neq 0 \qquad (5.3.25)$$

This requirement can also be justified graphically. It means that nodes of an equitemporal hyperplane may not be projected to the same PE. The schedule vector s is perpendicular to an equitemporal hyperplane. In the case where the identity (5.3.25) equals zero, the projection vector u is orthogonal to s, i.e. the projection vector is parallel to the equitemporal hyperplane.

This mapping using the matrix M (5.3.19) split as in (5.3.22) is applied to all nodes in the dependence graph, including the input and output nodes.

The same matrix also describes the mapping of the edges in the dependence graph since an edge in the dependence graph is a vector between two nodes.

$$\begin{bmatrix} \tau(e) \\ e \end{bmatrix} = \begin{bmatrix} s^T \\ \Pi^T \end{bmatrix} d \qquad (5.3.26)$$

Given that the edges of the dependence graph are combined to a matrix D, the edges of the signal flow graph are combined to a matrix E and that $\mathcal{T}$ is an ordered group of n elements representing the delay times corresponding to the edges e, (5.3.26) becomes

$$\begin{bmatrix} \mathcal{T} \\ E \end{bmatrix} = MD \qquad (5.3.27)$$

The presented projection method is to be applied to the example of sorting. As opposed to the algorithm in section 5.2, in particular Figure 5.2.4, the data sets are numbered from 0 to $N-1$ instead of from 1 to N. Systolic scheduling is assumed, i.e. each edge in the dependence graph is assigned a delay. Assuming minimal delay of one clock cycle for the vertical and horizontal edges, the equation

$$s = \begin{bmatrix} 1 \\ 1 \end{bmatrix}$$

holds.

Now the effects of the projections (horizontal, vertical and diagonal) on the resulting circuit structures are to be studied. The projection vector for a horizontal projection is

$$u = \begin{bmatrix} 1 \\ 0 \end{bmatrix}$$

The requirement (5.3.25) is easily verified. Thus, the given schedule vector is a valid projection vector.

$$s^T u = 1 \neq 0$$

One possible processor base is determined in accordance with (5.3.24). For

$$\Pi^T = [0 \ 1]$$

orthogonality is given to be met. Thus, the corresponding mapping matrix (5.3.22) is

$$M = \begin{bmatrix} 1 & 1 \\ 0 & 1 \end{bmatrix}$$

For the mapping of the nodes, it follows that

$$\begin{bmatrix} t \\ p \end{bmatrix} = Mc = M \begin{bmatrix} i_1 \\ i_2 \end{bmatrix} = \begin{bmatrix} i_1 + i_2 \\ i_2 \end{bmatrix}$$

This means that the processor index is identical to the vertical index and that the time of execution results from the sum of the horizontal and vertical indices. The data dependence matrix contains all data dependence vectors. Thus according to (5.3.27), the product of M and D yields all the nodes of the signal flow graph.

$$\begin{bmatrix} \tau(e_1) & \tau(e_2) \\ e_1 & e_2 \end{bmatrix} = MD = M \begin{bmatrix} 1 & 0 \\ 0 & 1 \end{bmatrix} = \begin{bmatrix} 1 & 1 \\ 0 & 1 \end{bmatrix}$$

The edge $e_1 = 0$ symbolizes feedback to itself. An edge $e_2 = 1$ in a 1-D array processor is a connection from node to node in the direction of positive indexing. Each edge is assigned unit delay. The signal flow graph resulting from projection is shown in Figure 5.3.9. The temporal specification of data input, output and multiplexer control can also be determined via the transformation matrix M. Data input in the dependence graph is carried out via the nodes

$$c = \begin{bmatrix} i_1 \\ 0 \end{bmatrix}$$

The transformation leads to

$$\begin{bmatrix} t \\ p \end{bmatrix} = \begin{bmatrix} i_1 \\ 0 \end{bmatrix}$$

This means that the positional index i_1 turns into a temporal index. The results can be extracted from the following positions in the dependence graph:

$$c = \begin{bmatrix} N \\ i_2 \end{bmatrix}$$

The transformation leads to

$$\begin{bmatrix} t \\ p \end{bmatrix} = \begin{bmatrix} N + i_2 \\ i_2 \end{bmatrix}$$

Thus, the results are available in parallel in accordance with the positional index i_2. The results have a basic delay of N cycles in addition to a temporal offset due to the further delay corresponding to the positional index i_2. The special input of the smallest negative value (symbolized by $-\infty$) is carried out in the dependence graph along the diagonal $i_1 = i_2$. The transformation yields

$$\begin{bmatrix} t \\ p \end{bmatrix} = \begin{bmatrix} 2i_2 \\ i_2 \end{bmatrix}$$

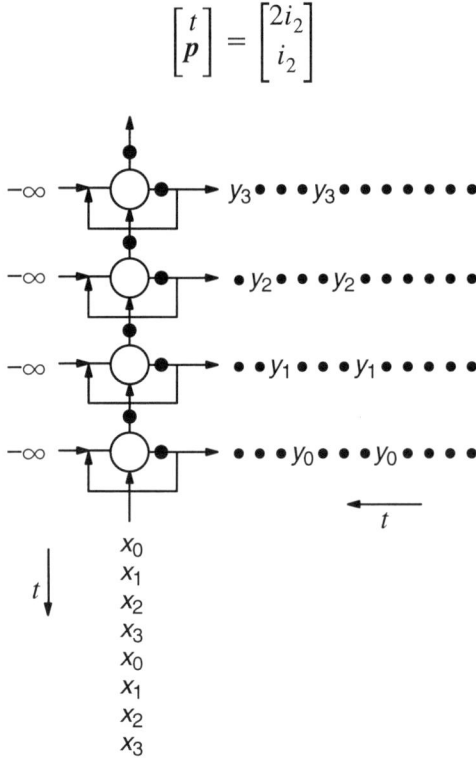

Figure 5.3.9: Signal flow graph for a sorter after horizontal projection

It thus follows that the multiplexer at node i_2 of the signal flow graph must be switched at time $2i_2$ such that the value $-\infty$ is input; at other times, the feedback loop is closed.

Figure 5.3.9 also shows how the data input and output can be periodically continued. The projection results in a mod N operation of the temporal index.

In a manner similar to horizontal projection, vertical and diagonal projection can also be carried out. The individual steps will not be thoroughly discussed anew. Instead, just the transformation matrix and the resulting signal flow graph will be shown.

For a vertical projection,

$$u = \begin{bmatrix} 0 \\ 1 \end{bmatrix}$$

The corresponding mapping matrix is

$$M = \begin{bmatrix} 1 & 1 \\ 1 & 0 \end{bmatrix}$$

The signal flow graph resulting from M is shown in Figure 5.3.10.

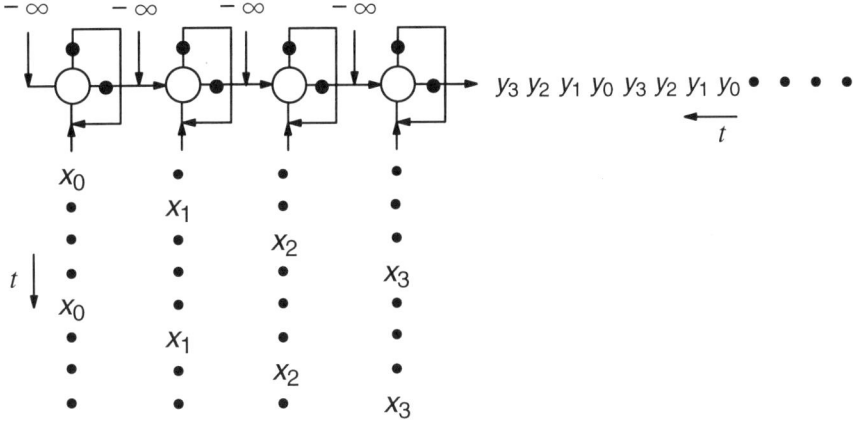

Figure 5.3.10: Signal flow graph for a sorter after vertical projection

The projection vector

$$u = \begin{bmatrix} 1 \\ 1 \end{bmatrix}$$

leads to projection along the diagonal of the dependence graph. One corresponding mapping matrix is

$$M = \begin{bmatrix} 1 & 1 \\ 1 & -1 \end{bmatrix}$$

The signal flow graph that then results is shown in Figure 5.3.11.

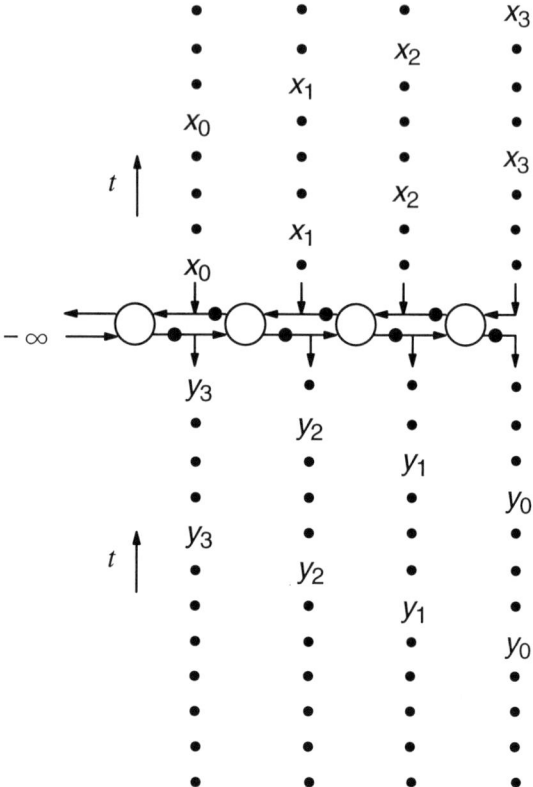

Figure 5.3.11: Signal flow graph for a sorter after diagonal projection

In all three cases, the projections carried out on the example of sorting lead to a signal flow graph with N nodes. The difference is the data input and output, which is serial in some cases and parallel in others. Due to the systolic implementation, a temporal offset results in parallel data input or output. Borrowing from known sorting algorithms [79], the results shown are labelled insertion, selection and bubble sorter [63].

A special case
The mapping of a dependence graph to a signal flow graph with the same number of nodes is a special case. In this case, the processor base is an identity matrix.

$$p = c \qquad E = D \qquad\qquad (5.3.28)$$

The schedule vector s is chosen such that the edges are given the desired delay. For systolic structures, each edge must have a delay larger than zero. This means that (5.3.23) must yield a delay larger than zero for all edges d_m.

5.3.5 Multiprojection

The projection method previously described only allows mapping of an n-dimensional dependence graph to an $(n-1)$-dimensional signal flow graph. If one wishes to further reduce the dimension of the target array, this can be achieved via projection applied to the signal flow graph. No nodes of the signal flow graph that are mapped onto one another may possess valid data at the same point in time. As explained in the introductory example of a matrix–vector multiplication, this can be achieved via a retiming of all nodes in the direction of projection such that no two are ever simultaneously active. Retiming and the following projection are to be explained for the case in which all nodes in the direction of projection are given a small temporal offset. The size of the temporal offset depends on the maximal number of nodes N_u in the direction of projection. Given τ, the delay in the original signal flow graph and τ', the delay in the new signal flow graph, the following condition for the basic clock period [63] must be fulfilled:

$$\tau_0 \geq \tau_0' + \tau_0' (N_u - 1)s^T u \qquad (5.3.29)$$

In this equation, s is the schedule vector of the given signal flow graph and u is the projection vector of the plan projection. The expression

$$(N_u - 1)s^T u$$

specifies the maximal time difference between the first and the last node in the direction of projection.

One wishes to retain the representation of the delay time as an integer multiple of a common clock cycle. Since τ_0' is the smaller time span, it will serve as a reference from now on, and the delay τ_0 will be represented as an integer multiple of τ_0'.

$$\tau_0 = K\tau_0' \qquad (5.3.30)$$

Thus, (5.3.29) yields

$$K \geq 1 + (N_u - 1)s^T u \qquad (5.3.31)$$

In most applications, an appropriate value for K is

$$K = N_u s^T u \tag{5.3.32}$$

Giving attention to the new time scale, a further projection is possible. The schedule vector to be used for the renewed mapping is given to be s'. The new timing is a combination of temporal scaling and linear projection. Thus the mapping matrix is

$$M' = \begin{bmatrix} K & s'^T \\ 0 & \Pi^T \end{bmatrix} \tag{5.3.33}$$

This mapping matrix is valid for transforming time, processor indices, edge delays and edges. Analogous to (5.3.19) and (5.3.27)

$$\begin{bmatrix} t' \\ p' \end{bmatrix} = M' \begin{bmatrix} t \\ p \end{bmatrix}$$

$$\begin{bmatrix} \mathcal{T}' \\ E' \end{bmatrix} = M' \begin{bmatrix} \mathcal{T} \\ E \end{bmatrix} \tag{5.3.34}$$

follows for the mapping.

The causality criterion must be upheld for all edges of the transformed signal flow graph. As in (5.3.23),

$$\tau'(e'_m) = K\tau(e_m) + s'^T e_m \geq 0 \qquad \forall e_m, e_m = col_m E \tag{5.3.35}$$

must hold. For feedback to itself ($e'_m = 0$) or systolic implementations, all edges must possess delay. Equality in (5.3.35) is not allowable in these cases.

The new schedule vector must be chosen such that the unambiguity formulated in (5.3.18) is met. This means that all nodes to be mapped onto one another in the direction of projection must be active at different times. Since the nodes of the signal flow graph are active at specific times due to previous projections and temporal scaling, requirement (5.3.25) is not sufficient in this case. Due to the linearity of the mapping, it is sufficient to test if any nodes in the direction of projection (excepting the first) are active at multiples of K in order to guarantee that no nodes are simultaneously active in the direction of projection. This is founded on the temporal scaling by K since data are then only input every K base clock periods. Thus

$$js'^T u \bmod K \neq 0 \qquad \forall j, j = 1, ...K - 1 \tag{5.3.36}$$

must hold. A suitable choice for the components of s' also follows from this equation. The components of s' that have contributions in the direction of projection should have the values

$$-K + 1, \; -1, 1, K-1, K+1$$

so that no multiples of K result. The first value leads to the smallest delay among the edges.

It is easily seen that this method of projection can be applied repetitively and that, through k-fold projection, a dependence graph can be converted to a signal flow graph of dimension $n - k$. However, the size of the design space is problematic since several valid vectors s and possible directions of projection u exist for each mapping. Several CAD programs are presented in the literature for supporting the design process [80], [81], [82], [83], [84].

The projection of a signal flow graph is to be explained using an example. Figure 5.3.12 shows the original signal flow graph. A few edges have no delay. The number of delay cycles is given by the weighting aside the corresponding edge. The schedule vector to be used is

$$s' = \begin{bmatrix} 0 \\ 1 \end{bmatrix}$$

The projection vector is

$$u = \begin{bmatrix} 0 \\ 1 \end{bmatrix}$$

The number of nodes in the direction of projection is $N_u = 4$, from which (5.3.31) yields the smallest value of K:

$$K = 4$$

The processor base

$$\Pi^T = [1 \; 0]$$

leads to the mapping matrix (5.3.33)

$$M' = \begin{bmatrix} 4 & 0 & 1 \\ 0 & 1 & 0 \end{bmatrix}$$

The edges of the given signal flow graph are

$$\begin{bmatrix} \mathfrak{T} \\ E \end{bmatrix} = \begin{bmatrix} 2 & 0 & 1 & 0 \\ 1 & 0 & 0 & -1 \\ 0 & 1 & -1 & 1 \end{bmatrix}$$

Transformation with M', results in the following edges for the new signal flow graph:

$$\begin{bmatrix} \mathfrak{T}' \\ E' \end{bmatrix} = \begin{bmatrix} 8 & 1 & 3 & 1 \\ 1 & 0 & 0 & -1 \end{bmatrix}$$

The signal flow graph shown in Figure 5.3.12 contains these edges and delays.

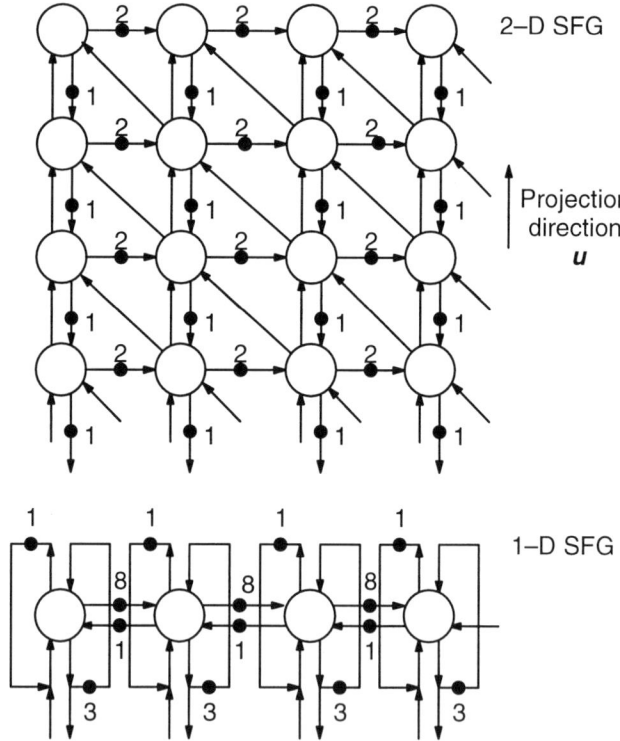

Figure 5.3.12: Example mapping of a 2-D SFG to a 1-D SFG (taken from [63])

A direct method

The disadvantage of the method presented is that the mapping to a SFG with the desired dimension is carried out in several individual steps. Beginning with a DG with an n-dimensional index space, an SFG is determined with dimension $n-1$. Giving attention to special retiming requirements, further projections then follow.

The equations (5.3.19), (5.3.22) and (5.3.27) also support direct projections to SFGs with an index space of arbitrarily reduced dimension. If the processor base Π^T is of dimension $n-k$, the schedule vector s^T must guarantee the unambiguity of the mapping for k dimensions of the dependence graph. In the given case, the mapping matrix M consists of $n-k+1$ rows and n columns. Due to the large degree of freedom that remains, general conditions beyond (5.3.17) and (5.3.18) are difficult to formulate.

However, for the special case in which the projections are carried out along the axis of the dependence graph, further statements about the mapping matrix can be made. It is assumed that the first k axial directions are mapped, temporally expanded, to the same processor node. The maximal number of nodes in the j-direction of the dependence graph is given to be N_j. Using $K = N_j$, the time scaling factor is set for the multiprojection in accordance with (5.3.32).

Under the given assumptions, the following holds for a projection to an SFG of dimension $n-1$.

$$M = \begin{bmatrix} s^T \\ \Pi^T \end{bmatrix} = \begin{bmatrix} 1 & s'^T \\ \mathbf{0} & \Pi'^T \end{bmatrix} \tag{5.3.37}$$

Here, s'^T and Π'^T represent the rest elements of s^T and Π^T with $n-1$ columns. For a projection with further reduction of the dimension, time scaling must be carried out corresponding to the maximum number of nodes in axial direction. so as to maintain unambiguity For a projection to an SFG of dimension $n-2$ in accordance with (5.3.33),

$$M = \begin{bmatrix} s^T \\ \Pi^T \end{bmatrix} = \begin{bmatrix} 1 & N_1 & s'^T \\ \mathbf{0} & \mathbf{0} & \Pi'^T \end{bmatrix} \tag{5.3.38}$$

Accordingly, a mapping to an SFG of dimension $n-3$ can be carried out with

$$M = \begin{bmatrix} s^T \\ \Pi^T \end{bmatrix} = \begin{bmatrix} 1 & N_1 & N_1 N_2 & s'^T \\ \mathbf{0} & \mathbf{0} & \mathbf{0} & \Pi'^T \end{bmatrix} \tag{5.3.39}$$

This can be continued until a mapping to a single node is described. Then

$$M = [s^T] = [1 \quad N_1 \quad N_1 N_2 \quad N_1 N_2 N_3 \quad ... \quad N_1 ... N_{n-1}] \tag{5.3.40}$$

Figure 5.3.6 shows an example of a mapping to a single node using an equation like (5.3.38). It is obvious that further alternative mapping matrices are formulated by changing the order in which the individual axial directions are evaluated.

5.4 Extensions of the Mapping Methods

5.4.1 Partitioning

The projection method presented in the previous section leads to architectures whose size is determined by the algorithm. The matrix–vector multiplication discussed can be implemented as a 2-D array, a 1-D array or a single PE. The size of the 2-D array is $n \times n$ PEs given an $n \times n$ matrix. The length of a 1-D array is n PEs or more, depending on the direction of projection. Mappings to arrays with preset size are desired. Partitioning is one way of mapping to problem-dependent, preset sizes of arrays.

The literature differentiates between two fundamental approaches [71], [85], [86]:

1. LPGS (locally parallel, globally sequential)
2. LSGP (locally sequential, globally parallel)

Besides these two, there are others such as geometric convolution and algorithm specific mappings. In the following, the two partitioning methods stated above will be further explained.

LPGS partitioning
The dependence graph or the signal flow graph of an algorithm is split into blocks of preset size, whereby the size of the blocks corresponds to the size of the target array. The left side of Figure 5.4.1 depicts such splitting of a 2-D graph. In this figure, it is assumed that a 4×6 graph is to be implemented by a 2×2 array. The nodes of the graph are split into blocks measuring 2×2. Each of these blocks is treated as a supernode. The given example has 2×3 such supernodes that are mapped to one supernode using the known mapping method. If all the edges possess delays even within the supernode, this must be taken into account in the temporal scheduling of data transfers and control. The result of the mapping after partitioning is shown on the right-hand side of Figure 5.4.1.

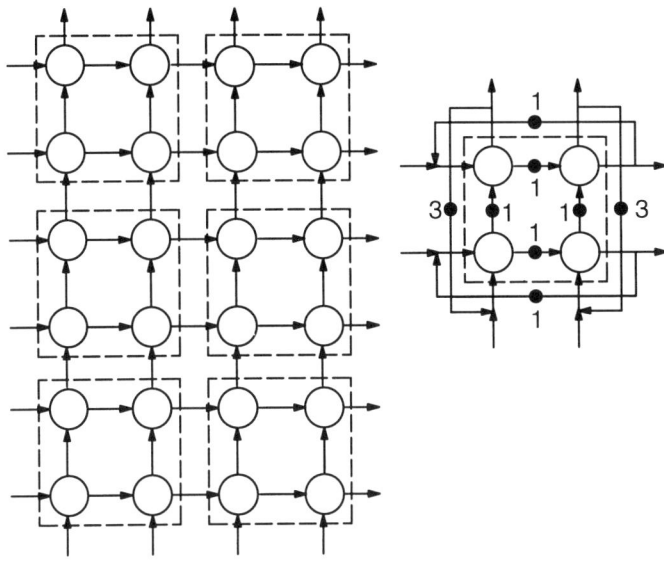

Figure 5.4.1: Example of LPGS partitioning

The characteristics of the method can be summarized as follows. The structure of the array is retained. The remaining array is a segment of the complete array. The structure and the timing in local sections within the block are unchanged. Additional external hardware (registers, multiplexers, control logic) is required. Furthermore, the data input and output are altered.

LSGP partitioning

In this method, the graph is split such that the configuration and number of blocks is identical to the target array. Neighbouring nodes within a block are mapped to one node (clustered) so that processing within a block is implemented sequentially. The left-hand side of Figure 5.4.2 shows such splitting of a 2-D graph. As before, it is assumed that a 4×6 graph is to be implemented through a 2×2 array. The nodes of the graph are split such that the number of blocks is 2×2. Each block then contains 2×3 nodes. Within each block, the nodes are mapped to one node using the known mapping method. Due to the data dependencies between the blocks, temporal offset between these must be taken into account. As the result on the right-hand side of Figure 5.4.2 shows, a configuration with 2×2 nodes remains.

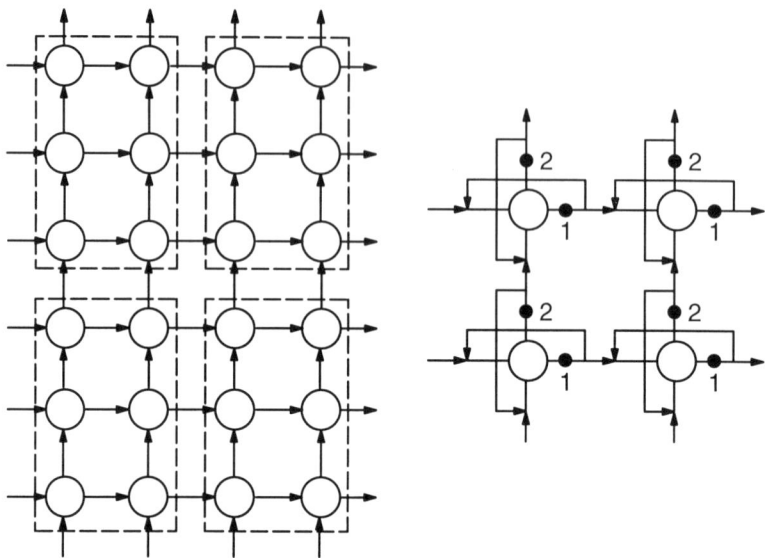

Figure 5.4.2: Example of LSGP partitioning

A characteristic of this method is that local structural changes are carried out with additional hardware (registers, multiplexers). The number of registers is specified by the size of the block. Here, feedback of the intermediate results is carried out locally, whereas in LPGS partitioning it is carried out globally.

Formal description of the mapping

For the present, a 2-D dependence graph is assumed that is to be converted to a 2-D signal flow graph of given size through partitioning. Partitioning through division into blocks is an extension of the dimension of the index space. A block index and a node index within the block result.

The nodes c are converted to the nodes c_{part} after partitioning. A set of parameters λ specifies the partitioning function

$$c_{part} = f(\lambda, c) \tag{5.4.1}$$

Each index point i_k of the original graph is described by two index points j_{2k-1} and j_{2k} in the partitioned graph. The mapping function is

$$i_k = \lambda_k j_{2k-1} + j_{2k} \quad \lambda_k \in \mathcal{N} \tag{5.4.2}$$

The value j_{2k-1} and j_{2k} are evaluated from i_k by dividing by λ_k. The first value is the integer quotient, the second is the remainder and is smaller than λ_k.

Given that the original dependence graph has $n_1 \times n_2$ nodes and the target array has $m_1 \times m_2$ nodes, a parameter μ_k can be derived from the quotient of n_k and m_k.

$$\mu_k = \left\lceil \frac{n_k}{m_k} \right\rceil \tag{5.4.3}$$

It is assumed that the original indexing is from 0 to $n_k - 1$. The following then holds for the two methods of partitioning:

$$\begin{array}{ll} LPGS: & \lambda_k = m_k \\ LSGP: & \lambda_k = \mu_k \end{array} \tag{5.4.4}$$

Using (5.4.1), a 2-D dependence graph is converted to a 4-D graph.

$$\begin{bmatrix} i_1 \\ i_2 \end{bmatrix} \Rightarrow \begin{bmatrix} j_1 \\ j_2 \\ j_3 \\ j_4 \end{bmatrix} \tag{5.4.5}$$

With the methods of projection and multiprojection from section 5.3, the new 4-D dependence graph is converted to a 2-D signal flow graph. In general, this transformation is

$$\begin{bmatrix} t \\ p \end{bmatrix} = M c_{part} \tag{5.4.6}$$

where M contains a combination of two local projections. As before, M can be split into two components.

$$M = \begin{bmatrix} s^T \\ \Pi^T \end{bmatrix} \tag{5.4.7}$$

For the two cases, LPGS and LSGP, possible solutions for the processor base Π and the schedule vector s can be determined.

LPGS mapping
In the case of LPGS, an $m_1 \times m_2$ array indexed with j_2 und j_4 remains after the projection. Accordingly, the processor base must be formed from the vector

$$[0 \; 1 \; 0 \; 1]$$

An obvious choice for the processor base is

$$\Pi^T = \begin{bmatrix} 0\,1\,0\,0 \\ 0\,0\,0\,1 \end{bmatrix} \tag{5.4.8}$$

All $\mu_1 \times \mu_2$ blocks must be projected onto one another. This means that temporal spacing must be created between them. If the temporal spacing is first created in the i_1 direction and then in the i_2 direction, the schedule vector is

$$s^T = [1\ 0\ \mu_1\ 0] \tag{5.4.9}$$

This corresponds to the solution that is shown in (5.3.38). If the order of the axial directions for the temporal spacing is exchanged,

$$s^T = [\mu_2\ 0\ 1\ 0] \tag{5.4.10}$$

is a valid schedule vector.

In the two schedule vectors shown previously, all nodes of the $m_1 \times m_2$ arrays created are located in the same time plane. If unit delays are to be present between the nodes of the created arrays, the delay between the blocks increases by m_1 or m_2 in accordance with the size of the block. The schedule vector of (5.4.9) is changed to

$$s^T = [m_1\ 1\ \mu_1 m_1\ 1] \tag{5.4.11}$$

Accordingly, the schedule vector of (5.4.10) becomes

$$s^T = [\mu_2 m_2\ 1\ m_2\ 1] \tag{5.4.12}$$

LSGP mapping
In the case of LSGP, an $m_1 \times m_2$ array is indexed with j_1 and j_3. Thus, the processor base for this case must be derived from the vector

$$[1\ 0\ 1\ 0]$$

A possible processor base is

$$\Pi^T = \begin{bmatrix} 1\ 0\ 0\ 0 \\ 0\ 0\ 1\ 0 \end{bmatrix} \tag{5.4.13}$$

All nodes within a block are to be projected to one result node. For this, a temporal division is necessary in which all nodes indexed by j_2 and j_4 are active at different times. Analogous to the solution in (5.3.38), the following two schedule vectors lead to the desired temporal spacing within a block.

$$s^T = [0 \ 1 \ 0 \ \mu_1]$$ (5.4.14)

$$s^T = [0 \ \mu_2 \ 0 \ 1]$$ (5.4.15)

However, the data dependency between the blocks is not taken into account. Each block measures $\mu_1 \times \mu_2$. Thus, depending on its axial direction, the next block cannot receive intermediate results from its predecessor until μ_1 or μ_2 cycles have elapsed. Accounting for these data dependencies, (5.4.14) and (5.4.15) must be changed to

$$s^T = [\mu_1 \ 1 \ \mu_1\mu_2 \ \mu_1]$$ (5.4.16)

$$s^T = [\mu_1\mu_2 \ \mu_2 \ \mu_2 \ 1]$$ (5.4.17)

Additional observations

It was previously assumed that 2-D dependence graphs are mapped to 2-D signal flow graph through partitioning. If the dependence graph is of dimension larger than 2, two indices are split as in (5.4.2); the others remain unchanged. In accordance with the equations above, projection is first carried out with the split indices. Using the method of multiprojection, projections are subsequently carried out using the remaining indices. In principle, it is possible to mix LPGS and LSGP partitioning. This means that LPGS projection is carried out in the one axial direction and LSGP projection is carried out in the others.

The partitioning methods can be understood as hierarchical structuring. A large, regular array is split into homogeneous blocks. These blocks, in turn, consist of regularly networked nodes. In signal processing, many procedures can be split into subprocedures. Each subprocedure can be individually formulated as a dependence graph and converted to an application-specific array processor through mapping. If all the subprocedures are treated independently of one another, the possibility of direct data transfers between the sub-modules cannot be guaranteed. In certain cases, the temporal order of the results of the predecessor does not fit the input for its successor. Through hierarchical representation and common mapping, these data transfer problems can be solved automatically for certain cases. Examples exist for which no linear, temporal division is possible over the entire configuration but only for each section. The special requirements for data transfer often then demand special configurations for reformatting the data stream.

A 2-D linear transformation will be discussed as an example of a signal processing procedure that can be split into subprocedures. Let

$$Y = A \, X \, A^T \qquad\qquad (5.4.18)$$

where A, X, Y are matrices with $n \times n$ elements, X are the input data and Y are the result data. Using the rules of matrix transposition, (5.4.18) can be converted to

$$Y = A \, (A \, X^T)^T \qquad\qquad (5.4.19)$$

This means that a 2-D linear transformation can be implemented through two-fold application of a 1-D transformation. First, n vectors of X^T are multiplied by A. The result is transposed and again multiplied by A. The subprocedures here are matrix–vector multiplication, transposition and matrix–vector multiplication. The transposition is implemented through reformatting memory. The cascade structure of the subprocedure is shown in Figure 5.4.3. During the projection of the matrix–vector multiplication parts, the data transfer criteria for the transposition memory change automatically.

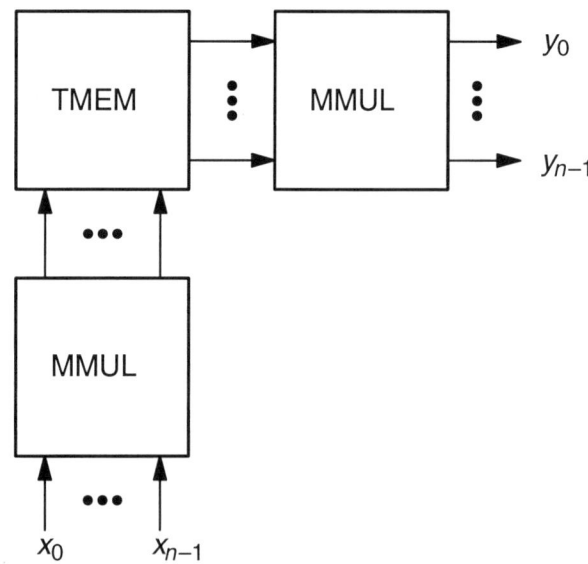

Figure 5.4.3:   Implementation of a 2-D transformation by cascading two 1-D transformations. MMUL: Matrix–vector multiplication. TMEM: Transposition memory

5.4.2 Projection of nodes with different operations

It was previously assumed that all nodes in the dependence graph carry out the same operation. The mapping to a signal flow graph only took the data dependencies into account. Through the use of delay elements (registers) and multiplexers in the data wiring, chronologically correct processing is guaranteed. In signal processing, however, many algorithms exist for which, in addition to the data dependencies, various operations are also to be executed. Multiplication at the bit level will be treated as an example.

The operations that must be observed for the multiplication are bit level multiplication (AND) and bit level addition (FA). The corresponding nodes are labelled M (multiplication) and A (addition). The multiplier implementations in section 3.3 show various configurations of these basic cells. In Wallace tree and Dadda multipliers, the M and A cells are assigned to separate regions and connected. In carry-save array multipliers, M and A cells are knitted to MA cells wherever they are structural neighbours (see Figure 4.1.10).

The computation of distance measure L_1 will be observed as an additional example of an algorithm with various basic operations. One such method is used for template matching in pattern recognition and for block matching in motion estimation. The distance between the two-dimensional blocks X and Y is to be evaluated. For rectangular blocks, (1.1.7) can be generalized as

$$D = \sum_{i_1=0}^{N_1-1} \sum_{i_2=0}^{N_2-1} |x(i_1, i_2) - y(i_1, i_2)| \qquad (5.4.20)$$

The basic operations are subtraction, absolute value and addition. Subtraction, absolute value and addition are taken to be implemented together in an AD cell. Addition is an associative operation, i.e. it can be implemented in any order. Due to this associativity, structures with various networks result. A configuration with horizontal evaluation of the partial sums and vertical summation of the partial sums is shown in Figure 5.4.4.

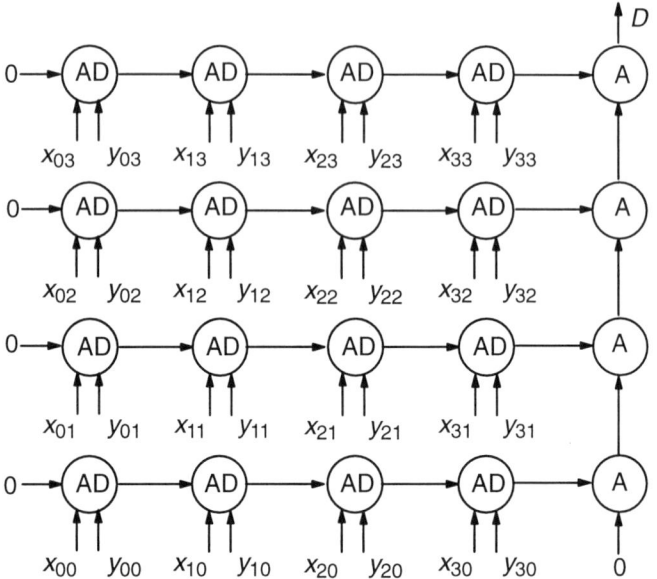

Figure 5.4.4: Dependence graph of a block matcher $(N_1 = N_2 = 4)$

When projecting the 2-D graph of Figure 5.4.4 to a 1-D signal flow graph, the various node types must be observed. In the case of vertical projection, only similar type nodes are mapped onto one another. A corresponding 1-D signal flow graph with systolic temporal processing is shown in Figure 5.4.5. In the case of horizontal projection, two types of nodes are mapped onto one another (see Figure 5.4.6). Thus, switchable processors are required for implementation. Addition is common to both operations. Thus, it is sufficient to implement only one adder per node.

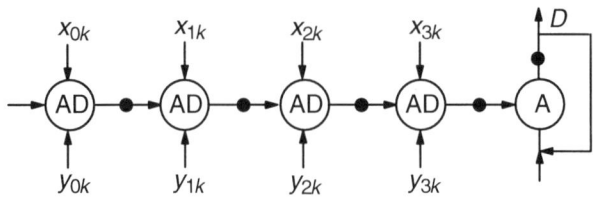

Figure 5.4.5: Signal flow graph of a block matcher after vertical projection

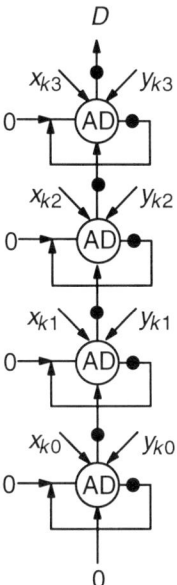

Figure 5.4.6: Signal flow graph of a block matcher after horizontal projection

As a further example of mappings with different nodes, the use of NOP nodes (no operation) for improving data input and output in systolic arrays will be treated. Through the projection of dependence graphs to signal flow graphs, it often happens that data must be input or output to inner rather than outer nodes. Often, dependence graphs can be extended using NOP nodes such that the data transfer problem is significantly diminished.

In the example of sorting, projection with

$$u = \begin{bmatrix} 1 \\ 1 \end{bmatrix}$$

leads to a signal flow graph with parallel data transfer (Figure 5.3.11). Through the use of NOP nodes, an implementation with serial data transfer can be achieved. The extended dependence graph with NOP nodes is shown in Figure 5.4.7. The NOP nodes are labelled N, the nodes for min/max selection are labelled M. The extension is carried out such that only the right-hand node has external data transfers for the given direction of projection. The nodes along the diagonal were also converted to NOP nodes since, due to the comparison with $-\infty$, the result is known in advance and this node steers the data from the horizontal to the vertical. The signal flow graph that results from projection is shown in Figure 5.4.8. The nodes must be cyclically switched in a particular temporal order between transfer mode (NOP) and active operation (compare) via an additionally created control signal.

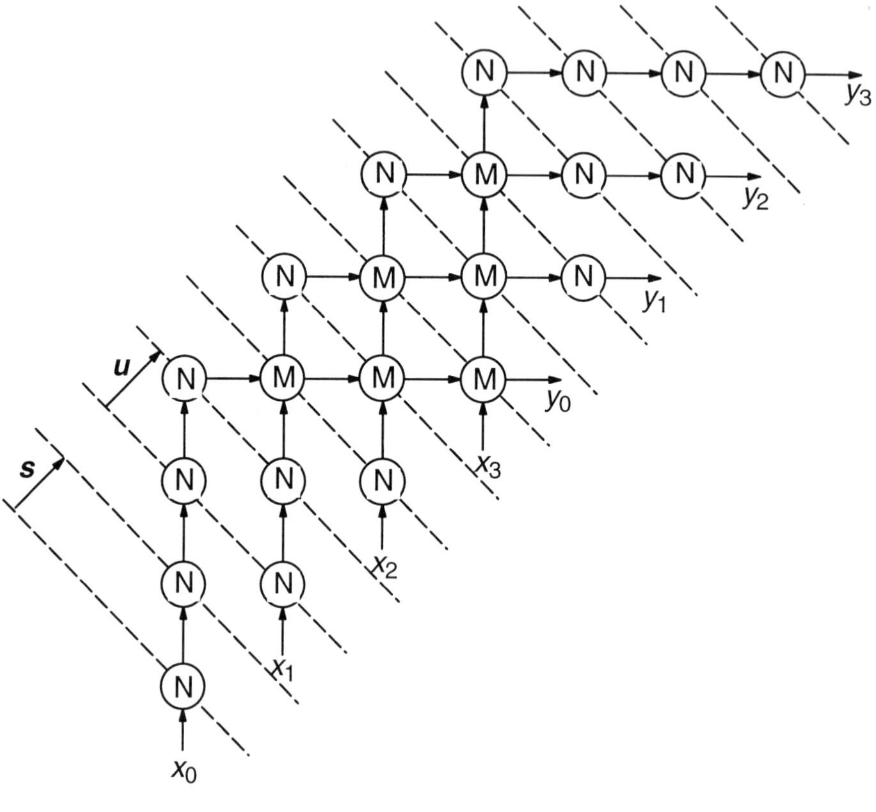

Figure 5.4.7: Extended dependence graph for a sorter indicating its equitemporal lines. M: Min/Max selector. N: NOP node

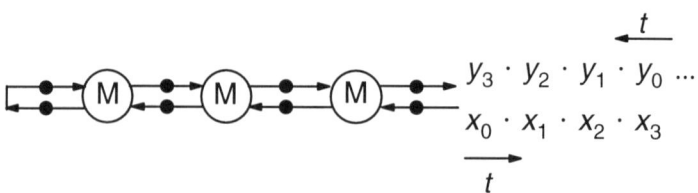

Figure 5.4.8: Signal flow graph of a sorter with switchable nodes

5.5 Implementation Aspects and Optimization

The steps previously covered pertain to the description of algorithms, the formulation of data dependencies in a dependence graph and mapping to a signal flow graph. To obtain a loop-free dependence graph, the algorithm must be formulated in single-assignment code. This dependence graph can be further adjusted to achieve local data dependencies (localization) and a preset array size (partitioning). Linear mapping methods were presented for assigning operations to a processor and an execution time. For the sake of completeness, it should be noted that non-linear mapping methods are also possible. However, such general solution schemes cannot be given for non-linear methods. In many such cases, it is necessary to make adaptations for the particular problem at hand.

Mapping between dependence graphs and signal flow graphs requires an extremely large solution space. These structures are of interest for an implementation in which the data transfer problem is also solved. In signal processing tasks, several sets of parameters are sequentially input anew, for which the particular results are computed. Only structures that account for this data input/output make best utilization of the processing elements.

From the signal flow graph, circuit structures for the processing elements and their networking can be easily derived. The operations in the nodes of the signal flow graph are implemented through dedicated modules for these operations. The delays along the edges of the signal flow graph are implemented through D-FF registers. The registers are assigned to the inputs or outputs of the processing elements. The time-dependent selection along the edges and the switching between various operations is implemented through multiplexers. The attainable throughput is determined by the maximal delay between two registers. If the demands on the throughput are higher, higher throughput must be achieved through additional pipelining. Through time scaling, the number of registers is increased, and through delay transfers, the portion of logic between two registers is reduced, as is its delay. So that all applicable connections are properly accounted for in the delay transfer, this should be carried out using the cut–set method.

The chronological sequence for the control signals can be directly derived from the mapping of the dependence graph to the signal flow graph. In periodic signal processing, the control signals are also periodic. Triggered by the maximal clock cycle, the control unit creates the required signals. The synthesis of such control units is described in the literature for designing sequential circuits [18], [20]. Figure 5.5.1 shows a special structure for a con-

trol unit particularly suited for the periodic control signals of array processors. The mod-N counter is a synchronous counter that sequentially counts
from 0 to $N-1$. The decoder consists of AND gates for recognizing particular
counter states. D-FFs are used at the output for the impulses that are one
clock period long. For creating longer impulses, RS-FFs are used.

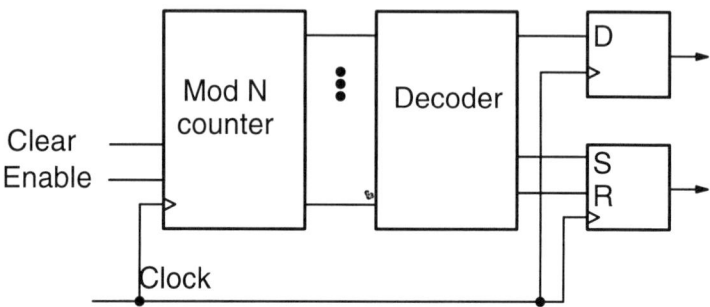

Figure 5.5.1:   Example of a control unit for application-specific array proces-
                    sors

In principle, it is possible to implement either central or local control
circuitry. Since the delays in the control sections are most often significantly
smaller than in the data path, central control of an array processor of moderate size is reasonable. For large, scalable arrays, more complicated local control is necessary. In this case, one obtains networks with local exchange of
data and control information. The design of array processors can be carried
out at different resolutions (granularities). In the previous sections, word level was assumed, i.e. samples, intermediate and final results were seen as a
whole. If the data are dissolved into bits and the mapping process takes inter-
actions between the bit planes into account, one speaks of bit level design
[87], [63]. For complex systems, further structuring via hierarchization and
insertion of partial processing levels is required. One then speaks of block
level design [75].

The design of array processors leads to many alternative solutions, in
which an optimal structure is sought. Criteria for formulating this optimum
are application-specific. A few criteria are listed as follows.

Computation time: The time interval between the start of the first computation and the end of the last computation for a set of data.

Clock rate: The maximal clock frequency for operating an array. The
clock rate is limited by the maximal delay time between two registers.

Block pipeline interval: The time interval between the start of two sequential data blocks.

Array size: This is a measure of the hardware costs. It can simply be the number of processing elements or, more exactly, the number of transistors or the area of silicon. For detailed specification of the hardware costs, the contribution of the control circuitry must be evaluated in addition to the data path.

I/O channels: The number of parallel data channels for inputting data or for transporting the results.

Throughput: The number of data sets processed per unit of time. The throughput is inversely proportional to the block pipeline interval.

The most common optimization criteria are smallest array size for given throughput or highest throughput for a given array size. As with the measure of efficiency from section 4.2, the quotient of throughput and array size can be used. The principal design flow for designing application-specific array processors is shown in Figure 5.5.2.

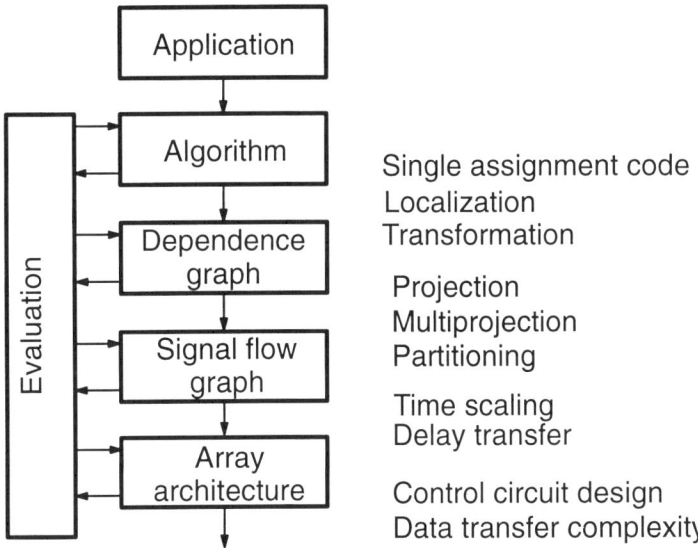

Figure 5.5.2: Design process for array processors

Beginning with the application, the algorithm to be used is described. Via dependence graph and signal flow graph representations, an array architecture is extracted. By adjusting the individual mapping steps, several alternatives are designed. Through evaluation of the architecture, an optimum is sought. Figure 5.5.2 lists several additional substeps that were treated. CAD tools have also been developed for supporting the design process. The references [80], [81], [82], [83], [84] are listed as representatives of such tools.

5.6 Exercises

1. Implementations for a matrix multiplication are to be found. As opposed to the representation in Figure 5.2.1, the indexing has been changed so that now

$$c_{ij} = \sum_{k=0}^{N-1} a_{ik} b_{kj}$$

Furthermore, systolic temporal ordering is assumed, i.e. a delay element can be found between all nodes in the dependence graph.

$$s = \begin{bmatrix} 1 \\ 1 \\ 1 \end{bmatrix}$$

The 3-D dependence graph is to be mapped to a 2-D signal flow graph via projection.

    a. The resulting 2-D array should be constructed such that the coefficients of the matrix A are permanently stored in the nodes, the coefficients of B are input along one side of the array and the results can be extracted from one side of the array. Determine the projection vector u and the data input and output. Sketch the array along with the chronological processing of the data for $N = 3$.

    b. Here, as opposed to a, a 2-D array is to be found in which the elements of A are also input along the edge.

2. Using systolic arrays, evaluate the cross correlation of two series of numbers $x(\cdot)$ and $y(\cdot)$.

$$u(k) = \sum_{i=0}^{N-1} y(i - k) \, x(i) \qquad \text{with} \ -k_0 \leq k \leq k_0$$

    a. The algorithm is to be described as a program. The program is to be constructed such that all statements are formulated as single assignment code and that only data access within the local neighbourhood occurs (localization).

b. Develop a dependence graph for the algorithm. Sketch the dependence graph for $k_0 = 2$ and $N = 4$. Note: The index space is to be created through the variables i and k.

c. Evaluate a schedule vector for systolic operation, i.e. all edges of the dependence graph are assigned a delay ≥ 1.

d. The 2-D dependence graph is to be mapped using the previously evaluated schedule vector to a 1-D signal flow graph such that the $x(i)$'s are input in parallel and the $u(k)$'s are output serially.

e. As an alternative to d, a 1-D signal flow graph is to be found in which the $x(i)$ and $y(i)$'s are input serially and the $u(k)$'s are output in parallel.

f. How can the solution from e be changed so that the autocorrelation functions of a series $x(i)$ are computed?

3. For an adaptive detector, a circuit is to be designed that counts the number of samples that are larger than a central value in an environment of length $(2N + 1)$. The samples of the input series are given to be x_i and v_i is the corresponding count.

$$u_{j,i} = \begin{cases} 1 & x_{i+j} > x_i \quad j = -N \dots N \\ 0 & \text{otherwise} \end{cases}$$

$$v_i = \sum_{j=-N}^{N} u_{j,i}$$

a. Construct a linear array of processing elements in which each PE consists of a comparison circuit (CMP) and an incrementer (INC) controlled by the result of the compare . Design shift register chains for the data input.

b. Design a systolic array with $(2N + 1)$ PEs using linear scheduling. Examine two solutions for the choice of processor index.

$$p_1 = N - j \qquad j = +N, \dots - N$$
$$p_2 = N + j$$

Change the shift registers to ensure the correct timing of the data input accordingly.

c. Map the linear array to a single PE and develop a shift register chain for data input. The schedule is given to be $t = p_1$.

4. Multiplier structures for the multiplication of positive numbers are to be developed. The operands are given to be

$$a_i \qquad i = 0, 1, \dots, n - 1$$

$$b_j \quad j = 0, 1, \ldots, n - 1$$

and the product bits are

$$p_k \quad k = 0, 1, \ldots, 2n - 1$$

For simplicity, $n = 4$ is assumed in the following.

a. A dependence graph is to be developed. All processing nodes should contain bitwise multiplication (AND) and a full adder. Unessential functions of the nodes should be eliminated through external switching. The carries from the full adder are to be passed in the direction of the i-axis. A regular array that makes all result bits available in the plane $j = n - 1$ is to be created through the introduction of NOP nodes.

b. The 2-D dependence graph is to be mapped to a 1-D signal flow graph, whereby a systolic schedule vector is to be assumed. In the resulting signal flow graph, the a_i's are to be input in parallel and the b_j's serially.

c. As opposed to b, the same mapping is to be carried out using a schedule vector that possesses a delay of 1 only in the j-direction. Compare the time-critical path of the solutions from b and c.

d. Due to the associativity of addition, the carries from the full adders can also be passed in the direction of the j-axis in the dependence graph. How do the solutions from b and c change for such a modified dependence graph?

e. A 1-D array from d is to be mapped to a single node through application of a multiprojection. Sketch the chronological course of input and output.

5. An implementation for the scalar product g of two vectors a and b is sought.

$$g = a^T b = \sum_{k=0}^{n-1} a_k b_k$$

The values of a_k and b_k are given to be positive, m-bit binary numbers. At the bit level, the scalar product is given by

$$g = \sum_{k=0}^{n-1} \sum_{j=0}^{m-1} \sum_{i=0}^{m-1} a_{ki} \, b_{kj} \, 2^{i+j}$$

A systolic implementation at bit level is sought.

a. Determine the 3-D dependence graph. Construct this dependence graph such that the individual product bits are available along the j-axis. Since the product $a_k b_k$ alone requires $2m$ bits, and n products are accumulated, $2m + \log_2 n$ nodes are require along the j-axis. In the i-axis, m nodes are required and n in the k-axis. Sketch the i-j plane and the j-k plane of the dependence graph for the particular values $m = 3$ and $k = 4$. Note that in the i-j plane, the carry bits run in the direction of j and the sum bits run along the diagonal . In the j-k plane, the sum bits should run in the direction of k and the carry bits in the direction of j. Use identical nodes comprising AND gates and FA cells. The required function is to be achieved through appropriate external switching. The index space is to be completely filled, i.e. non-occupied spaces are filled by NOP nodes.

b.   Find a schedule vector for a systolic scheduling.

c.   Determine a 2-D array through projection in the direction of i.

d.   Reduce the above 2-D array to a 1-D array through multiprojection in the direction of k.

6.   Determine a dependence graph for solving a system of equations. Through preprocessing, the matrix of coefficients has been brought to triangular form. Let

$$Ax = b$$

where x is a vector with n elements and A is a triangular matrix in which all the elements along the diagonal are non-zero and all the elements below the diagonal are zero. The recursive solution of the equation is

$$x_i = \frac{b_i - \displaystyle\sum_{j=i+1}^{n-1} a_{ij}\, x_j}{a_{ii}}$$

a.   Determine a 2-D depnendence graph using the two node types SD (sub-traction/division) and MA (multiplication/addition).

b.   Identify a schedule vector for systolic operation.

c.   Which are the time-critical elements that fix the throughput?

d.   The dependence graph is to be converted to a 1-D array through projection in which only nodes of the same type are mapped onto one another.

e.   The throughput of the 1-D array is to be increased through pipelining. The pipeline is to be constructed such that the limiting delay time is that of a carry-ripple adder or a subtractor. Approximately how high is the throughput (as a function of n) when the elements a_{ij} are m-bit, b_i are $2m$-bit and the division is to create m-bit results?

7.   With the aid of a matching method, determine the position in a data block y that bears the closest resemblance to a given pattern x. The L_1-norm is used

to evaluate the distance. The pattern x consists of n values, the data block y has $n + k$ values. The algorithm for the search can formally be described as follows.

$$u_j = \sum_{i=0}^{n-1} |x_i - y_{i+j}| \qquad j = 0,\ldots, k - 1$$

$$(v_j, w_j) = \begin{cases} (j, u_j) & u_j < w_{j-1} \\ (v_{j-1}, w_{j-1}) & \text{otherwise} \end{cases} \qquad j = 0, 1,\ldots, k - 1$$

with start value: $(v_{-1}, w_{-1}) = (0, \infty)$.

The first expression defines the calculation of distance for all possible positions j. The next expression defines the evaluation of the minimal distance w and the position v at which the minimal value occurs.

a. Develop a dependence graph for a 1-D matching method. Use two types of nodes. The two node types are labelled AD (addition of absolute values of differences) and M (evaluation of minima). The nodes M should be placed in the plane $i = n$ and all data y_j should be input in the plane $i = 0$.

b. The 2-D array is to be converted to a 1-D array with one M node and n AD nodes. With the aid of the known projection vector, find the processor base and a valid schedule vector.

c. A mapping to an array with 2 M nodes and $2n$ AD nodes is to be achieved using the partitioning method. Note: n is even. Determine the new indexing and the mapping matrix.

8. A mapping matrix is to be found for mapping a 4-D dependence graph to a 3-D signal flow graph. The data dependence matrix is given by

$$D = \begin{bmatrix} 1 & 1 & 0 & 1 \\ 0 & 1 & 0 & 0 \\ 0 & 0 & 1 & 1 \\ 0 & 0 & 1 & 0 \end{bmatrix}$$

a. Determine a schedule vector that creates a delay of 1 for all edges.

b. Specify the conditions for valid projection vectors based on the results of a.

c. Develop the corresponding processor bases for simple, valid projection vectors. Specify one projection vector and its corresponding processor base for a mapping to a feedback-free signal flow graph.

6 Filter Structures

Digital filters represent an essential part of digital signal processing [1], [2], [4]. Filters are employed in preprocessing for reducing noise and enhancing characteristic qualities. An example of this is edge enhancement in pictures. In source coding, filters are used for band splitting and interpolation [7]. Frequency multiplex transmission methods require filters for band limiting and band splitting. Filters are employed for pulse forming in digital modulators. Sigma-delta A/D and D/A converters use digital filters for decimation and interpolation. Furthermore, filters are used for echo suppression and as equalizers. This listing of their areas of usage is an indication of the importance and diverse use of digital filters.

Filters are categorized as either linear, non-linear or adaptive. Whenever possible, linear filters are preferred since their treatment can be completely put into theory. In particular, equivalent examinations in the time and frequency domains are supported by these filters. In their actual implementation, even linear filters have non-linear effects. However, the filters are implemented such that the non-linear effects are of subordinate importance.

In the following sections, first digital filters will be characterized, the design process will be briefly described and then alternative filter structures for implementing will be developed. Both non-recursive filters with finite impulse response and recursive filters will be treated. Among the non-recursive filters, special structures for two-dimensional filters, decimation filters and interpolation filters will be presented.

6.1 Characterization of Digital Filters

Linear digital filters belong to the special class of systems described as linear, time-invariant systems [1], [2]. One special characteristic of these linear, time-invariant systems is that they are completely described by their impulse response $h(\cdot)$. The impulse response $h(\cdot)$ is the reaction of a system to a unit step $\delta(\cdot)$.

$$\delta(k) = \begin{cases} 1 & \text{for } k = 0 \\ 0 & \text{otherwise} \end{cases} \qquad (6.1.1)$$

The reaction $y(\cdot)$ of a discrete, linear, time-invariant system to an input sequence $x(\cdot)$ is determined by convolution with the impulse response.

$$y(i) = h(i) * x(i) \qquad (6.1.2)$$

where $*$ is the convolution operator. This is a short notation for

$$y(i) = \sum_{k=-\infty}^{+\infty} h(k)x(i - k) \qquad (6.1.3)$$

Up to this point, filtering was formulated in the time domain. In addition to representations in the time domain, analysis in the transformed domain (frequency domain) is also essential for filtering. Analogous to the Laplace transform for continuous signals, the Z-transform is used for discrete signals. The Z-transform of the impulse response is

$$H(z) = Z[h(k)] = \sum_{k=-\infty}^{+\infty} h(k)z^{-k} \qquad (6.1.4)$$

The Z-transforms of the input sequence $x(\cdot)$ and the output sequence $y(\cdot)$ can be formulated in the same manner. A complete summary of the characteristics of the Z-transform can be taken from the relevant literature [1], [2], [88].

Two important characteristics in connection with filtering are the time-shift property

$$x(k - k_0) \quad \bullet\!\!-\!\!\circ \quad z^{-k_0}X(z) \qquad (6.1.5)$$

and that convolution in the time domain turns to a product in the frequency domain.

$$x(i) * h(i) \quad \bullet\!\!-\!\!\circ \quad X(z) H(z) \qquad (6.1.6)$$

Thus, the Z-transform of $h(\cdot)$ is the quotient of the Z-transform of the output sequence divided by the Z-transform of the input sequence.

$$H(z) = \frac{Y(z)}{X(z)} \qquad (6.1.7)$$

The function $H(z)$ is referred to as the system function. For analysis in the frequency domain, the discrete Fourier transform is applied. Through evaluation of the Z-transform along the unit circle, one arrives at connection to the discrete Fourier transform.

$$z = e^{j\omega} \qquad (6.1.8)$$

Substitution of (6.1.8) into the Z-transform of the impulse response yields the frequency response of the system

$$H(e^{j\omega}) = \sum_{k=-\infty}^{+\infty} h(k)e^{-j\omega k} \qquad (6.1.9)$$

Due to the specific characteristics of exponential functions with imaginary arguments, the system function is periodic with period $\omega = 2\pi$ or $f = 1$. It should be noted that a normalized frequency is assumed here. In principle, discrete sequences with normalized sampling intervals $T=1$ are handled. Scaling to real sampling intervals T' occurs using the relationship

$$\omega = \omega' T' \qquad (6.1.10)$$

According to their impulse response, filters are divided into those with a limited impulse response (FIR = finite impulse response) and those with an impulse response of infinite length (IIR = infinite impulse response).

For an FIR filter with an impulse response comprising N samples,

$$y(i) = \sum_{k=0}^{N-1} h(k)x(i-k) \qquad (6.1.11)$$

The corresponding system function is

$$H(z) = \sum_{k=0}^{N-1} h(k)z^{-k} \qquad (6.1.12)$$

According to (6.1.11), each output value is a weighted mean value of input data. Such a filter is thus labelled a moving average (MA) filter.

IIR filters are usually represented in the time domain by difference equations.

$$y(i) = \sum_{j=1}^{M} a(j)y(i - j) + \sum_{k=0}^{N-1} b(k)x(i - k) \qquad (6.1.13)$$

Using the rules of the Z-transform, it follows that the corresponding system function is

$$H(z) = \frac{B(z)}{A(z)} = \frac{\sum\limits_{k=0}^{N-1} b(k)z^{-k}}{1 - \sum\limits_{j=1}^{M} a(j)z^{-j}} \qquad (6.1.14)$$

In the case where $B(z) = 1$, an output value $y(i)$ is a linear regression of past output values. Such a filter is therefore called an autoregressive (AR) filter. The general function as in (6.1.14) is a combination of a moving average and an autoregressive filter. Thus it is called an autoregressive, moving average (ARMA) filter.

The frequency response of the system function as in (6.1.9) can be split into an amplitude (magnitude) spectrum and a phase spectrum. In many applications, filters with linear phase are desired.

$$\phi(\omega) = a\omega \qquad (6.1.15)$$

Filters with linear phase have constant delay in accordance with the slope of the phase spectrum. FIR filters with linear phase are characterized by a symmetric impulse response [1], [2].

$$h(k) = h(N - 1 - k) \qquad k = 0, 1, ... \, N - 1 \qquad (6.1.16)$$

The design process of filters can be divided into five sub-steps. These sub-steps are:

1. Specification of the filter characteristics

2. Approximation of the desired frequency response

3. Quantization

4. Architecture design

5. Circuit and layout design

The desired filter characteristics are often specified in the frequency domain. In this case, the pass and cutoff regions and their corresponding tolerances are specified.

In the case of FIR filters, the amplitude spectrum can be formulated as a polynomial. The best possible approximation to a desired frequency response is achieved using the methods of Chebyshev approximation. A known program after Parks and McClellan [89], [90] can be used here. Recursive filters are frequently extracted from analogue filters through transformations [1], [2], whereby bilinear transformations are used most often.

In the actual implementation, the filter coefficients and the signals must be represented by a limited number of bits. The quantization of the filter coefficients alters the frequency response. The effects of quantization and its treatment during design are described in the literature [1], [2]. One effect of limiting the bit count representing the signals is the creation of limit cycles in recursive filters. Furthermore, overflow effects must be considered.

After exact specification of the filter to be implemented, the next step is architecture design. In the following, special filter architectures and several alternatives are presented along with their particular characteristics. Filter implementations using general programmable processors are discussed in a later section.

6.2   1-D FIR Filter Structures

6.2.1   Signal flow graph of a 1-D FIR filter

An implementation can be directly derived from the definition of convolution in the time domain (6.1.11). These are termed direct form I structures in the literature. Figure 6.2.1 shows such a general filter structure. It comprises a filter arithmetic unit and a delay circuit. The filter arithmetic unit is an implementation of the data dependence graph. It contains multipliers and adders. The delay circuit furnishes the necessary data input for the filter arithmetic unit. It supplies several samples in parallel, the current sample and $N-1$ delayed samples, to be specific.

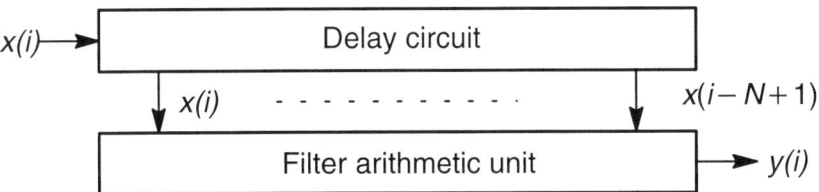

Figure 6.2.1: General FIR filter structure

The signal flow graph for direct form I is shown in Figure 6.2.2. Here, as is typical in the literature about filter structures, a delay of one clock interval is represented by z^{-1} and multiplication is symbolized by a triangle. This differs from the representations in Chapters 4 and 5.

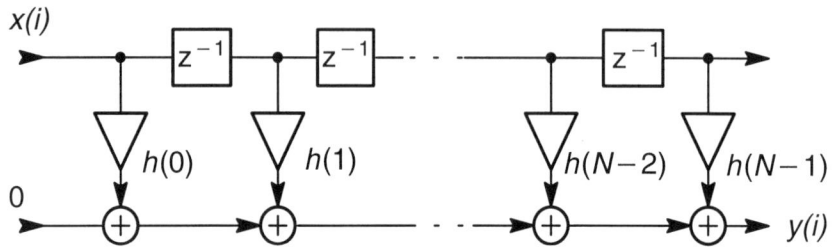

Figure 6.2.2: Signal flow graph of an FIR filter in direct form I

By exploiting the associativity of addition, a further structure for filter implementations can be extracted. It is called a direct form II structure. This structure can be derived from Figure 6.2.3.

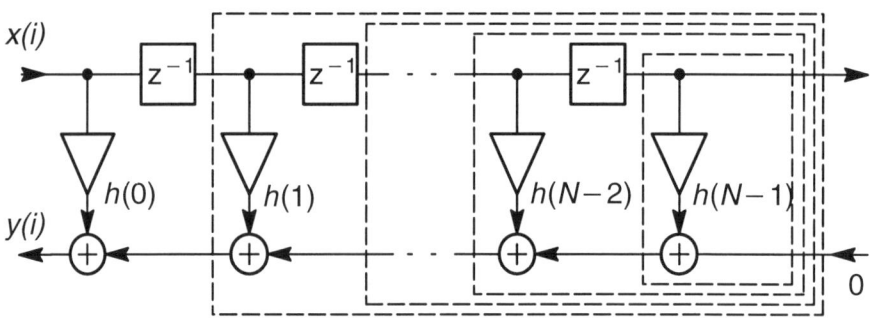

Figure 6.2.3: Shifting of the delay elements in an FIR filter using the cut–set method

The order of the addition for evaluating the output value is altered, and using the cut–set method, the delay elements are shifted from the upper path to the lower path. Input paths are given a negative delay, output paths a positive delay. The resulting signal flow graph in direct form II is shown in Figure 6.2.4.

x(i)

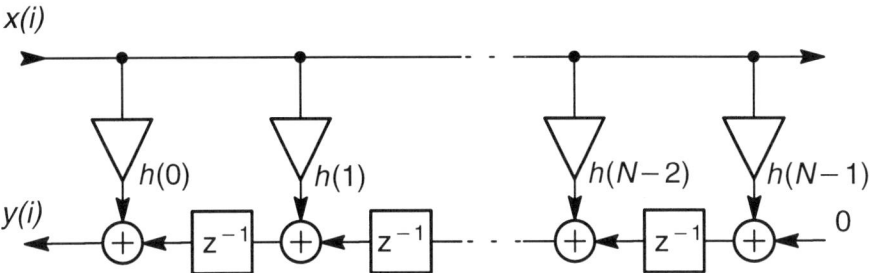

Figure 6.2.4: Signal flow graph of an FIR filter in direct form II

Direct form II can also be derived using the linearity of the function. The system function $H(z)$ is a polynomial in z^{-1}. Polynomials can be represented in two ways according to the Horner scheme. Both are used in the following.

$$Y(z) = \left[(...h(N-1)z^{-1} + ... + h(1))z^{-1} + h(0)\right] X(z)$$
$$= \left[h(0) + z^{-1}(h(1) + ... + z^{-1}h(N-1)...)\right] X(z) \qquad (6.2.1)$$

Interpreted from right to left, the first form designates that an input value is first delayed and then multiplied with a coefficient. The second form implies that an input value is first multiplied with a coefficient and that the resulting partial results are then delayed. If the delay operators and the order of the operations are properly observed, direct forms I and II can be derived from both lines of (6.2.1).

6.2.2 Implementations for low throughput

An FIR filter can be directly implemented in accordance with the signal flow graph of Figure 6.2.2. The time interval between two samples is given to be T_s. An implementation with this structure is possible whenever the delay of the time-critical path is less than the time T_s. The time-critical path comprises a multiplier, N adders and a register.

$$T_{D,max} = T_{D,MUL} + T_{D,NADD} + T_{D,REG} \qquad (6.2.2)$$

The following must hold for an implementation with this structure:

$$T_s > T_{D,max} \qquad (6.2.3)$$

If the time interval T_s is significantly larger than the delay time, the operative elements can carry out several operations in a row. If, for example,

$$T_s > N \cdot T_{D,max} \qquad (6.2.4)$$

all necessary operations can be mapped to one multiplier and one adder. Since all adders and multipliers in a time plane are active, it is necessary to force a temporal offset using the methods shown in section 5.3.5. Since N adders and multipliers are mapped onto one another, oversampling by a factor of N must be carried out according to (5.3.30) and (5.3.31). The scheduling vector must created a unit delay between the adders in the direction of projection. Figure 6.2.5 shows a signal flow graph after horizontal projection.

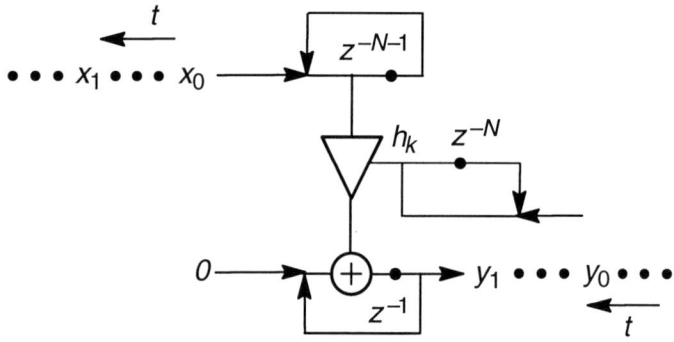

Figure 6.2.5:   Signal flow graph of an FIR filter with one MAC unit

The implementation requires two cyclic memories, one for the incoming samples and one for the coefficients. The cyclic memories consist of a shift register of appropriate length and a multiplexer. In audio processing, filters with an impulse response length N of 256 to 1024 are sometimes used. For such applications, it is reasonable to substitute the shift registers with RAM. The cyclic reads must be supported by an address generation unit in the memory unit.

Should the necessary throughput not be attainable through a single MAC unit, several remedial measures exist. First, the throughput can be raised with moderate additional expense by pipelining the multiplier as demonstrated in Example 5.3.1. A slightly less efficient implementation is a structure with several MAC units. Using the partitioning method of section 5.4.1, several alternative architectures can be determined. Figure 6.2.6 shows an example of an FIR filter with two MAC units. The throughput of this solution is twice that of Figure 6.2.5. This solution represents LSGP partitioning applied to one dimension. A further alternative solution can be extracted through LPGS partitioning. In addition, it is possible to operate two filter structures as in Figure 6.2.5 in parallel, offset by one input value.

It is to be noted that only the operative components (MUL, ADD) are reduced through projection and partitioning. The expenditures for memory and control circuitry increase.

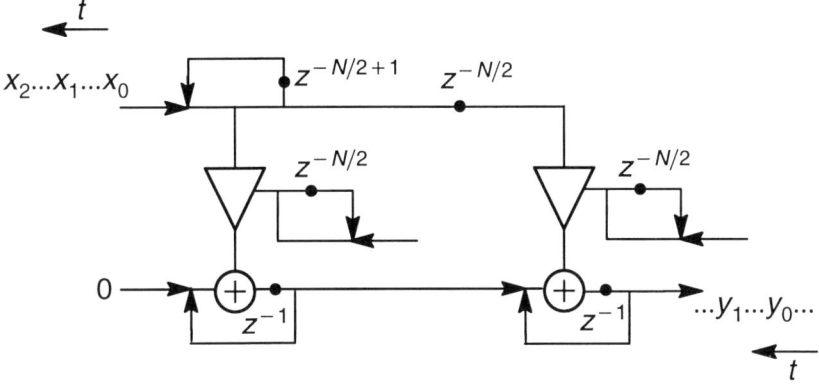

Figure 6.2.6: Example of an FIR filter with two MAC units (assumption: N is even)

6.2.3 Implementations for high throughput

Comparison of implementations in direct form
For high throughput, the direct implementation of signal flow graphs as in Figures 6.2.2 and 6.2.4 is quite practicable. It is not as obvious as it may seem which of the two structures is more favourable. The number of bits for representing the values is needed for a comparison. If the results are not truncated, then

samples $x(i)$: m_x bits
coefficients $h(k)$: m_c bits
results $y(i)$: $m_x + m_c + \log_2 N$ bits

If one compares the costs of direct forms I and II, it can be seen that direct form II requires more registers due to the larger word width in the sum path. The additional costs, measured in D-FFs, are

$$n_{D-FF} = N(m_c + \log_2 N) \tag{6.2.5}$$

The delay of the time-critical path is given by:

Direct form I:

$$T_{D,max} = T_{D,MUL} + T_{D,NADD} + T_{D,REG}$$

Direct form II:

$$T_{D,max} = T_{D,MUL} + T_{D,ADD} + T_{D,REG}$$

The delay of a chain of N adders is not N times the delay of a single adder since sum and carry signals simultaneously run through the configuration. The difference in delay of the two adder configurations is

$$T_{D,NADD-ADD} = (N - 1)T_{D,FA} \tag{6.2.6}$$

However, it must be noted that in direct form II, the input must drive N multipliers simultaneously compared to only one multiplier in direct form I. This increased load capacitance provides for a lesser time difference in the time-critical paths. The time difference in the time-critical path between direct forms I and II is roughly

$$T_{D,Diff} = (N - 1)(T_{D,FA} - m_c \tau_L) \tag{6.2.7}$$

Filters with high throughput are required, for example, for video signal processing. Here, 8-bit, 9-bit and 10-bit filter coefficients are generally used. For such coefficient word widths, filters in direct form II have a few advantages in terms of throughput, yet a higher number of D-FFs. Due to the less dominant advantages of direct form II, filters are normally implemented in practice in direct form I.

Systolic structures

A particular disadvantage of direct form filter implementations is the dependency of the delay time and thus the throughput on the number of filter coefficients N. Through the use of pipeline registers, systolic configurations can be created in which this does not hold. Such configurations can be extended without altering the throughput. This is called scalability. Pipeline registers can be threaded in with the aid of the cut–set method. A structure can be derived in which the input and the result values run in opposite directions. Figure 6.2.7 shows how this structure is extracted from Figure 6.2.4.

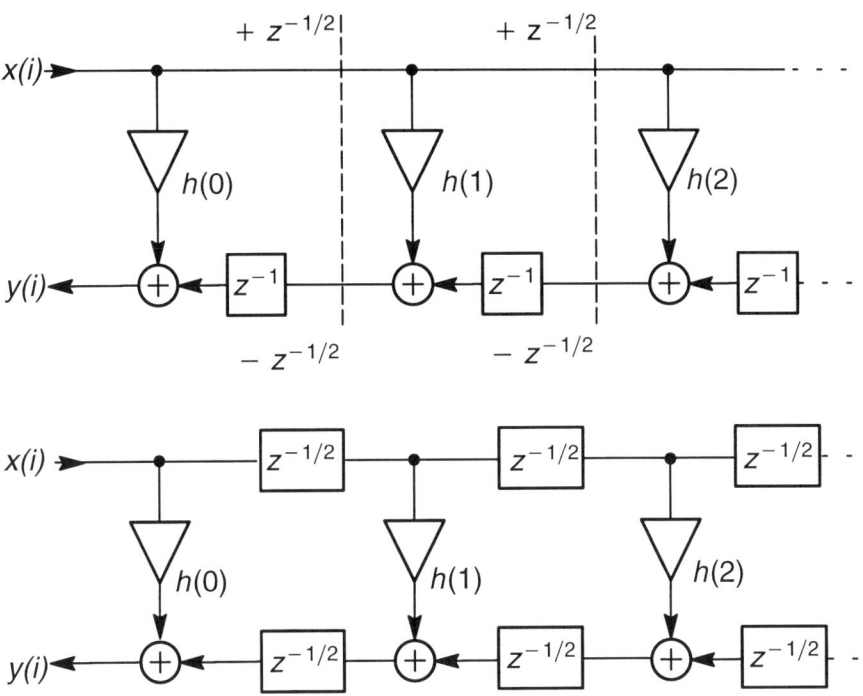

Figure 6.2.7: Systolic filter structure with counterrunning input and output signals

The corresponding period to $z^{-1/2}$ is half as large as that of z^{-1}. This means that filters are clocked twice as high as their incoming and outgoing data. Since the maximal clock rate is determined by the delay of the multipliers and adders, the throughput of filter structures of this kind is halved. Using a two phase clock and the implementation of clocked memory elements as in Figure 4.2.3, $z^{-1/2}$ delay elements can be implemented without halving the throughput.

If the input values and the results run parallel in the filter structure, registers are inserted in both paths. The existing register is doubled and one is inserted in the path without a register. Depending on the processing order of the filter coefficients, two different structures (see Figure 6.2.8) result. These structures are derived from the direct form I or II. The delay z^{-2} represents two clock registers.

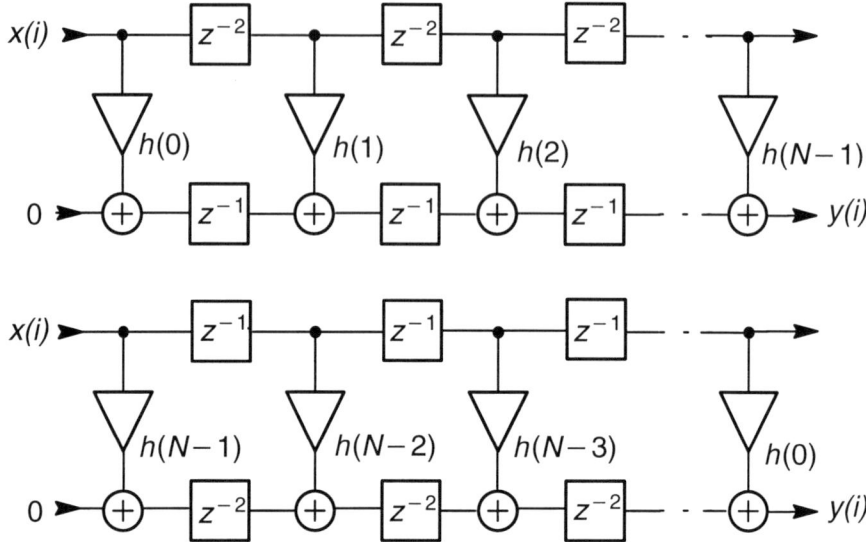

Figure 6.2.8: Systolic filter structure with parallel input and output signals

Filter implementations with fixed coefficients

In a filter, the multipliers require the most area. In cases where filters are to be implemented for fixed coefficients, the knowledge of these special coefficients can be used to reduce the hardware costs for multipliers. A CSD representation of the coefficients can be used uniformly for positive and negative coefficients and leads to a minimal number of non-zeros. The CSD representation of a coefficient $h(k)$ is given by

$$h(k) = \sum_{j=0}^{m_c - 1} h_j(k) \, 2^j \qquad h_j \in \{-1, 0, 1\} \qquad (6.2.8)$$

Through substitution into (6.1.11), it follows that

$$y(i) = \sum_{k=0}^{N-1} \sum_{j=0}^{m_c - 1} h_j(k) \cdot x(i - k) \, 2^j \qquad (6.2.9)$$

This relation is universally valid for various numeric representations. It can also be used with coefficients that have been coded with a modified Booth algorithm or that are given in two's complement representation. In all cases, the digits from the set $\{-1, 0, 1\}$ are to be used. In the case of two's complement representation, only the sign digit is weighted negatively. The equation (6.2.9) contains products using the digits of the coefficients. The result of a product with a digit of value zero is known in advance and can be cancelled. For multiplication by 1, the sample can be used directly. Multiplication by -1 can be implemented by a bitwise complement of the sample (one's complement) and the addition of 1 to the least significant bit of the sample. The sign digits of the intermediate results must be extended accordingly so that the following addition yields correct results. The necessary shift, corresponding to multiplication by 2, can be implemented by appropriate hardwiring of the bit planes (hardwired shift). Figure 6.2.9 shows a multiplier implementation of an FIR filter with fixed coefficients that takes the previous statements into consideration. All of the 1's of negative digits to be added are grouped in the term $g(k)$. As an alternative, due to the associativity and commutativity of addition, the individual adders can also be inserted into the horizontal result path [91].

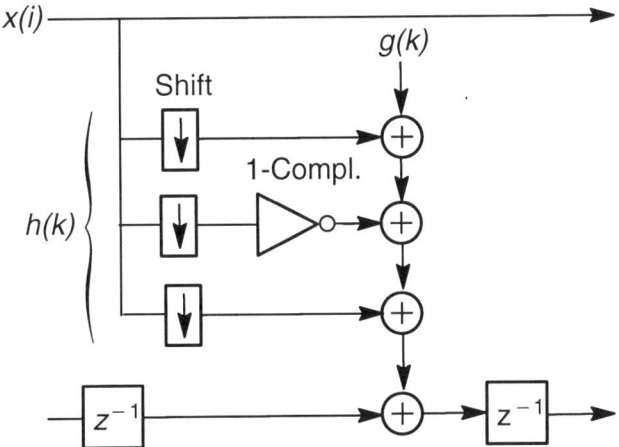

Figure 6.2.9: Multiplier implementation of an FIR filter with fixed coefficients

Bit-plane filters
In a filter implementation according to (6.2.9), the coefficients are broken up into individual bits, and the result is computed through continued addition of the samples with respect given to the weighting of the digits and the respective shift of the bit plane. In the case of an implementation of fixed coefficients, in particular the addition of zero-valued coefficient digits is suppressed. However, it is obvious that linear adder arrays are suited for the implementation of the result path of FIR filters, in particular in configurations with both variable and fixed coefficients [91], [92].

Due to the associativity of addition, the order of the summation in (6.2.9) can be exchanged. Then

$$y(i) = \sum_{j=0}^{m_c-1} 2^j \sum_{k=0}^{N-1} h_j(k)x(i-k) \qquad (6.2.10)$$

The inner sum has the typical structure of an FIR filter, yet only coefficients from the limited set of values $\{-1, 0, 1\}$ are used. The inner sum represents a filter for a bit plane of the coefficients. Therefore it is called a bit-plane filter. Thus, a bit-plane filter is given by

$$y_j(i) = \sum_{k=0}^{N-1} h_j(k)x(i-k) \qquad j = 0, ...m_c - 1 \qquad (6.2.11)$$

A complete filter must be implemented using m_c bit-plane filters. From (6.2.10), it seems readily apparent to create a parallel implementation that groups the m_c intermediate results in an adder tree.

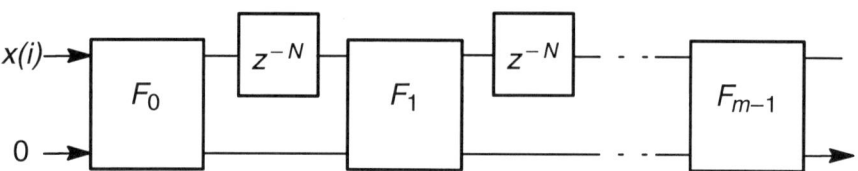

Figure 6.2.10: Cascade implementation of bit-plane filters. F_j is a bit-plane filter in direct form II for the bit plane j

One alternative is a cascade implementation. Here, however, attention must be paid to the delay between the input and output in the paths of the bit-plane filter. If the bit-plane filter is implemented in direct form II, the sample $x(i)$ must be delayed by an additional N clock cycles. Figure 6.2.10 shows the general architecture of a cascade implementation of an FIR filter with bit-plane structure (see also [91], [93]). Resolution down to the basic cells is shown in Figure 6.2.11. For simplicity, two coefficients with three bit-planes and a 3-bit representation of the input samples are assumed in this figure. Also, the particular treatment of negative digits is disregarded.

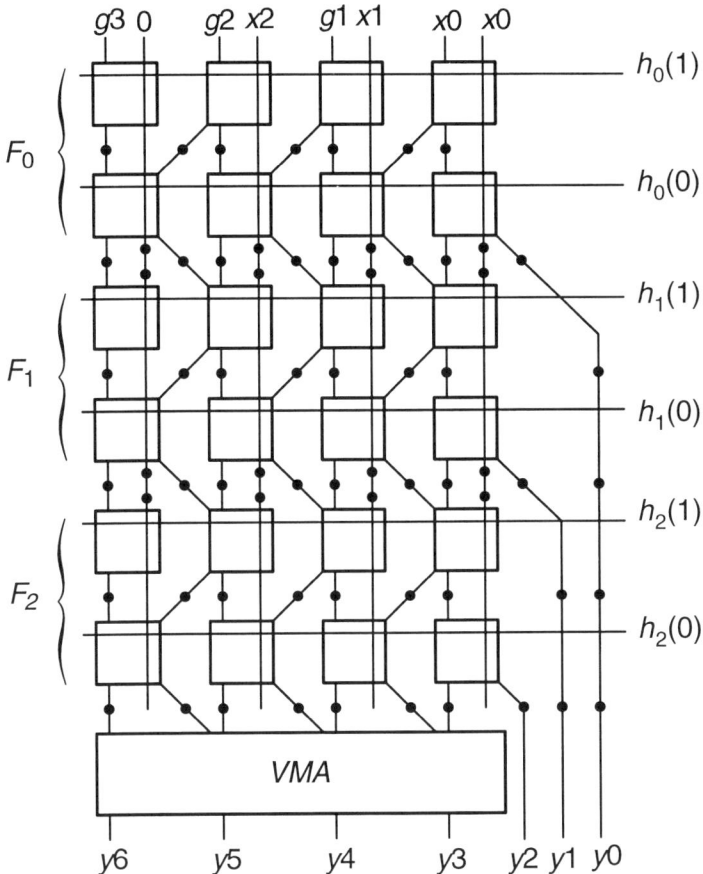

Figure 6.2.11: Pipelined carry-save bit-plane structure

Carry-save adder structures are particularly well suited for the implementation of multiple adders, as in this case. Because of the assumed three bit-planes, three bit plane filters were also implemented in this figure, whereby the lowest bit plane is the first. At the output of each bit-plane filter, the least significant result bit is already the final result bit and thus may be disregarded in further additions. It thus follows that the number of basic cells in a bit-plane architecture is smaller.

$$n_{cell} = (m_x + \log_2 N)m_c \cdot N \qquad (6.2.12)$$

The slowest element in the chain is the final vector-merging adder. Through intensive pipelining of this final adder, it can be achieved that the path determining the throughput comprises one full adder and one D-FF. This shows that, through pipelining, FIR filters for highest throughput can be implemented.

$$R_{T,max} = \frac{1}{T_{D,FA} + T_{D,FF}} \qquad (6.2.13)$$

Distributed arithmetic

For a bounded number of fixed coefficients, table-oriented implementation is possible. This is called distributed arithmetic [94].

As in (6.1.11), the filter function of an FIR filter can be seen as a scalar product of a vector h consisting of N filter coefficients with an equally large number of samples $x(i)$. The vector $x(i)$ is an N element section of the sequence $x(\cdot)$, whereby the index i signifies the first value. In vector notation, for an FIR filter,

$$y(i) = h^T x(i) \qquad (6.2.14)$$

Just as each sample can be broken up into its individual bit planes, this is also possible for the vector $x(i)$. In the case of an m-bit representation,

$$y(i) = h^T[x_{m-1} \; x_{m-2} \; \cdots \; x_1 x_0] \qquad (6.2.15)$$

where x_j is a vector that represents the bit plane j of x. It is assumed that each sample $x(\cdot)$ is given in two's complement representation. In this case, the value of $x(\cdot)$ is computed as in (3.1.12). Applied to (6.2.15), this leads to

$$y(i) = - h^T x_{m-1}(i) 2^{m-1} + \sum_{j=0}^{m-2} h^T x_j(i) 2^j \qquad (6.2.16)$$

The first term is evaluated negatively since the sign digit has negative weight in a two's complement representation. A function F is now defined that represents the scalar product of impulse response and bit-plane vector.

$$F[x_j(i)] = h^T x_j(i) = \sum_{k=0}^{N-1} h(k) x_j(i - k) \qquad (6.2.17)$$

The function F is found m times in (6.2.16) and has N bits as independent variables. If N is not too large, F can be implemented as a table in ROM. Figure 6.2.12 shows a corresponding implementation. The bit-plane vectors are input one after the other starting with the least significant bit. The function result according to (6.2.17) appears at the output of the ROM. The intermediate results are summed in a subsequent accumulator. The value for the sign digit is subtracted. The feedback loop via the shift register *SR* must be wired such that a shift by one bit plane is carried out, equivalent to multiplying by two. In each cycle, the least significant bit in the accumulator has its final value. After *m* cycles, the remaining result bits can be taken from the shift register. A parallel to serial converter and a register chain are necessary at the configuration's input to carry out the conversion from sequential data input to parallel bit planes.

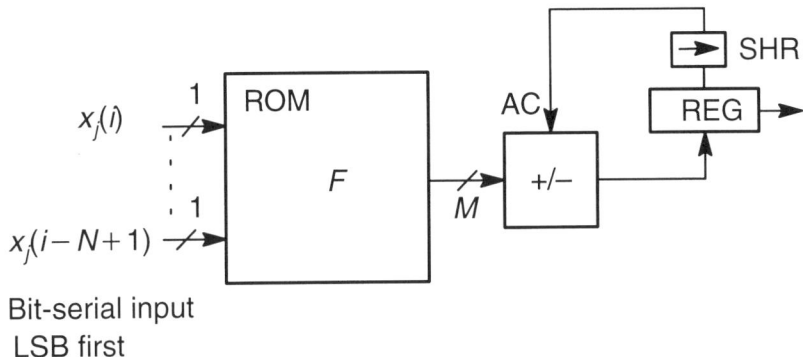

Bit-serial input
LSB first

Figure 6.2.12: Filter implementation with distributed arithmetic

The capacity of the ROM grows exponentially with the number of coefficients N. Theoretically, the number of output bits M depends on $\log N$ and the number of coefficient bits m_c.

$$M \leq m_c + \log_2 N \qquad (6.2.18)$$

The range of the ROM output is specified by the sum of all positive coefficients as the maximal value and the sum of all negative coefficients as the minimal value. Further it has to be considered that the magnitude of the filter coefficients decreases rapidly with distance from the centre of the impulse response, and that the impulse response is often so chosen that input and output have identical values for constant amplitudes. Under consideration of some additional rounding error M can be chosen to be significantly smaller than (6.2.18).

$$M \approx m_c + 1 \qquad (6.2.19)$$

The capacity of the ROMs is given by

$$N_{mem} = M \cdot 2^N \qquad (6.2.20)$$

In principle, it is possible to implement the distributed arithmetic using several parallel ROMs for large N. Each ROM then has a subset of the bit-plane vector and intermediate results computed in parallel must then be combined in an adder tree.

Linear phase FIR filters

Filters with system functions with linear phase have constant group delay. Filters with this characteristic are desired for video signal processing and communications applications. According to (6.1.16), such filters have symmetric impulse response. The filter multipliers normally represent the essential costs. By exploiting this symmetry, the multiplier costs can be nearly halved. Through mirroring of the signal flow graph in the point of symmetry of the coefficients, a modified structure for direct forms I and II can be derived (Figure 6.2.13). However, it is to be noted that the exploitation of symmetry in filters with high throughput (carry-save technique with extensive pipelining) only yields minor advantages [91].

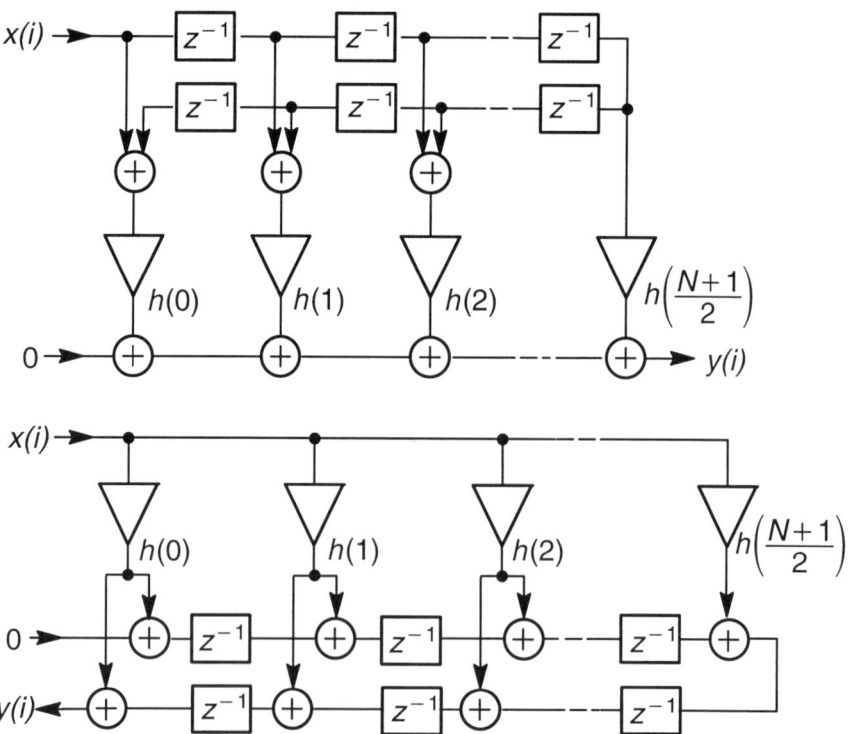

Figure 6.2.13: Modified signal flow graph for linear phase FIR filters (N odd)

6.3 2-D FIR Filters

In the area of image signal processing, filters are used that carry out filtering in both dimensions (horizontal and vertical) [3], [5]. Both image data and impulse response are doubly indexed accordingly. Thus, the convolution of an impulse response as in (6.1.11) must be extended to two dimensions.

$$y(i,j) = \sum_{l=0}^{M-1} \sum_{k=0}^{N-1} h(k,l)\, x(i-k, j-l) \qquad (6.3.1)$$

Its Z-transform must also be formulated in two dimensions. The 2-D Z-transform of (6.3.1) is

$$Y(z_1, z_2) = H(z_1, z_2)\, X(z_1, z_2) \qquad (6.3.2)$$

In this case, the system function is

$$H(z_1, z_2) = \sum_{l=0}^{M-1} \sum_{k=0}^{N-1} h(k,l)\, z_1^{-k}\, z_2^{-l} \qquad (6.3.3)$$

The indexing is chosen such that the first index is applied to the horizontal direction, the second to the vertical. The terms z_1^{-1} and z_2^{-1} represent delays of one sampling interval in the horizontal or vertical direction accordingly. Thus, in row-sequential data input, as is generally used in image signals, z_2^{-1} is the delay of one image row.

The inner sum of (6.3.3) can be interpreted as a 1-D filter for a fixed value of the variable l.

$$H_l(z_1) = \sum_{k=0}^{N-1} h(k,l)\, z_1^{-k} \qquad (6.3.4)$$

Substitution in (6.3.3) leads to

$$H(z_1, z_2) = \sum_{l=0}^{M-1} H_l(z_1)\, z_2^{-l} \qquad (6.3.5)$$

This means that the 2-D filter is implemented by M 1-D filters arranged vertically as given by the delay z_2^{-1}. Figure 6.3.1 shows the resulting filter structure.

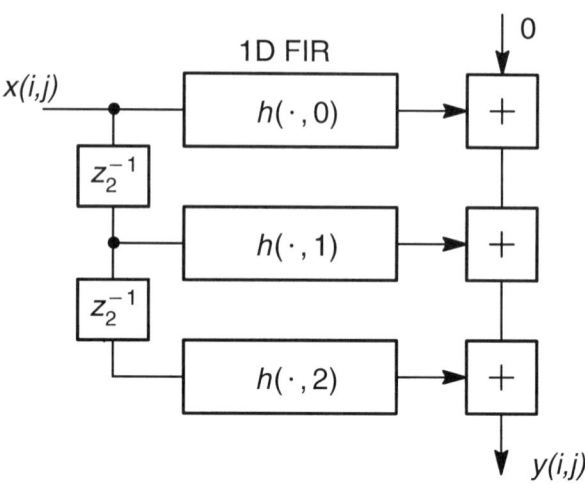

Figure 6.3.1:   Implementation of a 2-D filter through M parallel 1-D filters $(M = 3)$

In a 2-D filter, the filter arithmetic unit consists of $N \cdot M$ MA cells (multiplication/addition). Separable 2-D filters lead to a reduction in the filter arithmetic logic to $(N+M)$ MA cells. In separable filters, the 2-D impulse response can be split into the product of two 1-D impulse responses.

$$h(k, l) = h_1(k) \cdot h_2(l) \qquad (6.3.6)$$

The product of a horizontal and a vertical filter also holds for the system function.

$$H(z_1, z_2) = H_1(z_1) \cdot H_2(z_2) \qquad (6.3.7)$$

This means that the incoming data are first filtered by a filter H_1 in the horizontal direction and that the output data are then filtered by H_2 in the vertical direction. On account of linearity, the order of these two filters can also be exchanged. Figure 6.3.2 shows an implementation of a separable 2-D filter. In principle, the two filters have the same structure. Yet in row-sequential image data, each delay in H_2 is implemented by row memory.

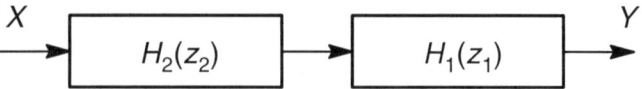

Figure 6.3.2:   Implementation of a separable 2-D filter as a chain of two 1-D filters

6.4 Polyphase Filter Structures

If the filter coefficients are split into several individual filters through sampling of the impulse response, the derived filters are termed polyphase filters [95], [96]. For a filter structure with M parallel paths, a given impulse response $h(k)$ is converted to $h_n(m)$ impulse responses M through division of the indices. Here n is the number of the partial filter.

$$h_n(m) = h(k) \quad \forall k \text{ with } n = k \bmod M$$
$$m = \lfloor k/M \rfloor \tag{6.4.1}$$

It should be noted that the index of h now designates partial filters and not bit planes as in (6.2.9). The sub-sampling of an impulse response is sketched in Figure 6.4.1 for $M = 2$. The convolution of the newly extracted impulse responses is given by

$$y_n(i) = \sum_{m=0}^{K_n} h_n(m)x(i - Mm - n) \tag{6.4.2}$$
$$K_n = \left\lfloor \frac{N - 1 - n}{M} \right\rfloor$$

The complete output sequence is the sum of all partial output sequences.

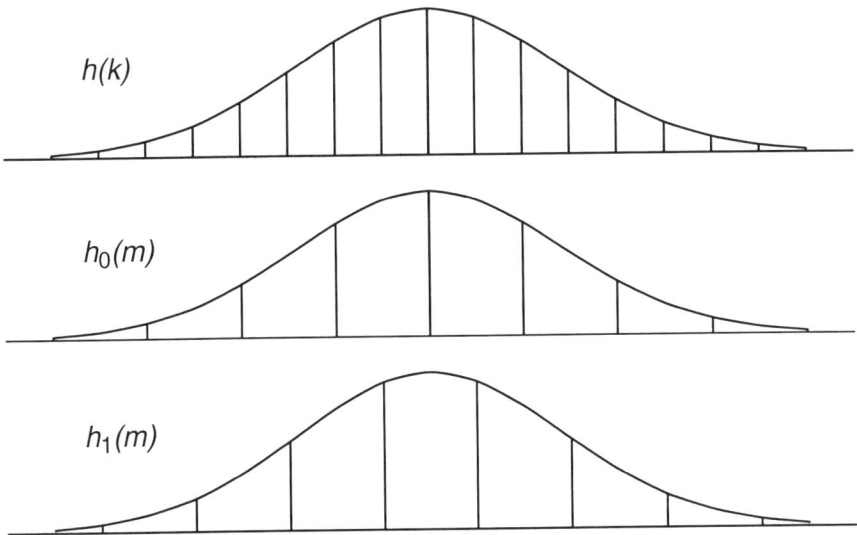

Figure 6.4.1: Splitting of an impulse response into two polyphase filters

$$y(i) = \sum_{n=0}^{M-1} y_n(i) \tag{6.4.3}$$

A corresponding equation for the system function can be derived through application of the Z-transform to (6.4.2) and (6.4.3).

$$H(z) = H_0(z^M) + z^{-1}H_1(z^M) + \ldots z^{-M+1}H_{M-1}(z^M) \tag{6.4.4}$$

This means that each partial filter is based on delay elements with an M-fold delay compared to the original structure. This allows a reduction in the clock rate by $1/M$ by replacing the M delays by one delay. The partial filters operate with mutual temporal offset among themselves. This offset corresponds to the start value during the convolution of the partial filters.

Polyphase filter structures can be used efficiently for the conversion of sampling rates. Filters for reducing the sampling rate are called decimation filters; those for raising the sampling rate are called interpolation filters. Decimation and interpolation filters are of particular importance for integer factors. Rational factors for sampling rate conversion are obtained through combination of these two filters.

6.4.1 Decimation filters

Sub-sampling of a sequence by a fixed factor M (decimation) cannot simply be carried out through suppression of original samples. As with any general sampling process, a periodic spectrum is created with a period corresponding to the sampling rate. To avoid aliasing through convolution of the spectra, a low-pass filter must be applied before sub-sampling.

Correlations and characteristic architectures are discussed in the following for a decimation factor of 2. The results can be applied to arbitrary integer factors.

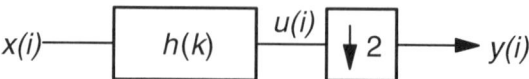

Figure 6.4.2:   Decimation filter for reducing the sampling rate by a factor of 2

The principle configuration for a decimation is shown in Figure 6.4.2. An input sequence x is filtered, and of the filtered sequence u, only every second value is used for the output sequence. The symbol used is meant to-represent sampling with a ratio of 2:1. The spectral representation of the three sequences x, u, y is shown in Figure 6.4.3. Here f_s is the sampling rate of the input sequence and f_s^* the sampling rate of the output sequence. The input sequence is bandlimited to $f_s/2$. After sub-sampling by a factor of 2, $f_s/2$ becomes the new sampling rate f_s^*. To avoid aliasing, the filter must carry out bandlimiting to $f_s/4$.

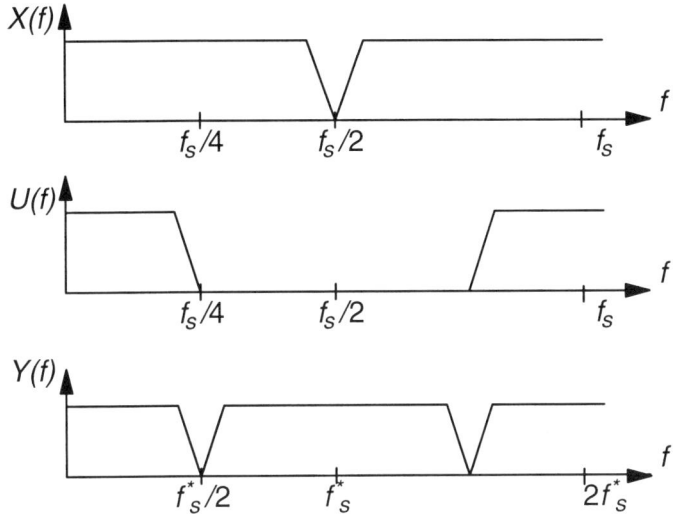

Figure 6.4.3: Spectral representation of decimation

The sequential index of x is given to be i, the sequential index of y is j and the even indices of x are to coincide with those of y, i.e. $i = 2j$. Substitution into (6.4.2) leads to

$$y(j) = \sum_{m=0}^{K_0} h_0(m)x(2j-2m) + \sum_{m=0}^{K_1} h_1(m)x(2j-2m-1) \quad (6.4.5)$$

The result is that h_0 and h_1 convolute disjunct subsets of x, i.e. the evenly indexed values of x are convoluted with h_0, the oddly indexed values with h_1. The corresponding division of the sequence x can be carried out by a switch controlled by the sampling clock (commutator, demultiplexer). The implementation of a corresponding switching structure is shown in Figure 6.4.4. The partial filters contain the impulse responses with half the number of coefficients derived through sampling.

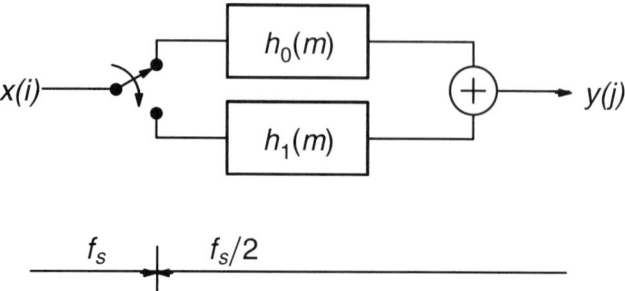

Figure 6.4.4: Decimation filter with polyphase structure and an input commutator

In the original structure of Figure 6.4.2, the sampler is at the output and only every second computed value u is used. The filter is driven by the sampling clock of the input sequence. In the structure in Figure 6.4.4, a commutator is used at the input and the two filter components are driven at half the clock rate. This raises the attainable throughput by a factor of 2. The cost in terms of the number of multipliers and adders is identical.

6.4.2 Interpolation filters

During interpolation with a fixed factor of M, $M-1$ new samples are computed between two existing samples. The computation of the intermediate values can be carried out by a weighted sum of the existing samples, i.e. a linear filter describes the interpolation process. As before, the particular example of $M = 2$ is to be observed.

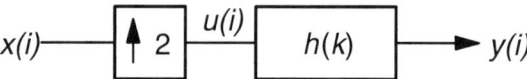

Figure 6.4.5: Interpolation filter for increasing the sampling rate by a factor of 2

Figure 6.4.5 shows a configuration for interpolation using a linear filter. The first block symbolizes the increase in the sampling rate by a factor of 2. The sampling rate is doubled through insertion of zeros into the positions to be interpolated. These samples represented by zeros are interpolated in a following filter. Figure 6.4.6 shows the interpolation process in the spectral domain. The spectrum is not changed through insertion of zeros, only the sampling rate is doubled, i.e. $f_s^* = 2f_s$. The spectrum $U(f)$ has period $f_s^*/2$ and not f_s^*. The period in the spectrum refers to the temporal distance between sequential samples. Since every second value of u is zero, the period in the spectrum does not change. Through elimination of the spectral region around $f_s^*/2$, a new spectrum is created with period f_s^*. The corresponding temporal sequence may not possess any more periodic zeros.

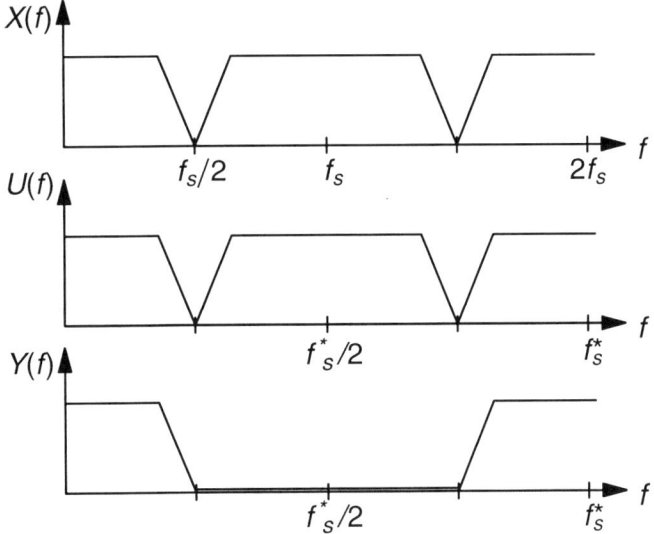

Figure 6.4.6: Spectral representation of interpolation

The sequence u created through insertion of zeros is to be indexed such that every evenly indexed sample leads to an input value and every oddly indexed sample leads to a zero.

$$u(2j) = x(i)$$
$$u(2j + 1) = 0$$

$$(6.4.6)$$

If the filter is now split into two parts as in (6.4.2), one part will always be convoluted with zeros due to (6.4.6). A sequence of zeros yields zeros even after filtering. Thus, the following equations hold for evenly and oddly indexed output values:

$$y(2j) = \sum_{m=0}^{K_0} h_0(m)u(2j-2m)$$

$$y(2j+1) = \sum_{m=0}^{K_1} h_1(m)u(2j+1-2m-1)$$

(6.4.7)

Together with (6.4.6), this means that the input sequence x must be convoluted with the two partial impulse responses in order to evaluate two sequential result values. Figure 6.4.7 shows the corresponding filter structure. The two partial filters are driven by the input clock, where in the original structure, the filter operated at the output clock rate. Thus, the filter is driven at a reduced clock rate in the new structure. The cost in terms of multipliers and adders is identical in both structures. In the original structure, half of the filter input values were zero, which lead to partial products, 50% of which could be ignored. Such partial products do not appear in the new structure.

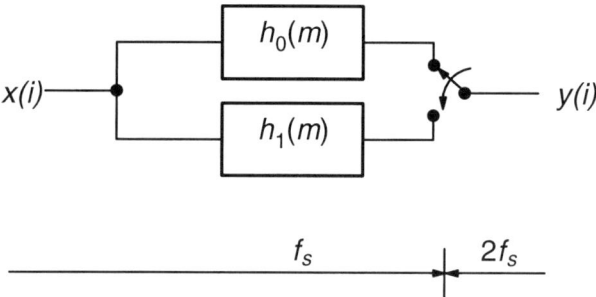

Figure 6.4.7:   Interpolation filter with polyphase structure and an output commutator

6.4.3   Filter banks

Through filter banks, an input signal is split into several parallel signals with distinct frequency bands. In accordance with the reduced bandwidth, the sampling rate of the output signal is reduced. A filter bank for spectral splitting is called an analysis filter. The reverse process, the merging of several parallel signals with distinct frequency bands to one signal, is called synthesis filtering. The applications for such filters are diverse [95]. Subband coding [7], for instance, requires such filter banks.

For simplicity, a filter bank with two bands, one low and one high, will be observed as in the previous section. The basic structures of an analysis filter and a synthesis filter are shown in Figure 6.4.8. As is typical in the relevant literature, the low-pass filter has the index 0 and the high-pass filter the index 1 [95]. The filters are to be constructed such that a circuit comprising an analysis and a synthesis filter bank in series reconstructs the original signal without error. This is called perfect reconstruction.

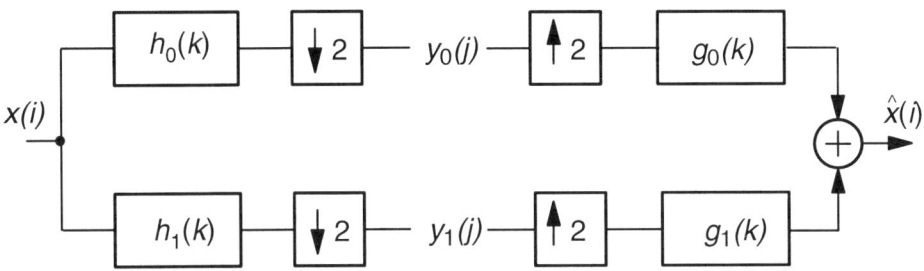

Figure 6.4.8: Two-channel analysis and synthesis filter bank

The filtering and the subsampling in the analysis filter can be described in the z-domain as follows [95]:

$$Y_0(z^2) = \frac{1}{2}[H_0(z)X(z) + H_0(-z)X(-z)]$$
$$Y_1(z^2) = \frac{1}{2}[H_1(z)X(z) + H_1(-z)X(-z)]$$

$$(6.4.8)$$

The factor 1/2 is caused by the decimation. The amplitude of the spectrum is halved. Through the sub-sampling with a factor of 2, a spectrum shifted by π is created. In the z-domain, this is symbolized by a negative sign. The argument z^2 of Y_0, Y_1 accounts for the scaling of the frequency axis between input and output due to the decimation.

The function of the synthesis filter can be formulated in the z-plane as follows:

$$\hat{X}(z) = G_0(z)Y_0(z^2) + G_1(z)Y_1(z^2)$$

$$(6.4.9)$$

Substitution of (6.4.8) into (6.4.9) yields the relationship between the output signal $\hat{X}(z)$ and the input signal $X(z)$.

$$\hat{X}(z) = F_0(z)X(z) + F_1(z)X(-z)$$

$$(6.4.10)$$

where

$$F_0(z) = \frac{1}{2} [G_0(z)H_0(z) + G_1(z)H_1(z)]$$
$$F_1(z) = \frac{1}{2} [G_0(z)H_0(-z) + G_1(z)H_1(-z)] \tag{6.4.11}$$

The function $F_0(z)$ describes the system behaviour. The alias components created are characterized by the function $F_1(z)$. For an alias-free filter bank, the following must hold:

$$F_1(z) = 0 \tag{6.4.12}$$

A filter bank with perfect reconstruction only shows delay characteristics. Then

$$F_0(z) = z^{-k} \tag{6.4.13}$$

In the literature, several methods for determining filter functions are known that fulfil the two conditions [95], [97], [98]. Requirement (6.4.13) is only roughly and not completely fulfilled in a few cases. The solution method using standard QMFs (quadrature mirror filters) in [97] is given by

$$H_1(z) = H_0(-z)$$
$$G_0(z) = 2H_0(z) \tag{6.4.14}$$
$$G_1(z) = -2H_0(-z)$$

Substitution into (6.4.11) leads to the system function

$$F_0(z) = H_0^2(z) - H_0^2(-z) \tag{6.4.15}$$

Methods for determining filter functions that fulfil requirement (6.4.13) are known [95]. For the implementation, it is important that the high-pass and the low-pass filter have a special relationship to one another. Both filters have coefficients whose absolute value is identical. The evenly indexed coefficients have the same sign; the oddly indexed coefficients have opposite signs.

Based on (6.4.4) the following holds for a polyphase filter with two partial filters:

$$H_0(z) = H_{00}(z^2) + z^{-1} H_{01}(z^2)$$
$$H_1(z) = H_{00}(z^2) - z^{-1} H_{01}(z^2) \tag{6.4.16}$$

Double indexing of the polyphase filter is now necessary since 0 and 1 are used to differentiate both the two partial filters and the low-pass from the high-pass filter. A relationship similar to (6.4.16) is also valid for synthesis filters. Substitution into the function for the synthesis filter leads to

$$\hat{X}(z) = G_{00}(z^2) \, [Y_0(z^2) - Y_1(z^2)] \\ + z^{-1} \, G_{01}(z^2) \, [Y_0(z^2) + Y_1(z^2)] \tag{6.4.17}$$

For the implementation of a two-channel filter bank, a cost efficient polyphase structure is derived using (6.4.16) and (6.4.17) [95]. The structure corresponds to a large degree to that of decimation and interpolation filters shown in Figure 6.4.9. For two-channel analysis and synthesis, however, only one subtractor is additionally necessary.

Due to the special relationship (6.4.14) between high-pass and low-pass filters, the implementation costs are practically halved. The delay requirements in terms of throughput are halved as in decimation and interpolation filters.

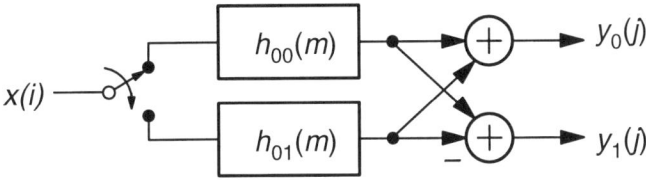

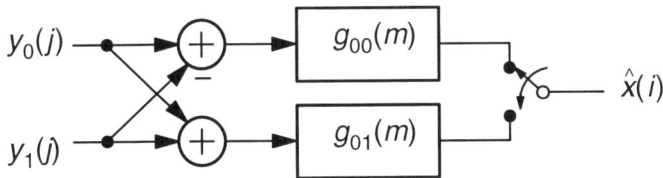

Figure 6.4.9: Two-channel analysis and synthesis filter bank with polyphase structure

To explain the principal relationships, two-channel filter banks were discussed. It is clear that by cascading two-channel filter banks, multi-channel filter banks can be implemented. The advantages shown for the implementations are then magnified.

6.5   IIR Filter Structures

In general, recursive filters can accomplish given selection tasks with a lower filter degree than non-recursive filters [1], [2]. The particular selection requirement can be defined by a transition from pass to cut-off region that is as steep as possible or by large cut-off damping. The lower filter degree of the recursive filter leads to fewer coefficients and thus also to fewer multipliers in a direct implementation. Due to their imperfect linearity, however, recursive filters often demonstrate stability problems. If particular values are stored in registers, periodic oscillations (limit cycles) occur when the input signal is zero [99]. These limit cycles can be limited or even entirely avoided through special measures. Wave digital filters show particularly favourable stability characteristics for simultaneously relatively low accuracy demands on the coefficients [100].

This section limits itself to a discussion of the throughput of simple recursive structures and the presentation of a few measures for increasing the throughput.

6.5.1   Recursive filters in direct form

According to (6.1.13), recursive filters can be represented by the sum of a purely recursive part and a non-recursive part. This means that a general recursive filter can be implemented by a non-recursive FIR filter and a purely recursive IIR filter (autoregressive filter AR) connected in series. This series circuit is admittedly not an implementation with a minimal number of storage elements, but it will be treated here in more detail for evaluation of the throughput.

The FIR filter at the front of the circuit has the coefficients $b(k)$ and an output sequence $u(i)$. According to (6.1.13),

$$u(i) = \sum_{k=0}^{N-1} b(k)x(i - k) \qquad (6.5.1)$$

For the autoregressive filter at the back of the circuit, it follows that

$$y(i) = \sum_{j=1}^{M} a(j)y(i - j) + u(i) \qquad (6.5.2)$$

For the special case in which the entire filter is autoregressive, the coefficients $b(k)$, with the exception of the first one, are zero and

$$u(i) = x(i) \qquad (6.5.3)$$

Figure 6.5.1 shows a direct form of an autoregressive filter according to (6.5.2). The attainable throughput of this structure is determined by the delay time of the time-critical path that consists of one multiplier, M adders and a register.

$$T_{D,\max} = T_{D,MUL} + T_{D,MADD} + T_{D,REG} \qquad (6.5.4)$$

The throughput is

$$R_T \leq \frac{1}{T_{D,\max}} \qquad (6.5.5)$$

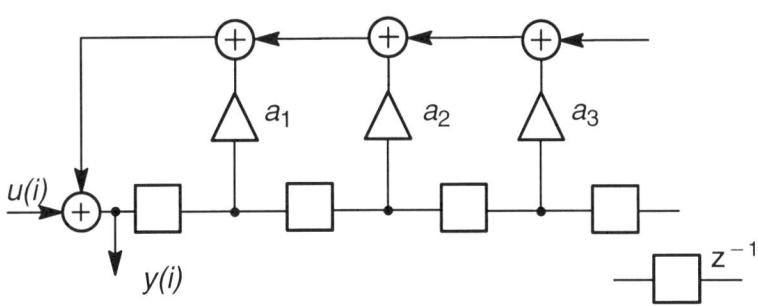

Figure 6.5.1: Autoregressive filter in direct form

Even the use of pipelining yields no significant improvement in throughput. Due to the counterrunning paths, the cut–set method only yields a splitting of the delay. A filter structure in which the delay element is split into two halves is shown in Figure 6.5.2. In the new structure, the adder chain is truly broken up, yet the filter must be driven at twice the clock rate. The delay along the time-critical path is then

$$T_{D,\max} = T_{D,MUL} + T_{D,ADD} + T_{D,REG} \qquad (6.5.6)$$

Due to the doubling of the clock rate, the throughput, however, is

$$R_T \leq \frac{1}{2T_{D,\max}} \qquad (6.5.7)$$

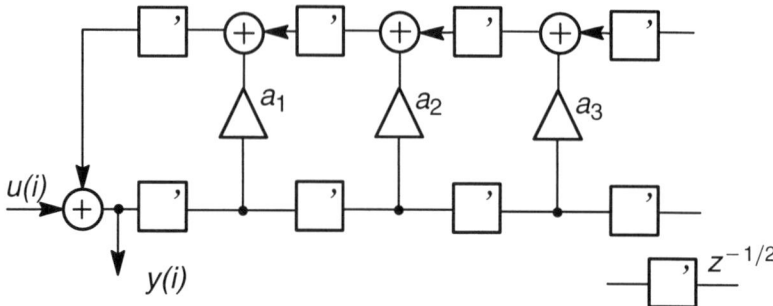

Figure 6.5.2:   Autoregressive filter after delay splitting using the cut–set
method

For moderate sizing of the filter degree M, this form of pipelining results not in an improvement but rather in a decline in throughput. It can thus
be seen that, as opposed to FIR filters, pipelining in recursive structures does
not lead to the desired increase in throughput. However, special measures
have been developed that also allow the throughput to be increased in recursive structures [91], [101], [102].

6.5.2 MSB first arithmetic

One measure for improving the throughput is to change the order of computation of the result bits. In general, this is done from LSB (least significant
bit) to MSB (most significant bit). Through computation of the result bits in
the opposite order and pipelining, the delay along the time-critical path is reduced. For the sake of simplifying its representation, this technique will be
presented using a first-order filter. For such a filter, the following equation
holds:

$$y(i) = a(1)\, y(i-1) + u(i) \tag{6.5.8}$$

Furthermore, in the following diagrams, the number of bits for $y(i)$, $a(i)$
and $u(i)$ will be limited to four. Both positive and negative representations
of numbers will be allowed.

$$
\begin{aligned}
y &= y_0 \cdot y_{-1}\, y_{-2}\, y_{-3} & -1 &\le y < 1 \\
u &= \quad .u_{-1}\, u_{-2}\, u_{-3}\, u_{-4} & -1/2 &\le u < 1/2 \\
a &= \quad .a_{-1}\, a_{-2}\, a_{-3}\, a_{-4} & -1/2 &\le a < 1/2
\end{aligned}
\tag{6.5.9}
$$

In a two's complement representation, the upper bit plane is negatively weighted (sign bit). Thus, in principle, the set of possible values for the result bits also includes −1.

In Figure 6.5.3, a recursive, first-order filter using a Pezaris array multiplier is shown. Due to the addition of u, the multiplier of Figure 3.3.17 had to be modified. For a negatively valued result bit, −1 must be added to the next higher level (see sign extension). In the figure one can also see that the multiplier yields twice as many bits as are used later in the recursion. This truncation is an underlying element of the non-linearity of recursive filters.

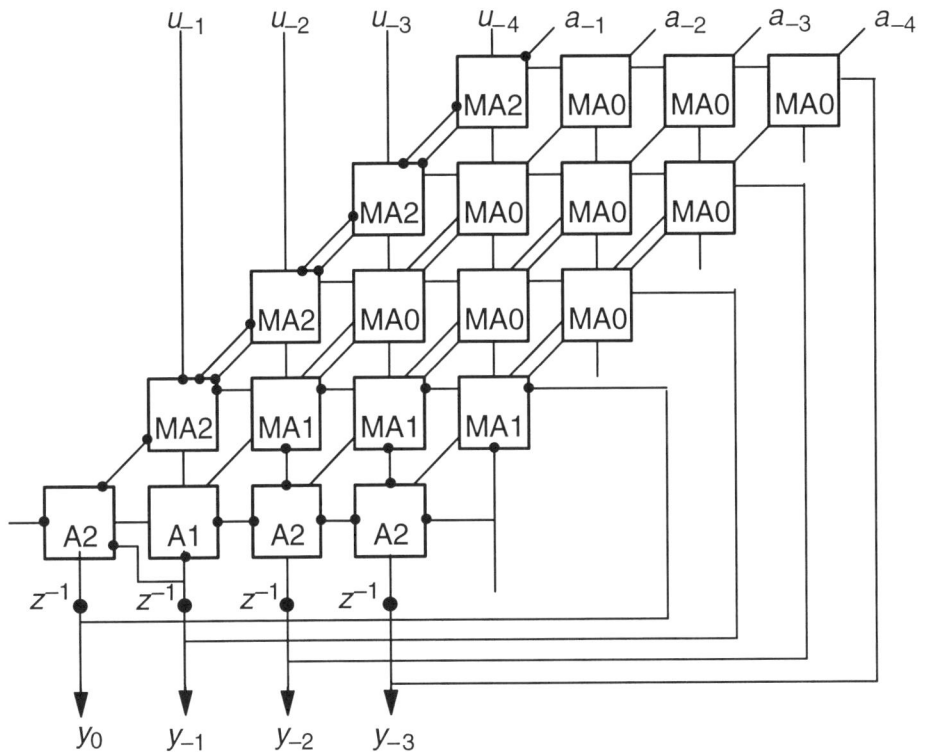

Figure 6.5.3: Recursive, first-order filter based on a Pezaris array multiplier

The time-critical path travels vertically through the middle of the array and then horizontally through the final adder. In the case of m bits, the delay along the time-critical path can be simply described as follows:

$$T_{D,\max} = T_{D,AND} + m(T_{D,FA,ss} + T_{D,FA,cc}) + T_{D,REG} \quad (6.5.10)$$

This, essentially, is the delay of the multiplier.

The order of addition of several numbers is arbitrary. Thus, the partial products from multiplication with the MSB can be added first and the partial products from the multiplication with the LSB can be added last. This changes the form of the rhombus, and the product of the highest valued bits is carried out first. However, the lowest valued bits can still contribute to the final value of the highest bits. This dependency can be split up through redundant representation of each bit plane. In the case of a two's complement representation, the set of values

$$y_j \in \{-1, 0, 1\} \tag{6.5.11}$$

is used for the redundant representation.

In this case, the redundant result of a bit plane must be coded by two bits. Several possibilities exist for coding the set of values in (6.5.11). Table 6.5.1 shows two of these. The left-hand table shows coding with sign (s) and magnitude (m), the right-hand shows coding with positive (p) and negative (n) values. The negative zero possible in the first table is disregarded.

Table 6.5.1: Coding of the value set $\{-1, 0, 1\}$

s	m	Value	p	n	Value
0	0	0	0	0	0
0	1	1	0	1	-1
1	0	-	1	0	1
1	1	-1	1	1	0

The product of the digits of the bit plane j with the coefficient a yields the set of values

$$y_j a \in \{-a, 0, a\} \tag{6.5.12}$$

These three numbers in the set of values are to be added under consideration of the position of the partial products. Instead of a subtraction of a, the two's complement of a may be added. This means that for negative digits, the bit-wise complement of a in addition to a one (sign s) must be added to the lower bit plane.

Figure 6.5.4 shows an implementation of a recursive filter using the MSB-first technique. Each bit plane is pn-coded. Through the altered order of multiplication, the rhombus form of the array has also been changed. The multiplication unit has become more complex through the introduction of negative digits. It is no longer an AND gate, but rather a controlled selection of a_j or $\bar{a}_j$. The second row of the array in Figure 6.5.4 is illustrated in more detail in Figure 6.5.5, in which the multiplication elements and the addition units have been separated.

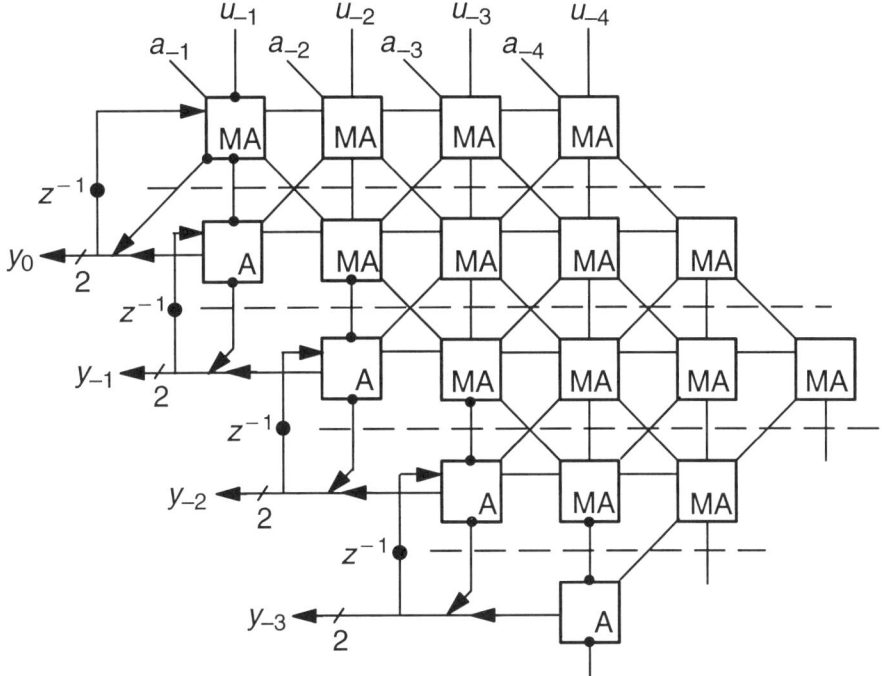

Figure 6.5.4: MSB-first filter structure for a recursive, first-order filter

Using the logic table in Table 6.5.1, the logic function of the multiplication unit is derived as being

$$w' = p\bar{n}a_k \lor \bar{p}n\bar{a}_k \tag{6.5.13}$$

The sign bit to be added is given by the function

$$s = \bar{p}n \tag{6.5.14}$$

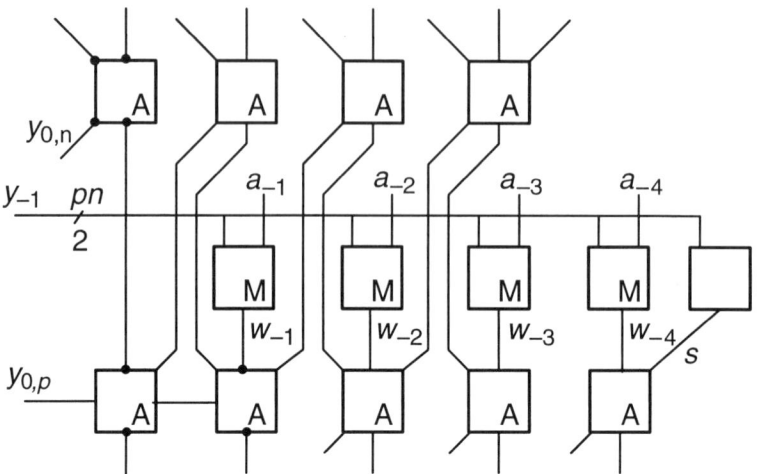

Figure 6.5.5: Close-up of the MSB-first filter structure

The logic function shown above holds for filter implementations with adjustable filter coefficients (programmable filters). In general, the coefficients a are fixed for a given filter to be implemented. In this case, depending on the value of a_k, the multiplication unit again simplifies to an AND gate.

$$w' = \begin{cases} \overline{p}n & a_k = 0 \\ p\overline{n} & a_k = 1 \end{cases} \qquad (6.5.15)$$

If each delay element is split using the cut–set method as already indicated by the dashed line in Figure 6.5.4, then the time-critical path consists of one multiplication unit, one full adder and a D-FF.

$$T_{D,\max} = T_{D,ME} + T_{D,FA,ss} + T_{D,FF} \qquad (6.5.16)$$

Due to the splitting up of the delay, $T_{D,max}$ defines the 50% throughput rate.

$$R_T \le \frac{1}{2T_{D,max}} \qquad (6.5.17)$$

The throughput of an MSB-first filter is no longer dependent on the word width of the multiplier. Relative to the structure as in Figure 6.5.3, the throughput is increased approximately by a factor of $m/2$.

In comparison to the original structure, the additional costs for implementing an MSB-first structure are modest. This means that this is a very efficient implementation since increasing the throughput by a factor of $m/2$ requires only modest hardware expenditures.

The results of an MSB-first filter are given in a redundant numeric representation. An additional carry-ripple adder is required for the conversion to a typical, non-redundant representation. This carry-ripple adder must be fully pipelined in order not to be a limiting element for the throughput. Besides the *pn*-coding of the signed digits used here, *sm*-coding (sign/magnitude) is also used in the literature [101], [102].

The representations above were limited to first-order filters. Connecting two first-order elements in series yields an autoregressive, second-order element (Figure 6.5.6). This can be implemented using the same technique as illustrated above. Complex filters are implemented more suitably by a cascade of biquad elements. A biquad element is a general, recursive filter whose numerator and denominator polynomials are of degree two or less. It has at most five coefficients a_1, a_2, b_0, b_1 and b_2.

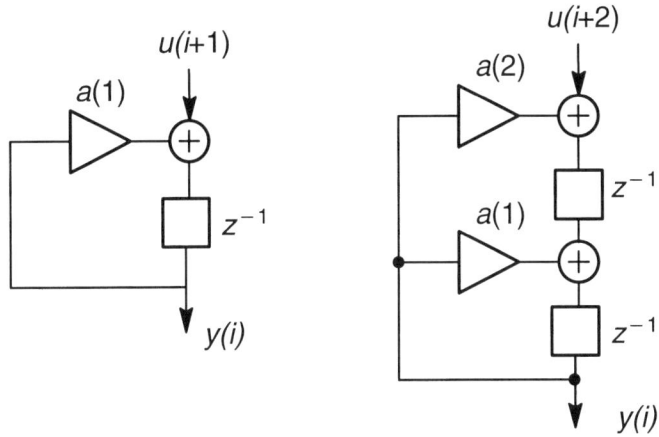

Figure 6.5.6: First- and second-order autoregressive filters in direct form II

6.5.3 Lookahead technique

As an alternative to the MSB-first technique discussed above, filter modifications that relax the temporal requirements on the recursive path are presented in the literature [103]. These are lookahead techniques and pole/zero compensations.

The lookahead technique inserts the general solution of previous output values into the recursive equations in order to derive a new recursive equation that does not depend on the preceding output values but rather on earlier output values.

From (6.5.8), a recursive, first-order filter yields the following equations for sequential output values:

$$y(i - 2) = a(1)y(i - 3) + u(i - 2)$$
$$y(i - 1) = a(1)y(i - 2) + u(i - 1)$$
$$y(i) \quad\;\; = a(1)y(i - 1) + u(i)$$

Substitution of the previous solution into the next one yields new differential equations.

$$y(i) = a^2(1)y(i - 2) + a(1)u(i - 1) + u(i) \qquad (6.5.18)$$

$$
\begin{aligned}
y(i) = a^3(1)y(i - 3) &+ a^2(1)u(i - 2) \\
&+ a(1)u(i - 1) \qquad\qquad (6.5.19) \\
&+ u(i)
\end{aligned}
$$

The recursive path of (6.5.18) has a delay of two clock cycles; that of (6.5.19) has a delay of three clock cycles. In general, a recursive path with an arbitrary number p of clock cycles of delay can be created. Using the cut–set method, the corresponding p registers can be split up and distributed to various levels of the multiplier.

The advantages of the recursive path must be paid for through the additional costs of a new, non-recursive section. The system function for (6.5.18) and (6.5.19) shows that this method creates additional poles and zeros that compensate for one another.

The system function of the original recursive filter is

$$H(z) = \frac{1}{1 - a\, z^{-1}} \qquad (6.5.20)$$

The two modified system functions are as follows:

$$H'(z) = \frac{1 + a\, z^{-1}}{1 - a^2\, z^{-2}} \qquad (6.5.21)$$

$$H''(z) = \frac{1 + a\, z^{-1} + a^2\, z^{-2}}{1 - a^3\, z^{-3}} \qquad (6.5.22)$$

In the end, these system functions must all be identical. This means that the division of the numerator polynomial by the denominator polynomial in (6.5.21) and (6.5.22) yields the nominator polynomial of (6.5.19) as its quotient.

For recursive filters of order higher than one, there are two principal approaches. The one is repeated substitution of earlier output values into the differential equation. This is called clustered lookahead. The other is the modification of each pole of the system function.

$$\frac{1}{z - a} \quad \Rightarrow \quad \frac{H_D(z)}{z^p - a^p} \tag{6.5.23}$$

This is called scattered lookahead.

One problem of lookahead techniques is that the pole/zero compensation is not perfect due to the finite accuracy in the implementation. This leads not only to a change in the system function but also to stability problems in many cases.

The high hardware costs for the implementation are one fundamental problem of lookahead techniques. In scatter lookahead, to increase the throughput by a factor of p, each pole is supplemented by an FIR filter with $(p-1)$ MAC units. If the additional costs for registers are also taken into consideration, one can see that the hardware costs increase disproportionately. Thus, lookahead techniques do not yield structures with high efficiency.

6.6 Exercises

1. N products are to be added in an FIR filter with N coefficients. This N-operand adder can be implemented in various ways.

    a. As in direct form I, the N-operand adder is to be implemented by a linear configuration of carry-ripple adders. Evaluate the costs and the delay for given word widths m_c and m_x. Give the costs in terms of the number of full adders.

    b. Carry out part a for a configuration using carry-save adders.

    c. The N-operand adder is to be implemented through a CSA tree. Evaluate the costs and the delay.

    d. The throughput is to be increased through pipelining. Which of the three configurations is to be preferred in terms of its costs and throughput?

2. An FIR filter with a fixed filter function is to be developed. The filter has four filter taps with the following coefficients:

    $$h = \left[\frac{1 + \sqrt{3}}{8}; \ \frac{3 + \sqrt{3}}{8}; \ \frac{3 - \sqrt{3}}{8}; \ \frac{1 - \sqrt{3}}{8} \right]$$

    Real-valued coefficients are quantized to five bits of accuracy for the implementation. The quantized filter coefficients are

$$h_d = \frac{1}{32} \ [11; \quad 19; \quad 5; \ -3]$$

   a. Formulate the coefficients in CSD representation.

   b. Carry-save adders are to be used for the basic arithmetic cells. Design an adder tree for implementing the filter function. The word width of the input signal $x(i)$ is 8-bit.

   c. The unused inputs of the adder are to be wired such that the one's necessary to implement the creation of two's complements are available (corresponding to the term $g(k)$ in Figure 6.2.9). Furthermore, the arithmetic unit is to add the constant value 0.5 to the output value through wiring of the unused inputs. This allows for rounding of the result to eight bits through truncation of the digits behind the period. It should be originally assumed that the output value lies in the range [0; 255].

   d. Specify the logic function of a limiter that limits the output value to the lowest (or highest) allowable value when the allowed range is exceeded.

3. FIR filters with various structures are to be designed and a comparison of their efficiency is to be carried out. Particular values are to be assumed for the comparison. The number of taps is given to be $N = 9$ and the word width of the input data and the coefficients is 8 bits. The input data and the coefficients are assumed to be positive numbers, whereby the sum of the coefficients is 512. It is taken that an estimation has shown that the necessary throughput can be achieved with three simultaneously active MAC units.

   a. Using the partitioning methods from section 5.4.1, filter structures are to be designed with the three MAC unit.

   b. Filters for parallel operation, each with one MAC unit as in Figure 6.2.5, are to be developed.

   c. As an alternative, pipelining is to be inserted into a structure like Figure 6.2.5.

   d. Compare the costs for the filter arithmetic units in all three solutions. Assumption: D-FF: 8 transistors, MUX-cell: 8 transistors, FA: 24 transistors, MUL-cell: 32 transistors.

   e. The data input necessary is to be determined for all three solutions. How large is the storage capacity needed in each case?

4. Subband coding is to be performed for data transmission. An analysis and a synthesis filter bank are to be designed. A QMF with 10 filter taps is to be used as the filter function. The low-pass filter function of the analysis filter has the coefficients

$$h_{QMF, LP} = \frac{1}{512} \ [8; \, -1; \, -43; \, 44; \, 248; \, 248; \, 44; \, -43; \, -1; \, 8]$$

a. Determine the filter function of the high-pass filter in the analysis filter bank and the filter function of the synthesis filter.

b. The partial band filters are to be implemented in a polyphase structure as in Figure 6.4.9. Determine the functions of the arithmetic blocks.

c. The word width of the signals $x(i)$, $y_0(i)$, $y_1(i)$, and $\hat{x}(i)$ is 8 bits. An arithmetic module is to be designed which can be used for the functions $h_{00}(m)$, $h_{01}(m)$, $g_{00}(m)$ and $g_{01}(m)$. How must the inputs of this module be connected?

5. A recursive, first-order filter is to be implemented in an MSB-first architecture. The starting point is the structure shown in Figures 6.5.4 and 6.5.5.

a. The output digits are to be *sm*-coded instead of *pn*-coded. Which logic function converts *pn*-coding to *sm*-coding? How must the logic function of the multiplication unit be changed? Compare the two solutions in terms of costs and throughput.

b. Configurations for converting redundant, *pn*- or *sm*-coded, numeric representations into a non-redundant, two's complement representation are to be determined.

c. The input value u and the coefficient a of a recursive, first-order filter are given to be positive. Determine an MSB-first filter structure based on a carry-save multiplier. A redundant numeric representation combining the carry and sum bits and using the set of values $\{0, 1, 2\}$ is to be used for the output digits.

6. An autoregressive, second-order filter is to be implemented using the lookahead technique. The coefficients $a(1) = \sqrt{3}/2$ and $a(2) = -1/4$ are to be assumed.

a. Using the clustered lookahead method and through continued insertion into the differential equations, new differential equations are to be determined that have a delay of two, three, and four clock cycles in the recursive part. Sketch the corresponding signal flow graphs and compare the hardware costs. Carry out delay shifting using the cut–set method and comment on the attainable throughput.

b. Carry out the same as in a using the scattered lookahead technique. Compare the results with one another.

7 Implementations of the Discrete Fourier Transform

The discrete Fourier transform (DFT) forms the basis for many methods of spectral analysis. Spectral analysis can be used for estimation of system parameters and for calculating the system function by filter models. Many applications from areas such as seismology, radar and sonar use spectral analysis methods based on the discrete Fourier transform. One important characteristic of the discrete Fourier transform is often exploited, namely that convolution in the time domain corresponds to multiplication in the frequency domain. In particular fast algorithms (FFT) support the quicker execution of convolutions.

In the following sections, several characteristics of the DFT will be discussed and theorems of the DFT will be presented. A further section treats the direct implementation of the DFT. Thereafter, the fast Fourier transform (FFT) will be explained and corresponding implementations will be derived.

7.1 Characteristics of the DFT

How the discrete Fourier transform can be derived from the standard Fourier transform is shown in the literature [1], [2]. The essentials will be briefly presented here following the formulation by Pearson [104].

Given a continuous signal $x(t)$ and its corresponding Fourier transform $X(jf)$, the effect of sampling $x(t)$ can be described as follows:

$$comb_T \cdot x(t) \quad \bullet\!\!-\!\!\circ \quad rep_{1/T} \, X(jf) \qquad (7.1.1)$$

Here, $comb_T$ is a sequence of Dirac delta functions a distance T apart and $rep_{1/T}(\cdot)$ stands for "repeat", i.e. the periodic repetition of the function $X(\cdot)$

spaced $1/T$ units apart and the summation of all these shifted functions. The essence of (7.1.1) is that continuous sampling at time intervals of T leads to a spectrum that is periodic with frequency $1/T$. The addition of several spectral components (aliasing) decreases when $X(jf)$ is band-limited to $\pm 1/(2T)$.

The Fourier transform is symmetric, i.e. the multiplication of the spectrum with a sequence of Dirac delta functions (sampling in the frequency domain) leads to a periodic time signal.

$$rep_T \, x(t) \quad \bullet\!\!-\!\!\circ \quad comb_{1/T} \cdot X(jf) \qquad (7.1.2)$$

This is the essence of a Fourier series representation. A periodic time-domain with period T has only spectral components at multiples of the frequency $1/T$. If $x(t)$ is a function extended over an arbitrary period of time, any components that overlap during repetition must be summed. Through multiplication with a windowing function (square windowing) of appropriate length, periodic continuation of an interval of length T of the time function is achieved. Then

$$rep_T \, [w_T(t) \cdot x(t)] \quad \bullet\!\!-\!\!\circ \quad comb_{1/T} \, [W_T(jf) * X(jf)] \qquad (7.1.3)$$

This means that discrete frequency lines in the spectrum are found that result from the convolution of the spectrum of the signal to be analysed with the spectrum of the windowing function.

The two procedures, sampling and periodic continuation, can now be combined. Using equations (7.1.1) and (7.1.3), it follows that

$$rep_{NT} \, [comb_T \cdot (w_{NT}(t) \cdot x(t))] \quad \bullet\!\!-\!\!\circ$$

$$comb_{1/NT} \cdot rep_{1/T} \, [W_{NT}(jf) * X(jf)] \qquad (7.1.4)$$

The periodic continuation of a sequence of samples leads to sampling of the periodic spectrum arising from the segment of the time signal. Due to the periodic spectrum, N spectral values suffice for the characterization of N periodically continued samples. It is to be noted that spectral values are calculated belonging to the evaluated time segment. This is the so-called short-time spectrum. Even when this segment of time behaves spectrally like a function extending indefinitely in time, the convolution with the spectrum of the windowing function will change the spectral values.

The discrete Fourier transform is the Fourier transform of a periodically continued input sequence of length N. The sampling interval is normalized to $T = 1$ and it follows accordingly that, for the periods in the spectrum, a frequency of $f = 1$ or rather an angular frequency of $\omega = 2\pi$ holds. To "denormalize", the frequency values must only be multiplied by $1/T$.

Given the values of the input sequence $x(\cdot)$ and the corresponding spectral values $y(\cdot)$ and taking into account equations for a Fourier series representation, the equations

$$y(k) = \sum_{m=0}^{N-1} w^{mk} x(m) \qquad k = 0, 1, ...N - 1$$

$$w = e^{-2nj/N}$$ (7.1.5)

$$j = \sqrt{-1}$$

follow for the computation of the spectral values [1], [2]. The inverse computation of the samples from the spectral values is carried out using

$$x(m) = \frac{1}{N} \sum_{k=0}^{N-1} w^{-mk} y(k) \qquad m = 0, 1, ...N - 1 \qquad (7.1.6)$$

7.2 Direct Implementation of the DFT

According to (7.1.5), the computation of the DFT can be seen as the product of a matrix and a vector. A vector with N elements of $x(\cdot)$ is multiplied by a matrix W with coefficients w^{mk}.

$$y = Wx \qquad (7.2.1)$$

The particular structure of this matrix–vector product for $N = 8$ is illustrated below.

$$
\begin{bmatrix} y(0) \\ y(1) \\ y(2) \\ y(3) \\ y(4) \\ y(5) \\ y(6) \\ y(7) \end{bmatrix} = \begin{bmatrix} 1 & 1 & 1 & 1 & 1 & 1 & 1 & 1 \\ 1 & w^1 & w^2 & w^3 & w^4 & w^5 & w^6 & w^7 \\ 1 & w^2 & w^4 & w^6 & 1 & w^2 & w^4 & w^6 \\ 1 & w^3 & w^6 & w^1 & w^4 & w^7 & w^2 & w^5 \\ 1 & w^4 & 1 & w^4 & 1 & w^4 & 1 & w^4 \\ 1 & w^5 & w^2 & w^7 & w^4 & w^1 & w^6 & w^3 \\ 1 & w^6 & w^4 & w^2 & 1 & w^6 & w^4 & w^2 \\ 1 & w^7 & w^6 & w^5 & w^4 & w^3 & w^2 & w^1 \end{bmatrix} \cdot \begin{bmatrix} x(0) \\ x(1) \\ x(2) \\ x(3) \\ x(4) \\ x(5) \\ x(6) \\ x(7) \end{bmatrix} \quad (7.2.2)
$$

In the above matrix, the periodicity of the complex e-function is exploited. Due to its periodic characteristics,

$$
w^{jN+i} = w^i \quad j \in \mathbb{Z} \quad (7.2.3)
$$

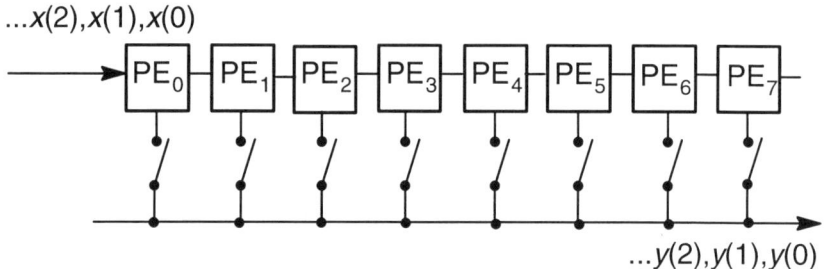

Figure 7.2.1: 1-D DFT array for $N = 8$

The implementation of a matrix–vector multiplication can be carried out with the methods shown in Chapter 5. Following Figure 5.3.4, an implementation with N^2 processor elements is possible. Figure 5.3.5 shows a structure as a 1-D array, and Figure 5.3.6 shows an implementation with one PE. Through the use of partitioning methods, one- and two-dimensional arrays that measure smaller than N are also possible. Figure 7.2.1 shows a 1-D array in which the results are accumulated in N parallel PEs. These results are available after a temporal offset and can be sequentially passed on via switches to a result bus. The switches in Figure 7.2.1 are switched cyclically via a pointer signal.

The processing element is shown in Figure 7.2.2. It consists of a multiplier and an adder. It should be noted that it is a time-dependent weighting function since the index m changes cyclically with each clock impulse.

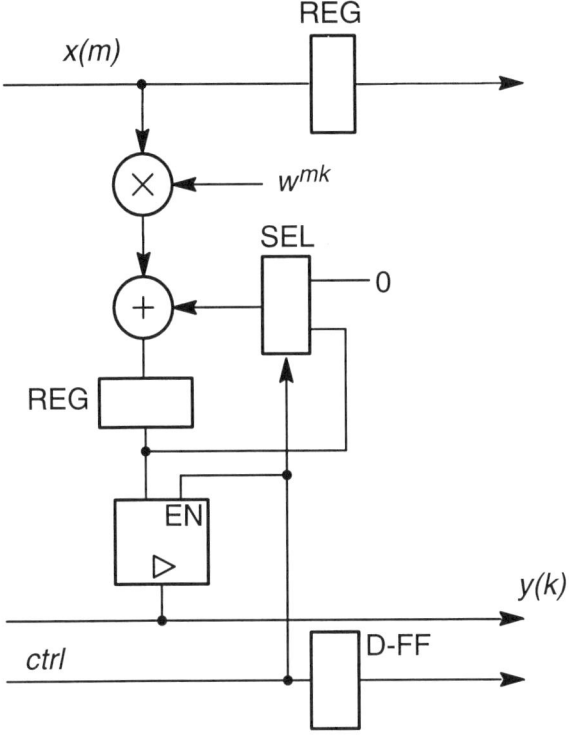

Figure 7.2.2: PE of a 1-D DFT array

Since w is complex, complex-valued results must be calculated in the PE. The multiplication of two complex numbers in component form is given by

$$x \cdot a = u$$
$$(x_r + jx_i)(a_r + ja_i) = (x_r a_r - x_i a_i) + j(x_i a_r + x_r a_i)$$

(7.2.4)

This means that the multiplication of two complex numbers can be implemented by the multiplication of four real-valued multiplications and two real-valued additions. Figure 7.2.3a shows the corresponding structure. If one also notes that a complex addition can be implemented by two real-valued additions, it follows that a DFT-PE comprises four multipliers and four adders. In comparison to a real-valued matrix–vector multiplication, this is four times as costly.

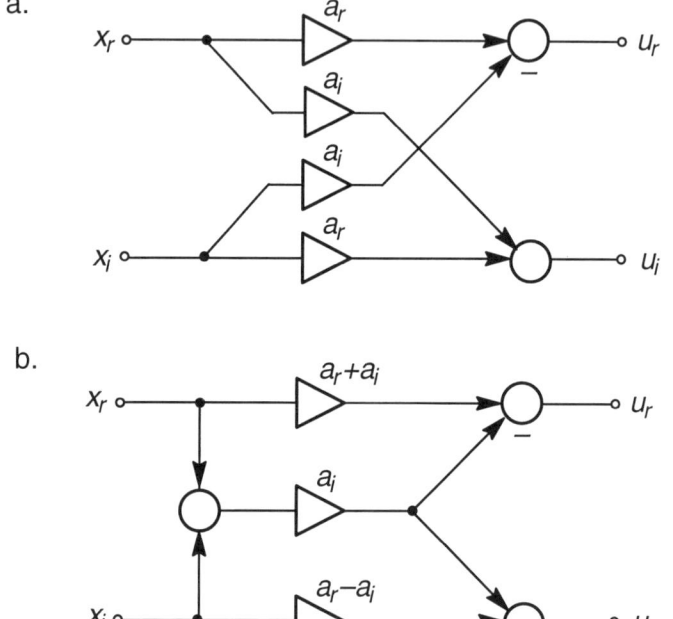

Figure 7.2.3:   Structures for implementing the complex multiplication $u = xa$.
a. Direct implementation with four real multipliers. b. Modified
structure with three real multipliers [106]

The DFT of real signal sequences $x(m)$ leads to a simplification with a total of two multipliers and two adders. Using equation (7.2.2), it follows for real $x(m)$, that $y(0)$ and $y(N/2)$ are real. Furthermore, using

$$w^i = (w^{N-i})^* \qquad 1 \le i < N \qquad\qquad (7.2.5)$$

one can derive that

$$y(i) = y^*(N - i) \qquad\qquad (7.2.6)$$

where $y*$ is the complex conjugate of y. Thus, instead of $2N$ independent components (real and imaginary parts), only N independent components appear in this case.

Complex multiplication has particular, inherent dependencies that can be used to reduce the circuit costs. Several possibilities exist for carrying out a complex multiplication with just three instead of four multiplications [105], [106]. For instance, equation (7.2.4) can be replaced by the following expression as an alternative.

$$u = [x_r(a_r + a_i) - (x_r + x_i)a_i] + j[x_i(a_r - a_i) + (x_r + x_i)a_i] \tag{7.2.7}$$

Figure 7.2.3b shows the corresponding hardware. This means that, in the end, a PE for a DFT can be implemented with three multiplications and five additions. The time-critical path consists of one multiplier and three adders. The delay of these stages determines the throughput. As already shown in Chapters 4 and 5, the throughput can be further increased via pipelining.

7.3 Fast Fourier Transforms

The fast Fourier transform leads to a reduction in computational operations through hierarchical division into transforms of shorter sequences. The abbreviation FFT is normally used for the fast Fourier transform. Two basic algorithms for the FFT exist. One is called decimation in frequency (DIF), the other decimation in time (DIT) [1], [2]. The DIF algorithm will be explained here as an example.

The general principle is a continuing segmentation into sequences of shorter length through division of the length by prime factors of N. Usually, N is a power of 2 so that a sequence on length N is halved several times. If the sum from (7.1.5) is split into two sums half as long, a common factor can be factored out of the second sum using the rules of exponents.

$$y(k) = \sum_{m=0}^{N/2-1} w^{mk}x(m) + w^{(N/2)k} \sum_{m=0}^{N/2-1} w^{mk}x(m + N/2) \tag{7.3.1}$$

Since $w^{N/2}$ corresponds to a rotation of $180°$, the factor of the second sum can be even further reduced.

$$w^{(N/2)k} = (-1)^k \tag{7.3.2}$$

Exploiting this special factor, both sums can be reunited.

$$y(k) = \sum_{m=0}^{N/2-1} w^{mk} \, [x(m) + (-1)^k x(m + N/2)] \qquad (7.3.3)$$

The division of k into even and odd values leads to the following two expressions:

$$y(2r) = \sum_{m=0}^{N/2-1} [x(m) + x(m + N/2)] \, w^{m2r}$$

$$y(2r + 1) = \sum_{m=0}^{N/2-1} [x(m) - x(m + N/2)] \, w^{m} \cdot w^{m2r} \qquad (7.3.4)$$

$$r = 0, 1, \dots \, N/2 - 1$$

Using the short notations

$$x_0' \, (m) = x(m) + x(m + N/2)$$
$$x_1' \, (m) = [x(m) - x(m + N/2)] \, w^{m} \qquad (7.3.5)$$

(7.3.4) leads to

$$y(2r) \quad = \sum_{m=0}^{N/2-1} x_0' \, (m) w^{m2r}$$

$$y(2r + 1) = \sum_{m=0}^{N/2-1} x_1' \, (m) w^{m2r} \qquad (7.3.6)$$

Through comparison with the original definition of the DFT in (7.1.5), one can see that a DFT of length N can be converted to two DFTs of length $N/2$. This requires, on the one hand, the summation of the samples spaced by $N/2$ and, on the other hand, the computation of the difference of the samples spaced by $N/2$ and its multiplication by w^{m}.

If N is a power of 2, the half-length DFTs can be further halved. This can be carried out until no further halving is possible. A signal flow graph is shown in Figure 7.3.1 that shows the operations required for $N = 8$. The continued halving after each stage is recognizable in that further operations can be carried out independently of each subset.

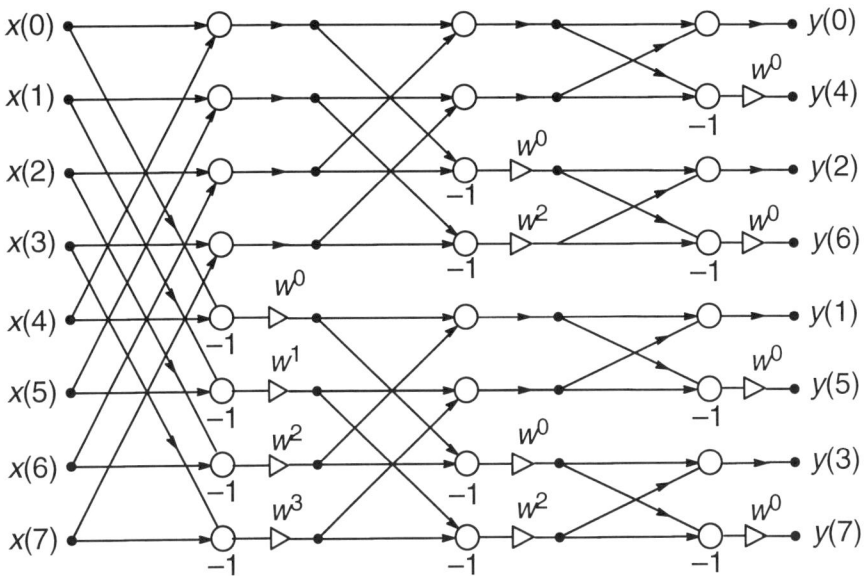

Figure 7.3.1: Signal flow graph of a DFT for DIF decomposition

Each stage of the signal flow graph requires $N/2$ complex additions, subtractions and multiplications. Since a total of only $\log_2 N$ stages is needed, the total amount of computation is clearly reduced. A comparison of the complex computations is shown in Table 7.3.1. The additions and subtractions are both treated as additions in this table. That no multiplication is necessary for factors such as w^0 and w^4 is also taken into consideration.

Table 7.3.1: Number of complex operations in a DFT and an FFT

DFT		FFT	
MUL	ADD	MUL	ADD
$(N-2)^2$	N^2	$N/2\log_2 N - (N-1)$	$N\log_2 N$

The FFT has a repetitive fundamental structure that is called a butterfly due to its particular shape. A butterfly PE is shown in Figure 7.3.2. The input comprises two complex data items and an exponential term consisting of a cosine and a sine term. Two complex data items are output as a result. As already demonstrated in the previous section, complex operations must be implemented through several real-valued operations. In accordance with the structure in Figure 7.3.2, a butterfly PE requires a total of three real multiplications and seven real additions. Since the costs for a multiplier are significantly larger than the costs for an adder, the implementation costs for the same throughput are reduced by approximately

$$a \approx \frac{2N}{\log_2 N} \qquad (7.3.7)$$

using the FFT.

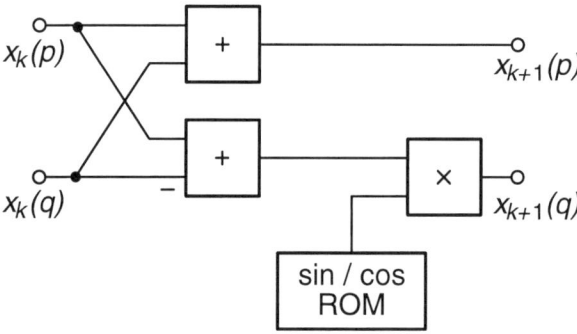

Figure 7.3.2: Butterfly PE

One particular problem of the FFT is its less regular data transport. Figure 7.3.3 shows the wiring of the butterfly PEs for $N = 8$. If the butterfly PEs are configured such that the PEs with lower exponents of w come first in each stage, a configuration results with identical communication networks between each stage. This communication network is called a perfect shuffle since each of the pairs

$$0, N/2$$

$$1, N/2+1$$

●

●

●

$$N/2-1, N$$

is created and fed to a PE. The effect of this rearrangement on the structure is shown in Figure 7.3.4.

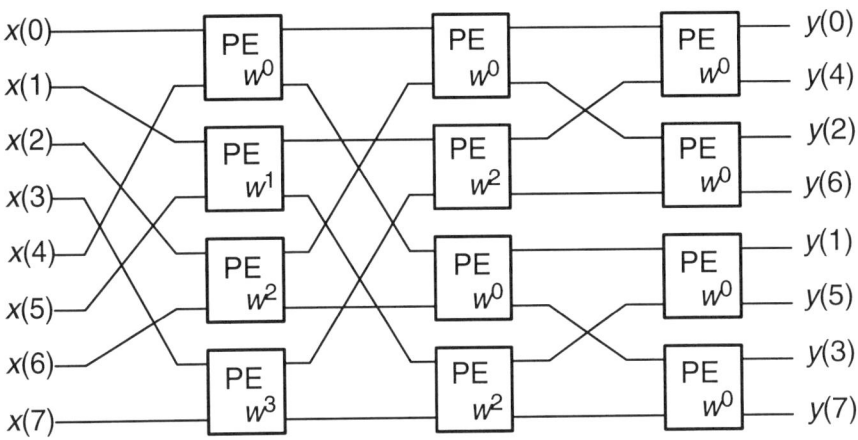

Figure 7.3.3: 2-D array of butterfly PEs for implementing an FFT

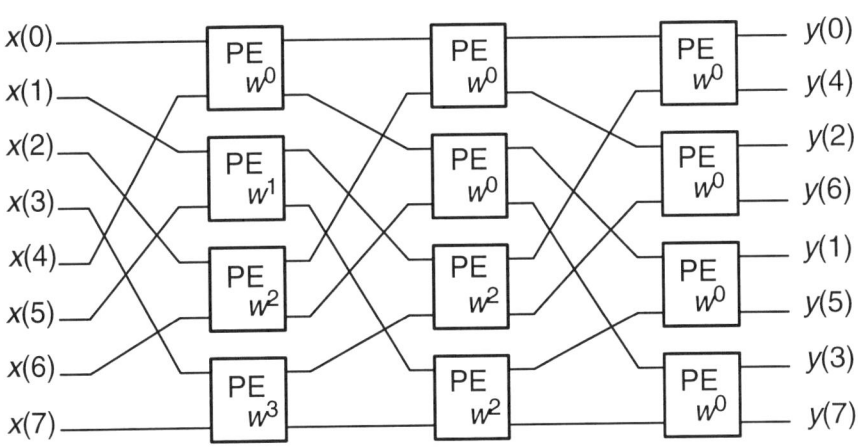

Figure 7.3.4: Modified 2-D array of butterfly PEs with identical communication networks between the stages

For implementations with less high demands on the throughput, it is possible to project to a configuration with a smaller number of butterfly PEs. Horizontal projection leads to $N/2$ PEs with a decrease in throughput by a factor of $\log_2 N$ in comparison to a 2-D array. Vertical projection yields $\log_2 N$ PEs and a decrease in throughput by a factor of $N/2$. A single butterfly PE must sequentially carry out the processing $(N/2)\log_2 N$ times. The essential data are summarized in Table 7.3.2.

Table 7.3.2: Number of butterfly PEs n_{PE} and factor α_R for throughput reduction for various structures

Structure	n_{PE}	α_R
2-D array	$(N/2) \log_2 N$	1
1-D array	$N/2$	$\log_2 N$
1-D array	$\log_2 N$	$N/2$
1 PE	1	$(N/2) \log_2 N$

As already implied previously, the FFT leads to complex communication structures. Figure 7.3.5 shows the configuration after horizontal projection. This structure can be directly derived from Figure 7.3.4. In this structure, the PEs do not have fixed coefficients, but rather they change after each cycle. In particular for large N, the global communication network is disadvantageous.

A disadvantage of the recursive structure in Figure 7.3.5 is that pipelining within the PEs does not allow a direct increase in throughput since the results of the current processing are required for the next processing step. However, sequential data blocks of length N can be processed independently of one another so that, in spite of the recursive structure, several data blocks can be processed by interleaving (see section 4.1). The increased number of intermediate registers in the recursive loop due to pipelining must correspond to the number of data blocks being processed by interleaving.

Due to the large number of PEs, horizontal projection is only applicable for moderate orders of N. For large values of N, partitioning methods are required. In this case, the $N/2$ butterfly PEs must be projected to M PEs, whereby M is also a power of 2 and $M < N/2$. On account of the particular data flow in an FFT, special registers for input data, intermediate results and result data are required. These registers must cyclically read and write a particular sequence of $2M$ pieces of complex data.

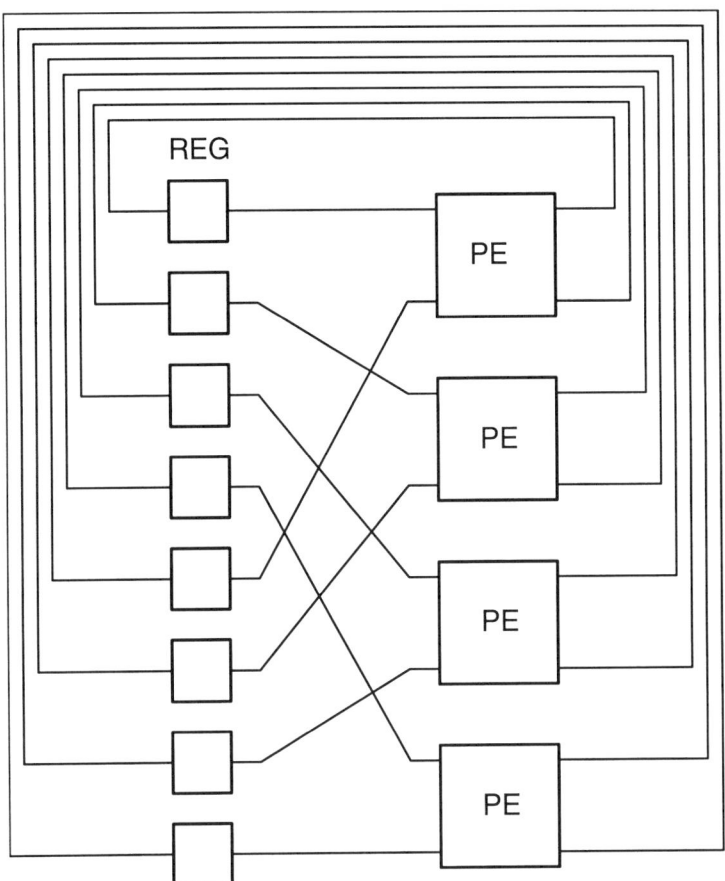

Figure 7.3.5: 1-D array of $N/2$ butterfly PEs after horizontal projection

The position of the PEs during vertical projection is obvious. However, special circuitry between the PEs is required to prepare the correct data input. From stage to stage, the length of the sequence onto which the FFT is applied is halved. Given that the previous stage led to a DFT of length $2n$, in accordance with the perfect shuffle, the sequence of length $2n$ must be halved, and the first and the $(n+1)$th values must be fed to the following PE. Then the second and the $(n+2)$th values are fed to it. The required data sorting is illustrated in Figure 7.3.6. In the illustration, the fact that the sequence must be delayed by n clock cycles in accordance with the position of the midpoint is also accounted for.

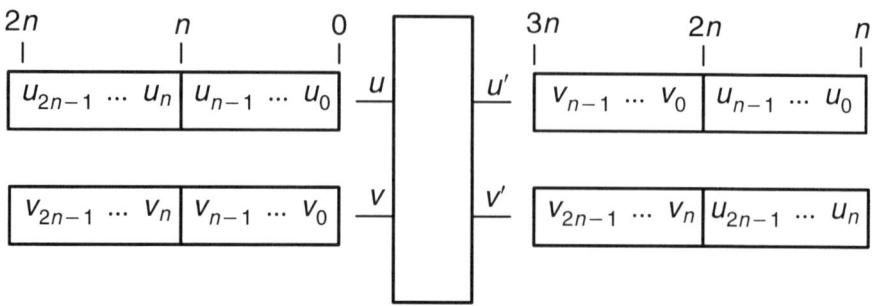

Figure 7.3.6: Specification of the data reformatting in linear FFT arrays

The corresponding circuit can be directly derived from the specifications in Figure 7.3.6. The block $u_{n-1} \ldots u_0$ must be delayed by n clock cycles. As soon as u_n is available, the values from the stream u must be fed to the new, lower stream v'. The values of u are input in parallel into the next butterfly stage for n clock cycles. Thereafter, the values of v are fed in parallel to the next butterfly PE for n clock cycles. This demands that the block $v_{n-1} \ldots v_0$ be delayed by $2n$ clock cycles and the block $v_{2n-1} \ldots v_n$ by n clock cycles. Figure 7.3.7 shows the resulting hardware structure. The multiplexer (MUX) serves to switch between the two data streams. It is switched every n clock cycles.

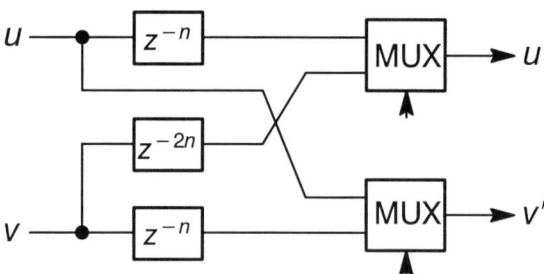

Figure 7.3.7: Registers for data reformatting in linear FFT arrays

By shifting n delay elements from the input to the output of the upper multiplexer and by merging the two parallel delay blocks in the input stream of v, the hardware structure can be further simplified (Figure 7.3.8). This new structure is called a delay commutator. In this configuration, the multiplexers connect the two data streams crossed or in parallel, alternating every n clock cycles.

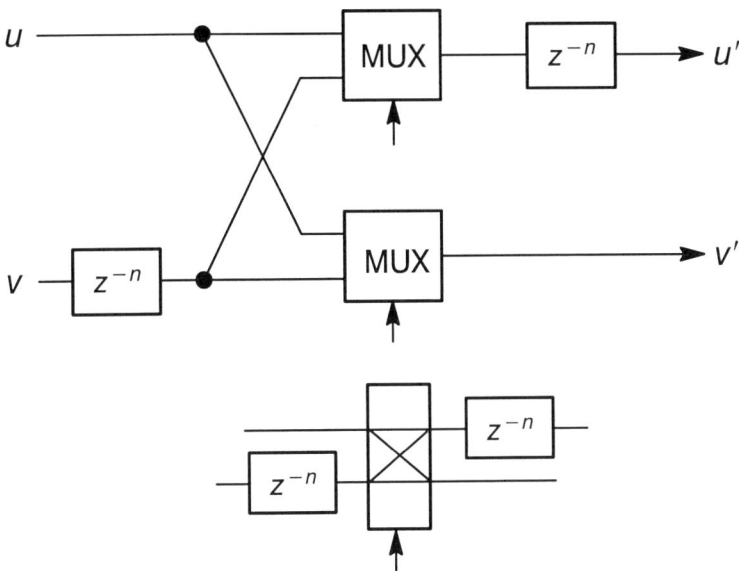

Figure 7.3.8: Delay commutator: principal construction and simplified symbol

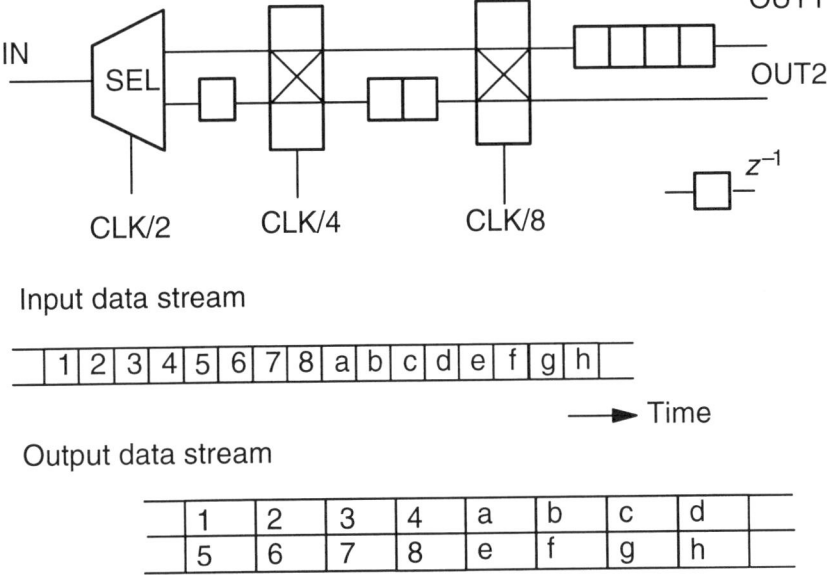

Figure 7.3.9: Division of a serial data stream in two phases with a synchronous FIFO register, illustrated for $N = 8$

A special circuit is necessary for the data input of the first stage. An incoming data stream of N pieces of data must be divided into two paths of $N/2$ pieces of data each. Due to division into two parallel paths, the clock rate must be halved. The circuit required for this is a demultiplexed followed by a FIFO register (FIFO-first in, first out). A special structure is shown in Figure 7.3.9 that carries out the splitting of the data with synchronous clocking and minimal storage capacity [107]. In the illustration, the delays are defined relative to the input clock. The illustration characterizes the general procedure. It does not specify the exact technical implementation.

The entire FFT processor resulting from vertical projection is shown in Figure 7.3.10. The configuration consists of N PEs. Delay commutators are located between the PEs. Due to the continued halving, their control signals are extracted using frequency dividers. This example shows clearly that solving the data input leads to considerable costs in term of registers and control circuitry.

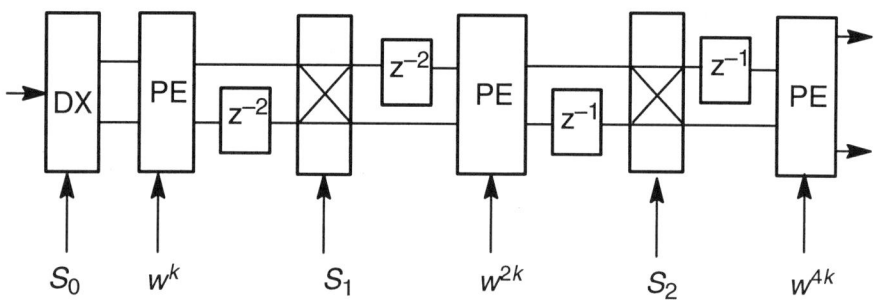

Figure 7.3.10: Linear FFT array based on butterfly PEs and delay commutators

When only one butterfly PE is used to carry out the FFT, a RAM for implementing the data input and output is a viable alternative. Figure 7.3.11 shows such a configuration. The PE requires two arguments. Thus, a dual port memory is suitable. As can be derived from the signal flow graph in Figure 7.3.1, memory with a capacity of N pieces of complex data is sufficient. The results of a butterfly operation can be written back to the same position in memory from which they were taken (in-place processing). The control ROM must supply N different coefficients and the control signals. The control signals essentially serve for memory addressing during the processing phase. Special control extensions make simultaneous reading of the results and storage of the new data possible at the end of each process.

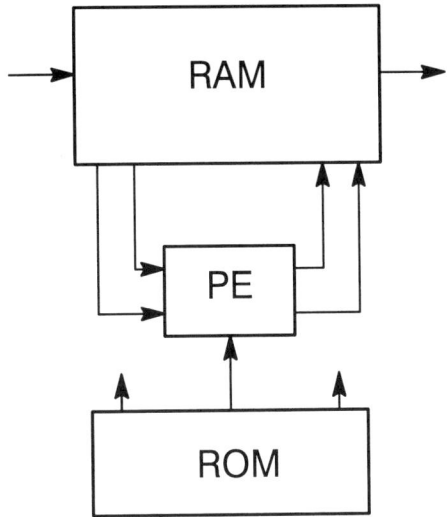

Figure 7.3.11: FFT implementation based on one butterfly PE

A comparison of the hardware structures for implementation of the FFT and the direct implementation of the DFT shows that direct implementation yields simple, regular structures with local connections and low control overhead. On the other hand, FFT structures are significantly more complex with highly complex wiring or special hardware costs for preparation of the data input. However, the total costs for an FFT are nonetheless significantly lower for a given throughput.

7.4   Exercises

1.   The achievable throughput of the various architectures from implementing the DFT is to be estimated. To simplify their evaluation, it is assumed that the delay characteristics of the multipliers and adders dominate and that the contributions to the delay from other components such as registers and multiplexers, for example, are negligible in comparison. The delay of an n-bit adder is given to be $n \cdot T_0$ and that of an $n \times n$-bit multiplier $3n \cdot T_0$. The input values and the coefficients are each given in component form with m_{bit} bits. The adder for accumulation is constructed such that the intermediate results can be represented with a maximum of $3m_{bit}$ bits. The necessary reduction to $3m_{bit}$ bits occurs via truncation.

   a.   The critical delays of a PE as in Figure 7.2.2 and of a butterfly PE as in Figure 7.3.2 are to be calculated for the general case.

b.   Which throughput can be achieved through use of one PE? The follow-
     ing particular values are to be assumed:

     $m_{bit} = 8$, $N = 64$, $T_0 = 0.5$ ns.

c.   How much does the throughput increase through use of 1-D arrays in
     accordance with Figures 7.2.1, 7.3.5 and 7.3.10?

2.   How the implementation of a DFT can be simplified by exploiting special
     boundary conditions is to be explored.

a.   By how much are the costs of a direct implementation as in Figure 7.2.2
     reduced when the input sequence is real?

b.   It is assumed that the input sequence is real and even over an interval
     of length N, i.e. $x(i) = x(N-i)$. How can the implementation costs of a
     direct implementation be further reduced using this knowledge? Hint:
     Equation (7.2.5) is to be observed.

c.   Can the special knowledge of the characteristics noted in a and b also
     be used in the case of an FFT?

3.   The achievable throughput of butterfly PEs is to be increased through pipeli-
     ning. It is assumed that the butterfly PE contains a total of four pipeline
     stages.

a.   In order for an increase in throughput to be possible in a recursive struc-
     ture as in Figure 7.3.5, four data blocks of N values are to be input di-
     rectly after one another and then processed together (interleaving).
     Sketch the processing steps for $N = 8$.

b.   Storage strategies for the structure in Figure 7.3.11 are to be developed
     for the case where the PE contains pipelining and the RAM possesses
     two ports (dual-port RAM).

8 Programmable Digital Signal Processors

Previously, special architectures adapted to a signal processing task were derived. In general, these dedicated structures have the advantage of minimal overhead for the desired throughput. Due to the adaptation of the architecture to the algorithm, one obtains a correspondingly patchworked hardware for a compound process. As shown in Chapter 5, the possibility exists of projecting various signal processing nodes onto one another. This, however, is only expedient for regular data transfers and similar operations. Otherwise, this only leads to duplication of individual nodes.

For signal processing tasks with moderate demands on the throughput, a flexible hardware is desired that makes sequential processing of the subtasks possible. In terms of flexibility, there are various categories: from a hardware structure controlled by few parameter lines that allows a limited number of different functions to a configuration programmed in a higher language that can implement practically all algorithms.

A programmable processor not only has the advantage of a uniform configuration for implementing compound processes, but can also be efficiently used for prototyping since the costs of developing software are generally lower than the development of dedicated hardware. In spite of technical progress, microprocessors do not offer sufficient performance for many signal processing tasks. Thus, special programmable processors have been developed for digital signal processing. These are called DSPs (digital signal processors).

In the following section, the architecture of DSPs is explained. It is shown by what means the signal processing performance is enhanced. Further, the architectural characteristics of a few commonly available DSPs are summarized. The architectural advantages are then demonstrated using programming examples for a microprocessor and a DSP. An FIR filter and a DFT serve as programming examples. The process of developing signal processing systems based on DSPs is explained and means of supporting the development of such systems are presented.

8.1   The Architectures of DSP Processors

8.1.1   The architecture of standard computers

A principal feature of the basic architecture of standard computers and microprocessors is their common memory for data and instructions. This is called a von Neumann architecture [108]. Through the common memory for data and instructions, one obtains flexible use of the memory since the space for data and instructions is mutually exchangeable. The position of data and instructions within memory can also be arbitrarily chosen. They can also be interwoven. In principle, it is even possible to dynamically change the program during execution.

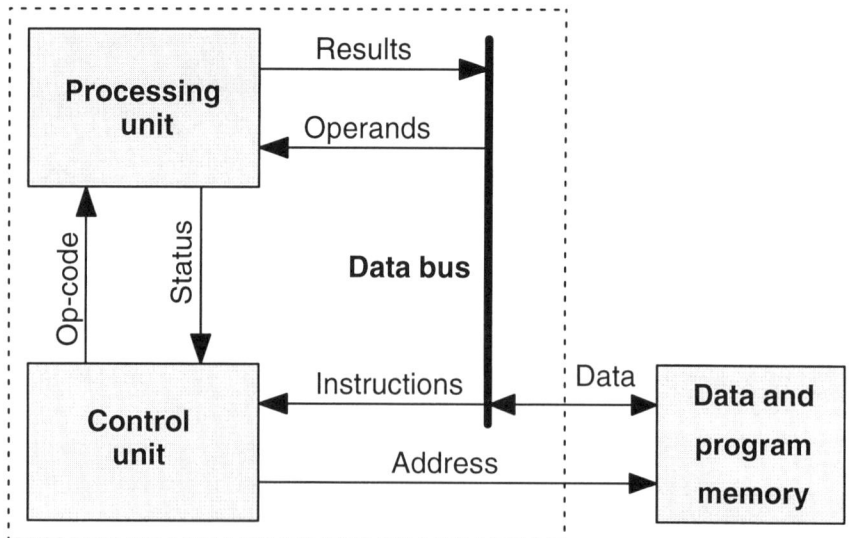

Figure 8.1.1: Architecture of a standard computer

Besides the data and program memory, a standard computer also contains a processing unit and a control unit. The basic structure of this type of computer is shown in Figure 8.1.1. An essential element of the computer is the arithmetic logic unit (ALU). An ALU can carry out all basic arithmetic operations and bitwise logic operations. The structure of an ALU using bit slicing is shown in Figure 8.1.2. The structure corresponds more or less to that of a carry-ripple adder (Figure 3.2.1). The operation code (op-code) is decoded, and the individual control bits (Ctrl) determine the function for the individual elements. Each element in the bit slice has the same function.

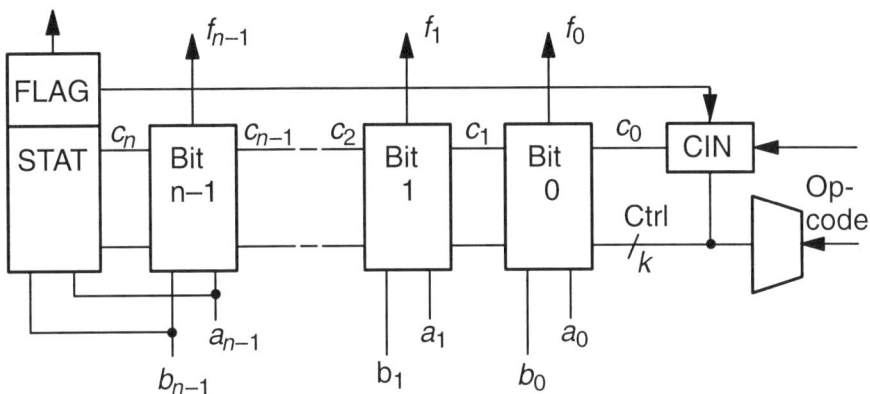

Figure 8.1.2: An arithmetic logic unit (ALU)

The basic element for the bitwise logical processing of two operands is a programmable logic function block. Figure 8.1.3 shows an implementation in switch logic. The corresponding logic function is

$$F = g_0\bar{a}_i\bar{b}_i \vee g_1a_i\bar{b}_i \vee g_2\bar{a}_ib_i \vee g_3a_ib_i \qquad (8.1.1)$$

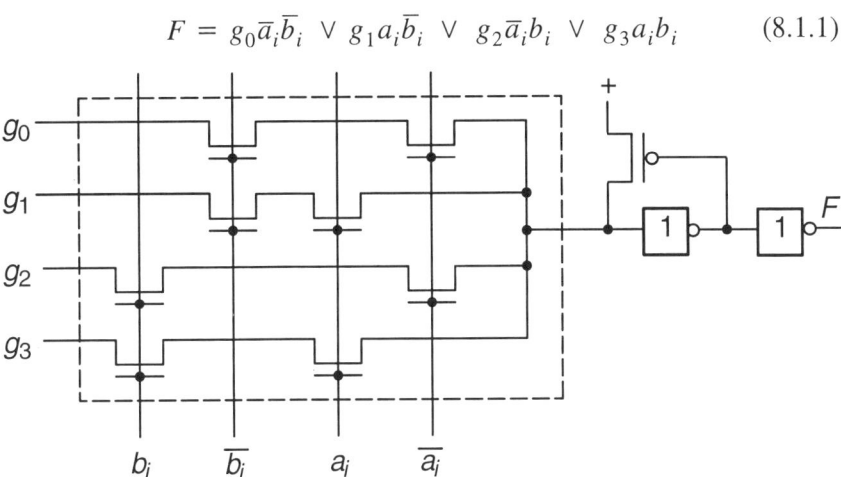

Figure 8.1.3: Programmable logic function block

The function contains four minterms that belong to the variables a_i and b_i. Via the special bit pattern of

$$G = (g_0, g_1, g_2, g_3) \qquad\qquad (8.1.2)$$

the minterms are activated or deactivated. From $G = (0, 1, 1, 0)$ one obtains an XOR operation. Accordingly, $G = (0, 0, 0, 1)$ is an AND operation, and $G = (0, 1, 1, 1)$ an OR operation. Through appropriate choice of G, all 16 possible logic functions for two variables can be set up.

For the implementation of arithmetic operations, the carry bit c_i must be taken into account as well as the operand bits a_i and b_i. In addition to the result bit f_i, the carry bit c_{i+1} must also be created. In the case of an addition, two XOR operations are required for the sum bit. This can be carried out by two logic function blocks as in Figure 8.1.3. One of the logic function blocks creates the propagate signal. For the carry signal c_{i+1}, a further logic block (kill signal) and carry logic as in Figure 3.2.7 is required [14]. Thus, in the end, a bit slice consists of three logic function blocks and carry logic [14].

Since each logic function block implements various functions through appropriate programming, it is obvious that together with additional logic circuitry CIN, with which c_0 is set, subtraction is also possible. Furthermore, A and B can be set to either 0 or -1, and thus an increase by 1 (increment) and a decrease by 1 (decrement) can also be implemented. If it is programmed such that an operator bit becomes the carry bit and the carry bits are mapped as result bits, the ALU carries out a shift by one position to the left (shift left). The ALU also contains additional logic for extracting status bits through evaluation of the uppermost bit slice. Status information such as positive, negative, overflow, etc. can be so derived. Thus, the ALU is in the position to carry out comparisons, too.

Multiplication and division can be reduced to additions and subtractions. For a multiplication, the addition is steered via a multiplier bit such that either the multiplicand or zero is added. In a division, the highest carry bit steers whether the divisor is subtracted or added in the next step (non-restoring division). These two operations require a total of n cycles for a word of length n bits. For such operations, additional registers for intermediate storage and a small control unit (microprogram control unit) for process control are required. Arbitrary shift and rotation operations also cannot be carried out in one cycle and must be steered via the microprogram control unit. Thus, one should remember that although an ALU completes most operations in one cycle, some operations exist that require several cycles. A few important basic operations are listed in Table 8.1.1.

Table 8.1.1: A selection of ALU operations

Operation		Description
Arithmetic operations		
$A + B$		Addition
$A + B + C_{in}$		Addition with carry
$A - B$		Subtraction
$A - B - C_{in}$		Subtraction with carry
$-A$		Two's complement
$A + 1$		Increment
$A - 1$		Decrement
$2 \cdot A$		Shift to left
$A + B$	Cond = 1	Multiplication element
A	Cond = 0	
$A - B$	Cond = 1	Division element
$A + B$	Cond = 0	
Bitwise logical operations		
$A \wedge B$		Logical AND
$A \vee B$		Logical OR
$A \oplus B$		XOR (antivalence)
$A \odot B$		XNOR (equivalence)
$\neg A$		NOT (one's complement)
A		Transport of operand A
0		Clear
1		Preset

To illustrate the program sequence of a standard computer with a structure as in Figure 8.1.1, a simple type of computer will first be assumed. It is presumed that all required jump addresses and operand addresses are listed following the actual instruction. This means that an addition occupies four positions in memory, namely the instruction, two operand addresses and the address for the result. Thus, the program segment for an addition is:

```
PC →      ADD
          # A
          # B
          # C
```

The corresponding program sequence could be described in pseudo-code as follows.

```
LOAD       (PC) +,      IR
DECODE
LOAD       (PC) +,      RADR
LOAD       (RADR),      RA
LOAD       (PC) +,      RADR
LOAD       (RADR),      RB
EXECUTE
LOAD       (PC) +,      RADR
STORE      RC,          (RADR)
```

The above instructions are defined as follows.

LOAD   Read a memory cell and store in a register of the control unit or ALU

DECODE   Decode the instruction in register IR

EXECUTE   Execute the operation

STORE   Store the contents of a register in a memory cell

PC   Program counter

()   Contents of the register are used as an address

(PC)+   After the suboperation, the program counter is incremented

In addition, RADR is a register that works on the address bus and RA, RB, and RC are registers for the ALU. The individual phases of the program sequence show that a total of seven read and write commands are necessary for an operation with two operands. This comprises the reading of the instruction, the addresses of the operand, the operand, the address for the result and the storage of the result. The illustrated example shows that, in order to carry out an arithmetic operation, a large overhead in terms of read and write operations is required. Thus, even in the first computer implementations, measures for reducing the number of read and write operations were introduced. Local registers were assigned to the ALUs. In the course of processing, it is then often possible to store the addresses of the operands and the result as pointers in local registers. In this manner, the nine cycles of the example are reduced to six, since three read operations for address can be dropped. Through enlargement of the instruction word width, it becomes possible to supply the instructions with address information of the operands. If the operands happen to be located in local registers and the result can also be stored locally, the number of cycles can be reduced to three (load, decode, execute) through corresponding structuring of the ALU. Here it is assumed that the contents of the register are fed via a multiplexer directly to the ALU or that the result is stored directly in an arbitrary register.

Instruction pipelining serves as a further measure for improving the computational performance [109]. During the decoding of an instruction, the next instruction is already read in advance. During execution of the instruction, the next instruction is decoded. The overlapping course of events in instruction pipelining is shown in Figure 8.1.4. Here it is assumed that the execution of an operation requires only one cycle. Instruction pipelining is particularly suited for computers with simple instructions that require one cycle for execution. These computers are termed RISC (reduced instruction set computer). On the other hand, computers with complex instructions steered via a microprogram control unit that require a differing number of cycles are termed CISC (complex instruction set computer).

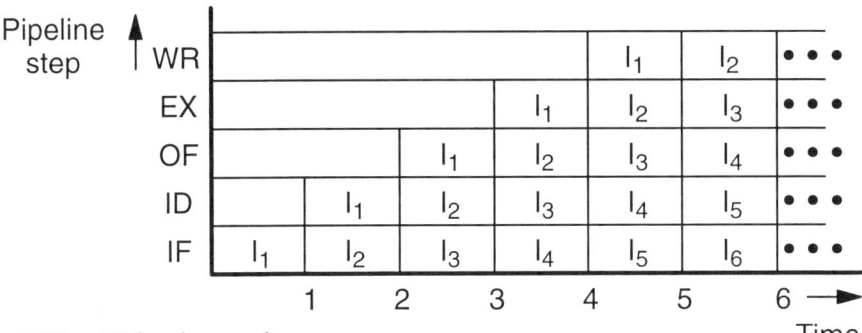

WR: Write the result
EX: Execute the instruction
OF: Operand fetch
ID: Instruction decode
IF: Instruction fetch

Figure 8.1.4:   Sequence of events in instruction pipelining

The storage cycles for a small RAM within a chip are significantly shorter than outside via a bus. Additional local storage is thus put on processor chips. This local storage can be arbitrarily addressable register fields or RAM. Instruction segments of data fields can be stored in this local storage.

Through appropriate programming, the local storage can be used efficiently and the number of memory calls to external storage can be distinctly reduced.

8.1.2   Architectural approaches for DSP processors

In the previous section it was shown that a great many read and write operations to the external storage media are necessary to carry out operations. Instructions and data, in particular, can only be read sequentially. Therefore, an architecture with separate storage for instructions and data was suggested for signal processing. This architecture is shown in Figure 8.1.5. It is called a Harvard architecture [110].

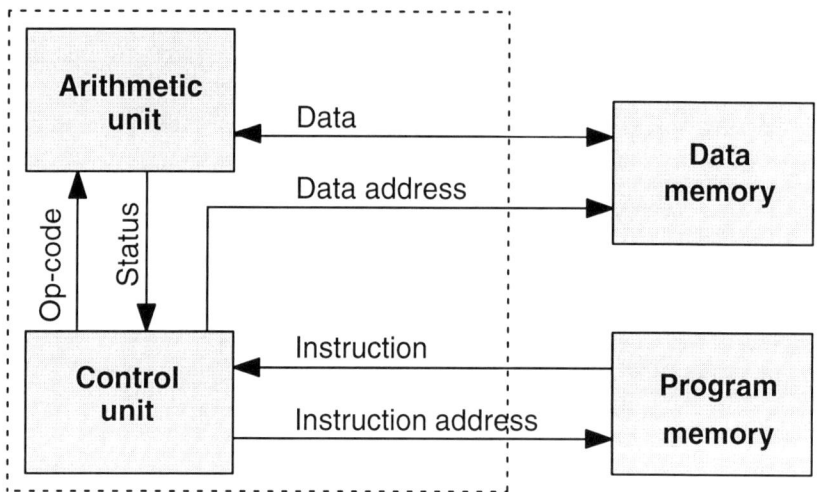

Figure 8.1.5: Harvard architecture

In order for a Harvard architecture to offer truly parallel instruction and data streams, the control circuitry must be constructed accordingly. This type of control sequence will be explained using Figure 8.1.6. An instruction is read from the program memory and written to the instruction register. The instruction is decoded and, via the control lines, the data path is set to carry out the corresponding operation. As is often typical, the processing unit is alternatively termed the data path. The data path contains registers and all the elements for carrying out the operation. During execution of the operation in the data path, the next piece of data can be read, controlled by an address generation unit in preparation for the next cycle. Such preparative reading of a piece of data is only possible in particular program segments. For this to happen, the address generation unit must be programmed via the instruction decoder before the end of each cycle. The instruction sequencer determines the address of the memory address to be read next. In general, this is the next address (increment). Since conditional and non-conditional jumps are contained in the instructions, the instruction sequencer is controlled by the instruction decoder. For the computation of relative jumps, increments (+1) and decrements (−1), the sequencer requires a small arithmetic unit. In addition, another stack register is required in which the return addresses can be stored in the case of subroutines.

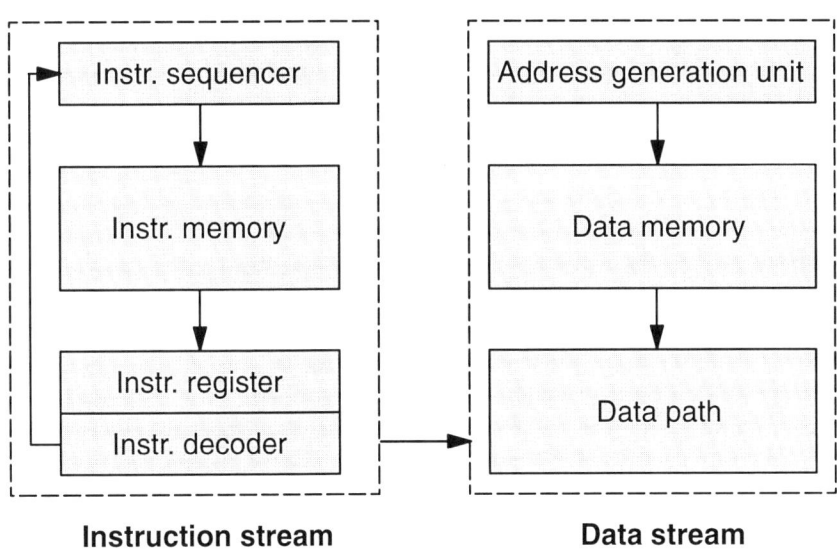

Instruction stream **Data stream**

Figure 8.1.6: Parallel instruction and data streams

Most of the operations to be carried out have two operands. If only one data memory is used, the instruction stream is interrupted since both operands must be written one after another to the input register of the data path. Furthermore, the address generation unit must manage two data pointers. Thus, for an efficient data stream, two data memories and two address generation units are appropriate (see Figure 8.1.7). Both operands can then be fed to the data stream in one cycle.

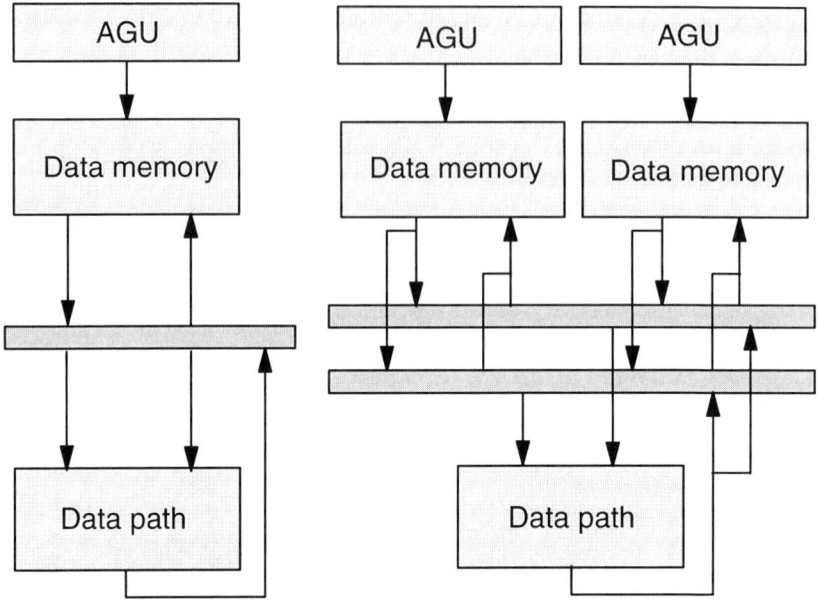

Figure 8.1.7: Data processing with one and two buses

For further increase of the performance of a processor the concurrency of operations should be increased. This could be reached by several parallel operating functional units. A very long instruction word (VLIW) is formed to specify the operations and operands for the multiple units [111], [112]. Different fields of the very long instruction word carry the individual instructions of each unit. A typical VLIW architecture is illustrated in Figure 8.1.8. The multiple functional units share a common large multi-ported registerfile. Parallel random access is supported by a crossbar network (see Figure 9.1.5). All partial instructions are simultaneously executed by the functional units. The functional units may be a collection of ALUs, MAC units, shifters and other special operational units. Concurrent with the synchronous operations of the functional units, load/store operation of data between a RAM memory and the register file is performed. A typical instruction format is displayed in Figure 8.1.8. In contrast to superscalar processors that dynamically exploit instruction level parallelism at run time, the long instructions of a VLIW architecture are completely specified at compile time. Thus, VLIW architecture relies on static operation scheduling at compile time.

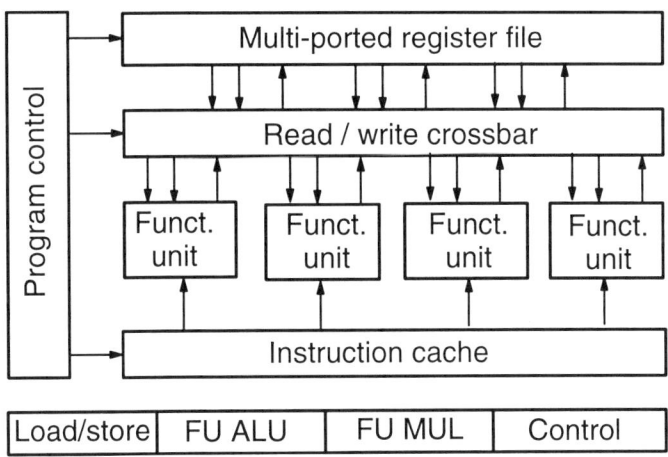

Figure 8.1.8: VLIW architecture and typical VLIW instruction

Performance gains achieved with VLIW architectures depend on the degree of exploitable instruction level parallelism in the algorithm and the selected functional units. A prerequisite of instruction parallelism is independent operations. Signal processing algorithms contain in many cases short inner loops which do not directly support instruction level parallelism. In order to increase the pool of operations that can be scheduled into a single instruction word, sophisticated compiler techniques are applied, such as loop unrolling, software pipelining, trace scheduling, or guarded execution [113].

Performance limitations of VLIW processors may arise from the growing hardware expense for multi-ported register file and crossbar switches when the number of functional units increases. Moreover, the long instructions result in high bandwidth requirements for instruction supply from an external memory. This problem can be tackled by compressing the instruction words in memory and expanding them after loading into the processor.

Up to now the signal processor architectures have been discussed on a highlevel considering the instruction stream and data stream. Also the possible parallelism has been described on the instruction level including concurrency of data access. For performance evaluation the data path has to be investigated in more detail.

Many signal processing methods contain convolution-like algorithms at their crux. In this case, two operands at a time must be continually multiplied and the results of this multiplication must be accumulated. The ALU presented in the previous section requires n cycles for the multiplication of two n-bit words. To accelerate the multiplication, array multipliers from section 3.3 can be employed. The Booth array multiplier is a suitable choice in this case. It supports two's complement multiplication and has a favourable product of area and delay.

A further element that is often required for processing is a shifter. An ALU can only carry out arbitrary shifts in several cycles. For fast execution of a shift, special hardware configurations should be employed. A barrel shifter is a configuration with which shifts set via control lines can be quickly carried out [14], [15] (see Figure 3.5.8).

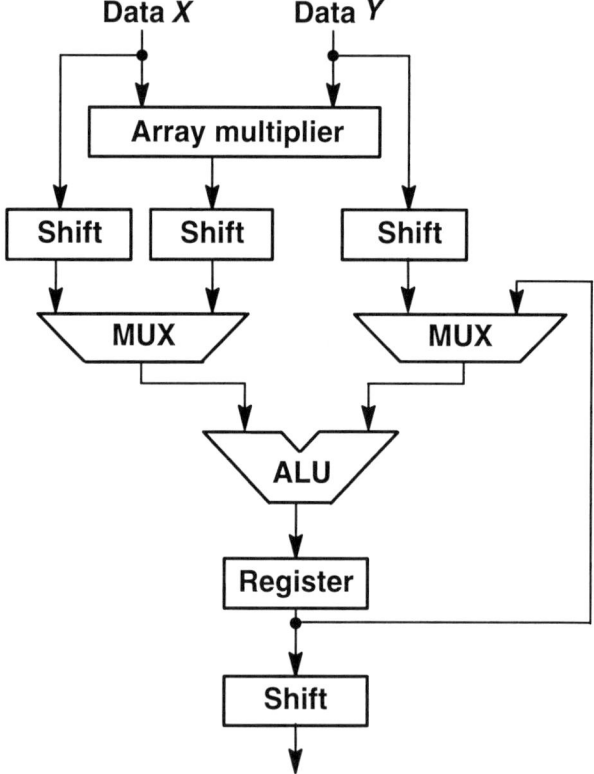

Figure 8.1.9: Data path based on a combination of MUL and ALU

It follows from the previous observations that the data path, in addition to an ALU, should also contain an array multiplier and a hardware shifter. A data path is shown in Figure 8.1.9 in which an ALU is supplemented by these units. Through appropriate control of the multiplexers, direct input of the two operands X and Y into the ALU can be achieved. If the left multiplexer is switched, the result of the multiplication from the array multiplier appears at the input of the ALU. The ALU can be set to a transparent mode, so that the result of the multiplication is then directly available at the output register. The ALU can also work as an accumulator. It is to be noted that the result of the multiplication is twice as long. If k products are to be accumulated without limiting effects, the ALU requires $2n + \log k$ bits. Not all of the shifters shown in Figure 8.1.9 must carry out arbitrary shift operations. This is generally only expected of the output shifter. The others only support particular shifts such as 2, 8 and 16, for example. In the case of a larger accumulation result, truncation to n and $2n$ bits must be supported for storing the results in the data memory. For $2n$ bits, two write cycles are necessary.

Figure 8.1.10 shows an alternative for implementing the data path. The array multiplier is implemented separately, together with the accumulator. Derived from multiplication and accumulation, this is also called a MAC unit. Here, the ALU only needs to support the word width of the input data X and Y.

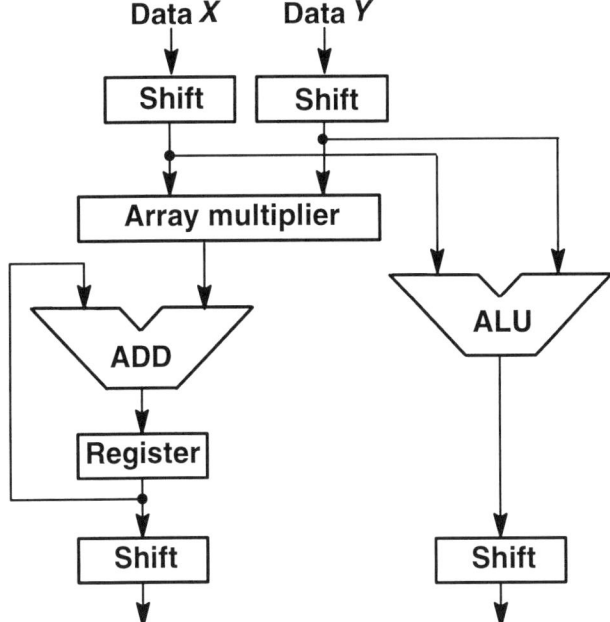

Figure 8.1.10: Data path with MAC unit and separate ALU

The data path displayed in Figures 8.1.9 and 8.1.10 increases the performance for algorithms involving largely multiplication, accumulation and arbitrary shifts. Hardware extensions of the data path with array multipliers, shifters and multiplexers reduce the number of clock cycles for a specific set of basic operations. In the particular case the execution of convolution- type core operations is accelerated. Thus, the data path is adapted to a class of algorithms where these basic operations occur frequently. It should be recognized that the frequency of occurrence of the specific hardware supported operations influences the performance increase.

As an extension of the idea formulated above further specialized instructions are included in signal processors for dedicated application fields. The basic idea is to incorporate additional logic units in order to reduce the number of execution cycles for operations from target algorithms. The decision which specialized instructions to implement involves a tradeoff between additionally required hardware effort and probability of their use. The benefit from specialized instructions is not universal. Generally, only a subset of algorithms will experience an acceleration. A few examples will be discussed below.

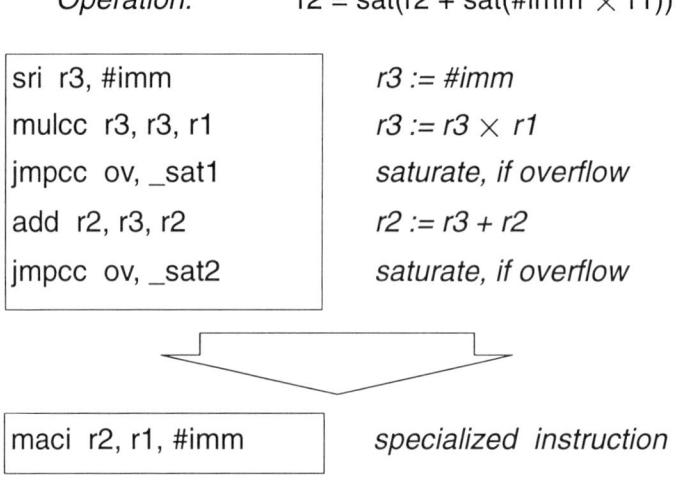

Operation: $r2 = \mathrm{sat}(r2 + \mathrm{sat}(\#imm \times r1))$

sri r3, #imm	*r3 := #imm*
mulcc r3, r3, r1	*r3 := r3 × r1*
jmpcc ov, _sat1	*saturate, if overflow*
add r2, r3, r2	*r2 := r3 + r2*
jmpcc ov, _sat2	*saturate, if overflow*

maci r2, r1, #imm	*specialized instruction*

Figure 8.1.11: Specialized instruction for multiply–accumulate with saturation. A single specialized instruction replaces a longer sequence of standard instructions. The branching to subroutines for saturation is eliminated

In fixed point signal processing overflow and underflow of a predefined range of data may occur. This overflow/underflow has negative side effects from wrapping around from large to small numbers and vice versa. Saturation is employed to avoid this effect. An instruction sequence for applying saturation on multiply–accumulate is shown in Figure 8.1.11. It is assumed that both the product and the sum have to be tested for overflow and saturated. By design of additional hardware the sequence of several instructions could be replaced by one instruction. It should be noted that the specialized instruction is not a macro for the original instructions. It is a new instruction with fewer cycles.

The hardware for a limiter consists of a comparator for comparison of the argument with the ceiling value and a multiplexer for setting the specified saturation value in case the argument is outside the defined range. The logic for the saturation block can be minimized considering predefined ceiling values as powers of 2.

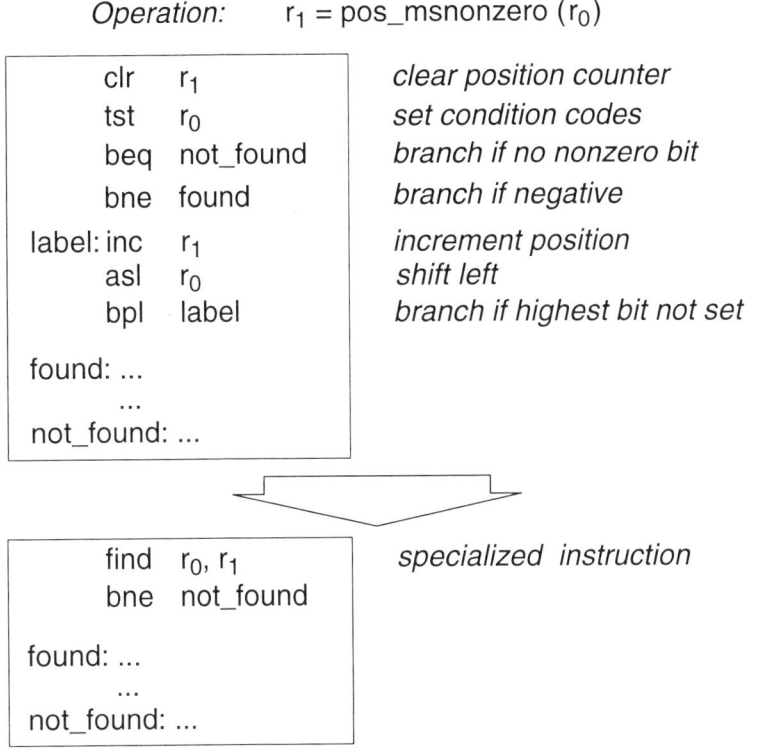

Operation: $r_1 = \text{pos\_msnonzero}(r_0)$

clr	r_1	*clear position counter*
tst	r_0	*set condition codes*
beq	not_found	*branch if no nonzero bit*
bne	found	*branch if negative*
label: inc	r_1	*increment position*
asl	r_0	*shift left*
bpl	label	*branch if highest bit not set*

found: ...
...
not_found: ...

find	r_0, r_1
bne	not_found

found: ...
...
not_found: ...

specialized instruction

Figure 8.1.12: Specialized instruction for the search for the most significant non-zero bit. A new "find" instruction and a branch instruction replace a sequence of standard instructions

Range control of intermediate results is frequently needed in fixed point signal processing. For this purpose the position of the most significant non-zero bit has to be investigated. A usual instruction sequence for the search for this bit is listed in Figure 8.1.12. This search routine contains some initial instructions plus a loop of three instructions. On average $3n/2$ cycles are needed for the loop with n as the word width. By insertion of specialized hardware the complete search can be accomplished in two cycles.

The hardware for finding the position of the most significant bit may be implemented in a tree structure. At first each of two neighbouring bit pairs is investigated for occurrence of a 1. As a result the occurrencebit for a 1 and the position code of the leading 1 are provided. The result of the first level is then further processed in the next level by amalgamating two neighbouring results to a new occurrencebit and a new position code. This hierarchical structure is continued until one position code for the complete argument is reached. The logic can be minimized by considering the a priori knowledge about the meaning of the intermediate results. This minimization results in a logic of 75 gates for a 32-bit argument.

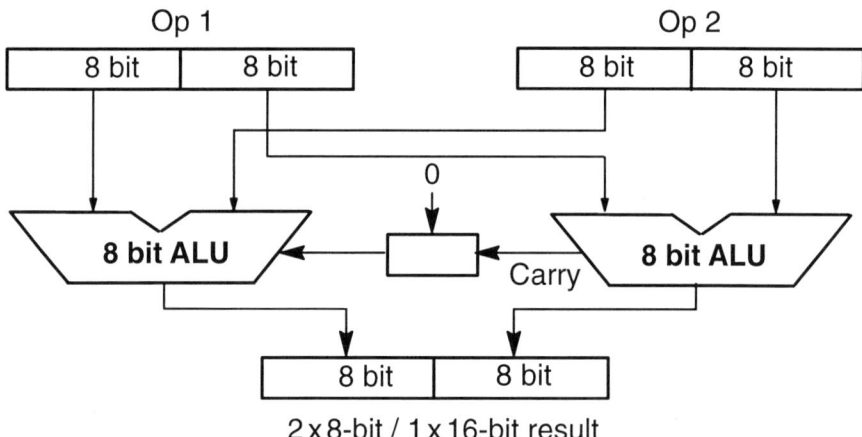

2 x 8-bit / 1 x 16-bit result

Figure 8.1.13: Split-ALU implementation. Two 8-bit ALUs can either operate independently or be combined to a 16-bit ALU by propagating the carry signal from the lower to the higher part

Besides specialized DSP instructions, performance increase of the data path is possible with the split-ALU concept which targets data level parallelism. This concept, also referred to as subword parallelism [114], involves processing of several lower precision data items on a single ALU of higher word length in parallel. For example, a 32-bit ALU can execute a single operation simultaneously on four 8-bit data items. The implementation of a split ALU requires only minor hardware extensions. Basically, the carry signals have to be prevented from being propagated across the boundaries of separate data items. A possible split-ALU implementation is shown in Figure 8.1.13.

The main benefit from a split-ALU is obtained for highly regular low-level algorithms involving identical operations being executed on large data volumes. The degree of exploitable data parallelism depends on the precision required for an operation. Video signal processing is a very demanding signal processing task based on operations with low-precision (8-bit) pixel data. Thus, subword parallelism can effectively be exploited to enhance the performance for these tasks. Audio processing generally has higher precision demands and does not benefit from split-ALU to the same extent. By providing different split-ALU instructions for various data formats, the achievable parallelism can gradually scale with the precision needs.

The idea of a split-ALU can be extended to a packed arithmetic which supports additional instructions in a subword parallelism [114], [115]. One example may be multiply–add on subwords.

$$A = a_3 \mid a_2 \mid a_1 \mid a_0$$
$$B = b_3 \mid b_2 \mid b_1 \mid b_0$$
$$C = (a_3b_3 + a_2b_2) \mid (a_1b_1 + a_0b_0)$$

Another example may be the sum of the magnitude of differences:

$$A = a_3 \mid a_2 \mid a_1 \mid a_0$$
$$B = b_3 \mid b_2 \mid b_1 \mid b_0$$
$$C = (\mid a_3 - b_3 \mid + \mid a_2 - b_2 \mid + \mid a_1 - b_1 \mid + \mid a_0 - b_0 \mid)$$

Functions associated with memory addressing are supported by an address generation unit. An address generation unit for the data must support direct addressing, i.e. the address is supplied by the instruction. The other two important modes are increment and decrement. In addition, a modulo arithmetic is required in the case of periodic repetitions. For fast Fourier transforms, bit reversal logic is also important, in which the bit levels are mutually exchanged. Figure 8.1.14 shows an address generation unit that supports the address modes mentioned.

If convolution is carried out, the processor must carry out identical operations in accordance with the length of the data to be convoluted. Only the address generation units are active and change the data pointers. To support such an operation, a loop counter is required in the control unit. Only a counter must be implemented that is set at the beginning and decremented with each cycle.

The particular measures for increasing the computational performance for signal processing are to be briefly summarized. A Harvard architecture with separate data and program memory is to be used. The operations in the data path are to be carried out overlapping with the read and write cycles of the memory. Two-operand operations, in particular, are supported by doubling the data memory and expanding the corresponding bus system. Instruction level parallelism may be supported by VLIW architectures. To accelerate the operations in the data path, a dedicated implementation of multipliers and shifters is to be employed. Further additional hardware extensions of the data path offer specialized DSP instructions with reduced number of cycles. Subword parallelism could be implemented with the split-ALU concept which increases the concurrency of operations. Furthermore, a special implementation of the data address generation unit and a hardware loop counter are to be designed. To accelerate memory access, local storage for data and instructions is to be present on the chip. A hierarchical storage system with small memory units on the chip and large, external memory improves the effective memory bandwidth due to the various access rates.

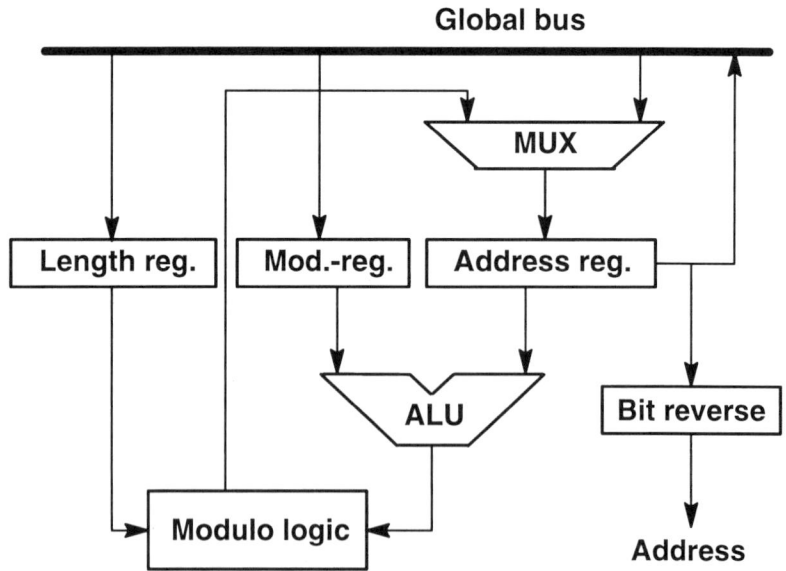

Figure 8.1.14: Address generation unit for a signal processor

8.1.3 Characteristics of available DSPs

This section limits itself to single processor systems. Their general structure is termed SISD (single instruction stream – single data stream) after Flynn [116]. Following the description and abbreviations of Madisetti [117], in Figure 8.1.15 a schematic of an SISD processor with Harvard architecture is shown. The continuous flow of instructions from the instruction memory IM forms the instruction stream IS. The data exchange between the data memory DM and the data path DP forms the data stream DS. The data path is often alternatively termed a data processor or processing unit. The control unit here is generically termed the instruction processor (IP). This is to show that complex address computations are carried out in the control unit. An alternative name often used is instruction sequencer. The DSPs commercially available are modifications of the basic structure of Figure 8.1.15.

One of the first DSP chips on the market was Texas Instruments' TMS 320C10 that appeared in 1979 [118]. This processor corresponded to a large degree to the basic structure noted above. On the chip it contained a program memory of 1.5k words and a data memory of 144 words. The data path consisted of a 16×16-bit multiplier, a 32-bit ALU, a 32-bit accumulator and a 16-bit barrel shifter. The structure of the data path corresponded to Figure 8.1.9.

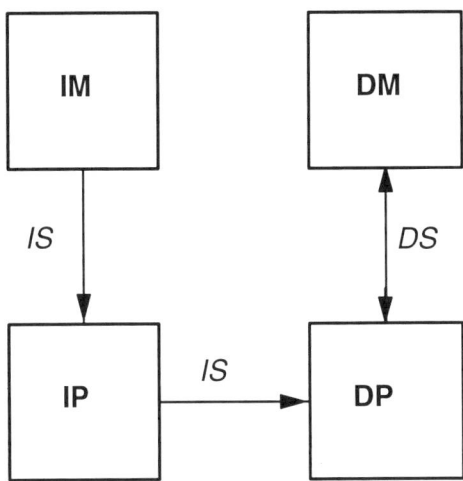

Figure 8.1.15: SISD processor with Harvard architecture

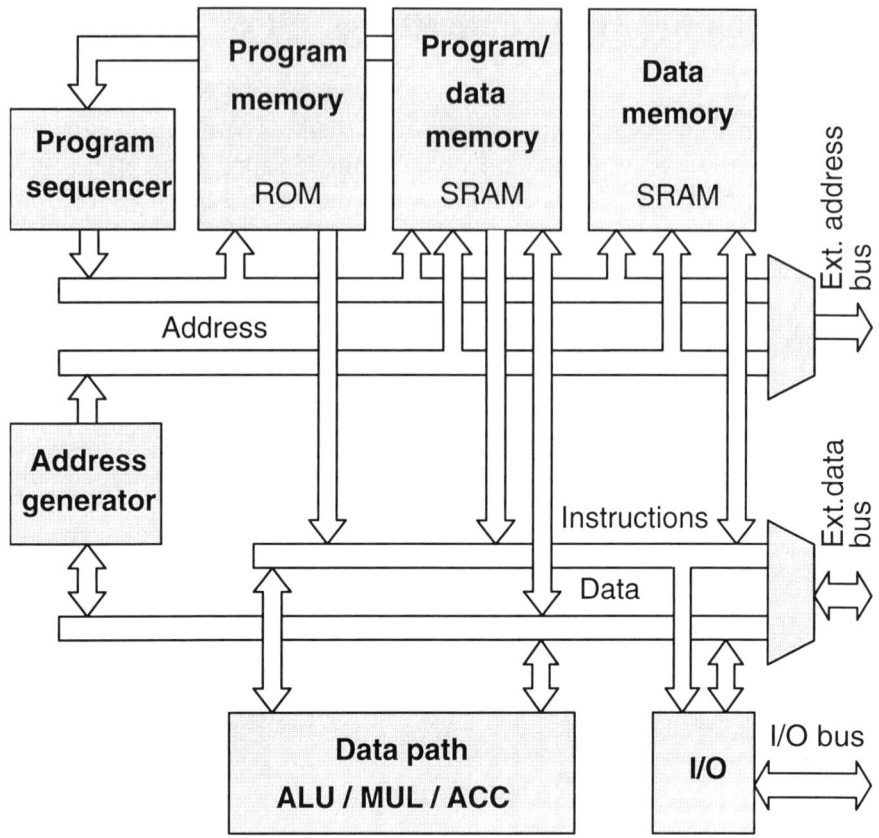

Figure 8.1.16: Block diagram of the TMS 320C25

This first DSP was further developed in subsequent years. The structure of the TMS 320C25 is shown in Figure 8.1.16 [119]. The integrated memory is more than twice as large and the cycle time was more than halved. The instruction memory can also be used as data memory, for example, for filter coefficients. To avoid conflicts between data and instruction access, the instruction processor was equipped with a mini-cache for one instruction. New variants of the TMS 320 family have clearly increased integrated memory, a significantly enlarged address space for external memory and further reduced cycle times. Whereas the first TMS 320s supported a processor clock rate of up to 5.5 MHz, new versions (C50) enable a processor clock rate of 50 MHz [119].

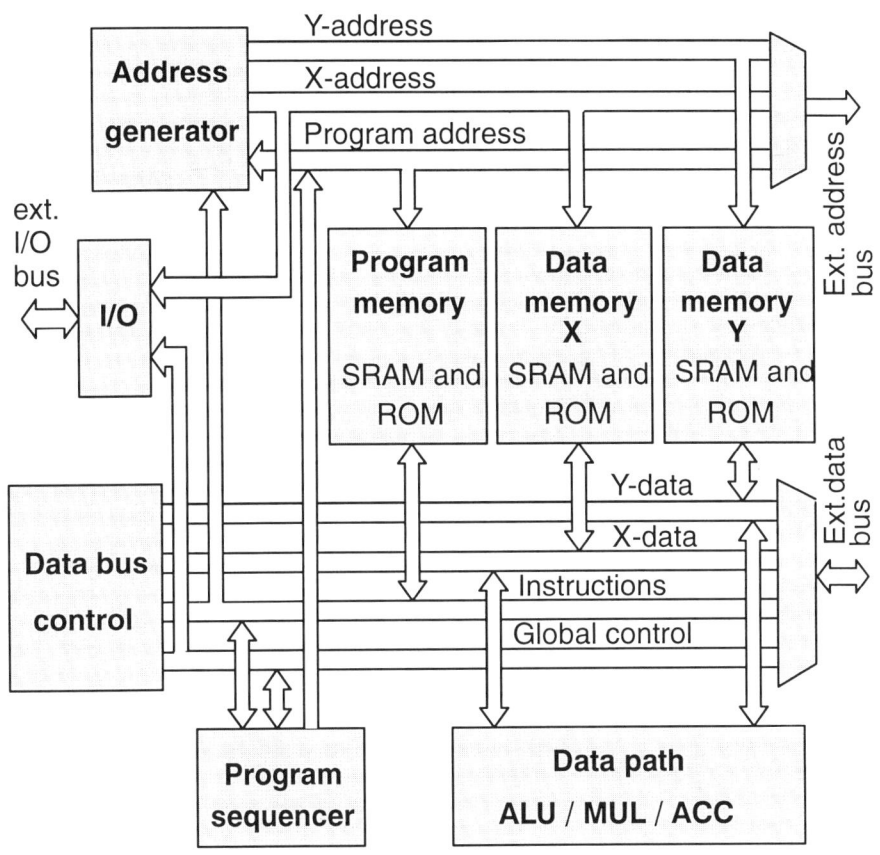

Figure 8.1.17: Block diagram of the DSP 56000

To avoid access conflicts during joint usage of internal memory for instructions and data, DSPs are also implemented with two data memories as in Figure 8.1.7. One example is Motorola's DSP 56000 [120]. Figure 8.1.17 shows a block diagram of this processor.

Deviating from others, this processor is based on a word width of 24 bits. The data path has a structure as in Figure 8.1.10. The processor possesses a complex address generator and a program sequencer with a hardware loop counter. Due to the large number of internal memories, several internal buses are used that can be connected via a special switching network.

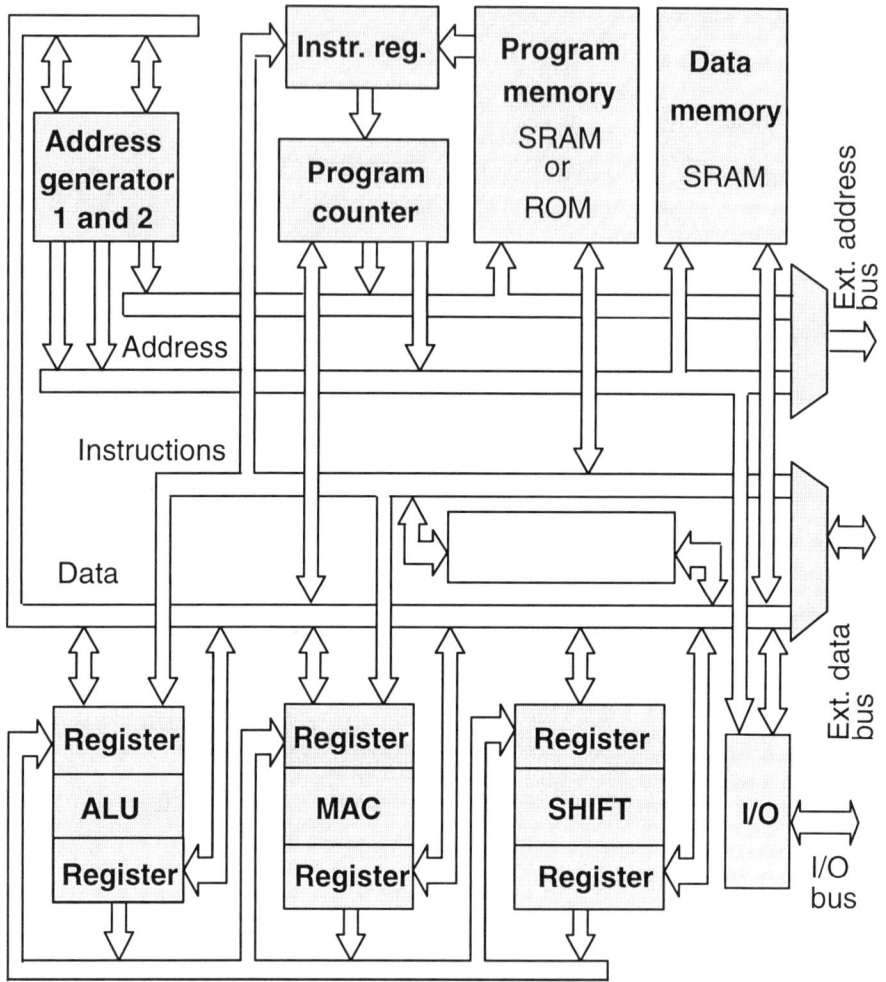

Figure 8.1.18: Block diagram of the ADSP 2101

As a further example, Analog Devices' ADSP 2101 is to be examined
(Figure 8.1.18) [122]. The data path consists of three independent logic
units. Each of these logic units has its own input and output register. This
avoids access conflicts between source and target registers. The results of a
logic unit can be directly transferred without intermediate storage in internal
RAM via a result bus to a following unit. Through these measures, the
throughput rate for operations is increased. The ADSP 2101 possesses two
address generators for two address spaces. To solve access conflicts between

instruction and data access to the program memory, the ADSP 2101 is equipped with a cache memory for 16 instructions.

Newer DSPs contain additional modules for special functions [121], [122]. A typical element of this kind is an integrated timer that can be used for clock signal generation for external modules and periodic interrupts. The DSP56156 and DSP56166 contain additional A/D and D/A converters according to the delta–sigma principle. In addition, DSP chips exist with special modules for accelerating Viterbi decoding. Besides the previously treated DSPs with fixed-point arithmetic, processors with floating-point arithmetic are also available. In general, these have a 32-bit word width with a 24-bit mantissa and an 8-bit exponent in accordance with the IEEE format 754. The extended IEEE standard with 11-bit exponents is only supported by a few processors. A summary of the characteristic data of a selection of available DSPs is listed in Tables 8.1.2 and 8.1.3.

Table 8.1.2: Selection of DSPs with floating-point arithmetic

Manufacturer/ type ID	Cycle time (ns)	Word width (bit)	Integrated memory (bit)		External memory (bit)
			Data	Program/data	
Texas Instruments					
TMS320C30-40	50	32/8	2k RAM	4k ROM	16M × 32
TMS320C31-60	33	32/8	2k RAM	Boot loader	16M × 32
TMS320C32-60	33	32/8	16k RAM	Boot loader	16M × 32
TMS320C40-60	33	32/8	2k RAM	4k ROM	4G × 32
TMS320C44	50	32/8	2k RAM	4k ROM	128M × 32
Motorola					
DSP96002	50	32/11	32k RAM 64k ROM	32k RAM 2k ROM	2 × 4G × 32
Analog Devices					
ADSP-21020	33	32/8		1.5 k cache	4G × 40
ADSP-21060 (Sharc)	25	32/8		4M RAM	4G × 48

Table 8.1.3: Selection of DSPs with fixed-point arithmetic

Manufacturer/ type ID	Cycle time (ns)	Word width (bit)	Integrated memory (bit)		External memory (bit)
			Data	Program/data	
Texas Instruments					
TMS320C10-25	180	16	144 RAM	1.5k ROM	8×16
TMS320C15-25	160	16	256 RAM	4k ROM	6×16
TMS320C16	114	16	256 RAM	8k ROM	8×16
TMS320C25	80	16	544 RAM	4k ROM	16×16
TMS320C50	20	16	10k RAM	2k ROM	$64k \times 16$
TMS320C52	20	16	1k RAM	4k ROM	$64k \times 16$
TMS320C53	20	16	4k RAM	16k ROM	$64k \times 16$
Motorola					
DSP56000	50	24	12k RAM 12k ROM	12k RAM 768 ROM	$64k \times 24$
DSP56001	30	24	12k RAM 12k ROM	12k RAM 768 ROM	$64k \times 24$
DSP56002	25	24	12k RAM 12k ROM	12k RAM 768 ROM	$64k \times 24$
DSP56156	33	16	32k RAM	32k RAM 1k ROM	$64k \times 16$
DSP56166	33	16	64k RAM	32k RAM 1k ROM	$64k \times 16$
AT & T					
DSP1610	25	16		128k RAM 8k ROM	$64k \times 16$
DSP1617	20	16		64k RAM 384k ROM	$64k \times 16$
DSP1618	20	16		64k RAM 256k ROM	$64k \times 16$
Analog Devices					
ADSP-2101	50	16	16k RAM	48k RAM	$16k \times 24$
ADSP-2161	60	16	8k RAM	192k ROM	$16k \times 24$
ADSP-2171	30	16	32k RAM	48k RAM	$16k \times 24$
ADSP-2181	30	16	256k RAM	384k RAM	$16k \times 24$

8.2 The Programming of Signal Processing Algorithms

The increases in throughput through the architectures of signal processors are to be illustrated in two examples here. A program for an FIR filter is presented as a favourable example for a DSP. A second example is the programming of a butterfly element for an FFT. The DSP 56000 from Motorola [120] is used as a DSP and the MC 68000 microprocessor [123], [124] from the same company is used as a CISC processor for the comparison. The MC 68000 was first presented in 1979. In the mid-1980s, it was the standard processor for PCs besides the Intel 8086. First, a comparison of DSPs and CISC processors that were available in 1986 is carried out. Further development has taken place since this time. Thus, the changes that the new CISC processors have brought with them are also shown.

The example programs are programmed in assembler. A development of the programs in detail is beyond the scope of this book. The programs are presented in their final forms, and the influence of the architectural characteristics on the program is explained.

8.2.1 An FIR filter program

The filter algorithm is given by Equation (6.1.11) alone. It is assumed that all filter coefficients $h(\cdot)$ are stored in the processor's local memory. Furthermore, a segment of the input sequence $x(\cdot)$ is also to be held in the local memory. The number of stored elements $x(\cdot)$ is assumed to be larger than the length N of the impulse response so that no external data must be loaded during the course of the algorithm. The reloading of input data and the reuse of the result data are not shown. Only the central process of filtering is shown.

A filter program for the MC 68000 is listed in Figure 8.2.1. The program comprises three sections. The actual filter loop with multiplication and accumulation follows the initialization process with its loading of start addresses and the initialization of the loop counter. To finish, the result is stored and the start of a new cycle is prepared. The meaning of the individual instructions is explained by comments. The number of processor cycles necessary is also given. Through addition of the processor cycles for a loop, it follows that

$$M_{CYC, Loop} = N \cdot 100 \qquad (8.2.1)$$

The large number 100 is due, in particular, to the large number of cycles for the multiplication. But even the addition and the loading of the coefficient require clearly more than one cycle. Neglecting the remaining cycles of the rest of the program, the maximally attainable sampling rate that could still run in real-time is to be determined from the loop data.

$$f_{S,max} = \frac{f_{CLK}}{M_{CYC,Loop}} \tag{8.2.2}$$

Under the assumption of an impulse response with $N = 128$ and a processor clock frequency of 16 MHz, a maximally processable sampling rate of 1.2 kHz results. This means that the MC 68000 processor is not even in a position to carry out the filtering of speech signals (typical sample rate of 8 kHz).

	Instruction	Operand	Processor cycles	Comment
	MOVE.L	D0,A0	4	load data start addr.
	MOVE.L	D1,A1	4	load coeff. start addr.
	MOVE.W	#N,D2	8	init. loop counter
	CLR.L	D3	6	clear accumulator
LP:	MOVE.W	(A1)+,D4	8	load coefficient
	MULS	(A0)+,D4	74	multiply with data
	ADD.L	D4,D3	8	accumulate
	DBRA	D2,LP	10/14	loop end ???
	MOVE.L	D3,(A2)+	12	store result
	ADDQ.L	#2,D0	4	new start addr. data

Figure 8.2.1: FIR filter program for an MC 68000

The corresponding filter routine for the DSP 56000 processor is shown in Figure 8.2.2. The typical structure of an assembler program with an instruction field and an operand field is extended here by X-bus and Y-bus transports since data can be loaded via buses to input registers of the ALU simultaneously with the execution of operations. In addition, the decreased number of cycles stands out. This is achieved through optimization of the processor logic and avoidance of a microprogram control unit. The actual filter loop consists of one MAC command since the next data can also be simultaneously transported. The hardware loop counter runs unseen in the background. It is initialized through the previous REP command. The processor cycle time for the loop in the above case amounts to

$$M_{CYC, Loop} = N \cdot 1 \qquad (8.2.3)$$

This shows the clear advantage of the DSP architecture. The maximally attainable throughput for real-time processing using, as before, $N = 128$ and an effective processor clock rate of 10 MHz evaluates to 75 kHz. The processor itself is fed a clock rate of 20 MHz, yet an operation is carried out in only two clock phases. Consequently, the result shows that the processor can be effectively used for the filtering of audio signals (sampling rate 48 kHz).

Instr.	Operands	X-Bus	Y-Bus	Cycles	Comment
MOVE	#D_ADR,R0			1	R0 points to data
MOVE	#K_ADR,R4			1	R4 points to coeff.
NOP		A1,X:(R1)+		1	store prev. result
CLR	A	X:(R0)+,X0	Y:(R4)–,Y0	1	clear accumulator
					load operand
REP	#N			2	init loop counter
MAC	X0,Y0,A	X:(R0)+,X0	Y:(R4)–,Y0	1	multiply & accumulate
RND	A			1	round the result

Figure 8.2.2: FIR filter program for a DSP 56000

The Intel P5 (Pentium) is currently the standard processor for PCs [125]. The signal processing gains that are possible through modified architectures and modern semiconductor technologies are to be demonstrated using this processor. The FIR filter program for a P5 is listed in Figure 8.2.3. The comments explain the function of the individual instructions. The P5 has two internal data paths that are designated U and V. For the evaluation of its time requirements, the division between the data paths and the scheduling of the inner loop must be taken into account. Figure 8.2.4 shows the inner loop with scheduling and the required number of cycles. The number of processor cycles for the loop amounts to

$$M_{CYC, Loop} = N \cdot 13 \qquad (8.2.4)$$

FIR1xN:

```
              ;         general register initialization
              ;
              lea       edi, [data]
              lea       esi, [result]

              ;         outer loop
              ;
block:        mov       edx, edi
              mov       ecx, –4*N              ; loop index
              sub       edx, 4
              xor       eax, eax               ; clear intermediate result
              xor       ebx, ebx               ; clear accumulator

              ;         inner loop
              ;
loop:         add       ebx, eax               ; Σ data*coeff.
              add       edx, 4                 ;increment pointer to input
                                                 data
              mov       eax, [ecx+4*N+coeff] ; load coefficient ...
              imul      eax, [edx]             ; ... multiply with data
              add       ecx, 4                 ; increase loop index
              jnz       loop                   ; back to inner loop

              add       ebx, eax               ; Σ data*coeff
              mov       [esi], ebx             ; store result
              add       edi, 4                 ;increment pointer to input
                                                 data
              add       esi, 4                 ;increment pointer to result
                                                 data

              j..       block                  ; back to outer loop
```

Figure 8.2.3: FIR filter program for an Intel P5

Under the assumption of a processor clock frequency of 133 MHz and an impulse response of $N = 128$, a maximally processable sampling rate of 79 kHz follows. The significant gains of modern processors can be split into two components. The number of cycles for the inner loop could be reduced from 100 to 13. This can be interpreted as an architectural improvement by a factor of roughly 8. The increase in processor clock rate from 16 MHz to

133 MHz also corresponds to a technological gain by a factor of roughly 8. Thus, the total gain amounts to a factor of approximately 64.

	U-pipeline	V-pipeline	Cycles
loop:	add ebx, eax	add edx, 4	1
	mov ea x, [ecx+4*N+coeff]		1
	imul eax, [edx]		10
	add ecx, 4	jnz loop	1

Figure 8.2.4: FIR filter program loop with scheduling for a P5

In the same timespan, simple DSPs have also further developed. The DSP 56002 is currently offered with a processor clock rate of 40 MHz. DSPs also exist with a clock rate of 50 MHz. Consequently, one has a technological gain by a factor of roughly 2.5. Direct architectural gains for an FIR filter are minor. Observed as a whole, the distance between CISC processors and DSP processors is clearly reduced.

8.2.2   A DFT program

The algorithm for carrying out a DFT is given by Equation (7.1.5). As shown in section 7.3, a fast algorithm leads to a significant reduction in the required computation performance. Thus, such an algorithm is to be used here. The central element of an FFT is the butterfly element. Therefore, only the program segment for a butterfly will be presented. Besides the FFT explained in section 7.3 based on a DIF (decimation in frequency) algorithm, FFTs based on the DIT (decimation in time) algorithm exist. A program for the DIT algorithm is shown in Figure 8.2.6.

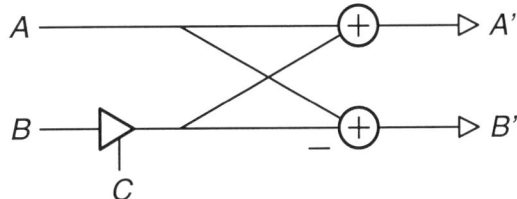

Figure 8.2.5: Radix 2 butterfly for a DIT algorithm

Figure 8.2.5 shows the corresponding butterfly element. It is a transposed form of the butterfly element of a DIF algorithm. One of the input operands is multiplied by a complex exponential term, then the sum and the difference are created and stored. Figure 8.2.6 shows a corresponding assembler program for the MC 68000. The number of cycles for a butterfly amounts to

$$M_{CYC, FFT-PE} = 489 \qquad (8.2.5)$$

Instruction		Operands	Processor cycles	Comment
	MOVE.W	#N,D7	8	loop counter
LP:	MOVE.W	(A0),D5	8	
	MULS	(A1)+,D5	74	Re{B} * Re{C}
	MOVE.W	(A0)+,D6	8	
	MULS	(A1),D6	74	Re{B} * Im{C}
	MOVE.W	(A0),D4	8	
	MULS	(A1),D4	74	Im{B} * Im{C}
	SUB.L	D4,D5	8	D5 = BrCr − BiCi
	MOVE.W	(A0),D4	8	
	MULS	−(A1),D4	74	Im{B} * Re{C}
	ADD.L	D4,D6	8	D6 = BrCi+BiCr
	CLR.L	D2	6	
	MOVE.W	(A2)+,D2	8	D2=Re{A}
	MOVE.L	D2,D3	4	D3=Re{A}
	ADD.L	D5,D2	8	D2=Re{A'}
	MOVE.L	D2,(A4)+	8	store
	ASL.L	#1,D3	10	D3=2*Re{A}
	SUB.L	D2,D3	8	D3=Re{B'}
	MOVE.L	D3,(A5)+	8	store
	CLR.L	D2	6	
	MOVE.W	(A2)+,D2	8	D2=Im{A}
	MOVE.L	D2,D3	4	D3=Im{A}
	ADD.L	D6,D2	8	D2=Im{A'}
	MOVE.L	D2,(A4)+	8	store
	ASL.L	#1,D3	10	D3=2*Im{A}
	SUB.L	D2,D3	8	D3=Im{B'}
	MOVE.L	D3,(A5)+	8	store
	DBRA	D7,LP	10/14	loop

Figure 8.2.6: Radix 2 butterfly for an MC 68000

Limited to the pure processing time of the butterfly PE and neglecting the computation time for pointer initialization, etc., the maximal sampling frequency for an FFT still executable in real-time evaluates to

$$f_{S,max} = \frac{2f_{CLK}}{\log_2 N \cdot M_{CYC,FFT-PE}} \qquad (8.2.6)$$

Under the assumption of a block length $N = 128$, a maximally executable sampling rate of 9.4 kHz follows for the MC 68000 ($f_{CLK} = 16$ MHz).

A modern P5 shows significant additional gains here. Programming is possible that leads to 52 cycles for a butterfly PE. Taking the distinctly higher processor clock rate of 133 MHz into account, a maximal sampling rate of 370 kHz is processable.

Figure 8.2.7 shows an assembler program for the DSP 56000. An efficient implementation with few cycles is visible. The cycle count in this case is

$$M_{CYC,FFT-PE} = 9 \qquad (8.2.7)$$

Under the assumption of a processor clock rate of 20 MHz, a maximally attainable sampling rate of 320 kHz results for a block length of $N = 128$.

Instr.	Operands	X-Bus	Y-Bus	Cycles	Comment
MOVE		X:(R1),X1	Y:(R6),Y0	1	data and coeff.
MOVE			Y:(R0),B	1	load
MOVE		X:(R0)+,A		1	
DO	#N,END			3	init. loop counter
MAC	Y0,X1,B	X:(R6)+,X0	Y:(R1)+,Y1	1	Re{B} * Im{C}
MACR	X0,Y1,B	A,X:(R5)+	Y:(R0),A	1	Im{B} * Re{C}
SUBL	B,A	X:(R0)+,B	B,Y:(R4)+	1	2*Im{A} – Im{A'}
MAC	X0,X1,B	X:(R0),A	A,Y:(R5)+	1	Re{B} * Re {C}
MACR	–Y0,Y1,B	X:(R1),X1	Y:(R6)+,Y0	1	Im{B} * Im{C}
SUBL	B,A	B,X:(R4)+	Y:(R0)+,B	1	2*Re{A} – Re{A'}
END					
MOVE		A,X:(R5)+		1	store result

Figure 8.2.7: Radix 2 butterfly for the DSP 56000

A comparison between CISC processors and DSPs on the basis of components available in 1986 shows clear architectural gains for the DSPs. Modern CISC processors have significantly decreased this discrepancy. The cycle count only differs by a factor of approximately 5. Since CISC processors with very high processor clock rates (133 MHz and above) are commercially available, the difference for various processor clock rates is further decreased.

The execution of a multiplication with CISC processors requires distinctly more than one cycle. Thus, in particular an FFT can be somewhat more efficiently implemented when the complex multiplication is implemented with three real multiplications instead of four (see Figure 7.2.3). Such a modification does not yield an advantage for DSP processors since both the multiplication and the addition require one cycle.

8.3 Architecture Optimization via Simple Models

There is a desire to examine quantitatively the influence of architectural parameters. Due to the complex dependencies, it is extremely complicated to account for all influencing factors exactly. Therefore, the evaluation is to be carried out using simple modelling.

As presented at the end of section 8.1, several architectural measures lead to the high performance of DSPs. Thus, to simplify the evaluation, the influencing factors will be examined individually, i.e. one parameter will be varied while the others have a fixed value. The programming examples in section 8.2 show that the algorithm to be implemented also has an influence. For this reason, the algorithms must also be modelled. The algorithms are described by a statistical distribution of characteristic areas or instructions for this evaluation.

The throughput is used as the measure of performance for an architecture. Another important criterion is the architecture efficiency. In the following, the influence of instruction pipelining and the improvement in performance through multipliers, barrel shifters and local memory are observed.

8.3.1 Instruction pipelining

Several steps are necessary to carry out an instruction in a processor. These are:

    a.   Address and read from the instruction memory and read the instruction into the instruction sequencer (IF: instruction fetch)

b. Decode the instruction (ID: instruction decode)

c. Address and read the operands from the data memory (OF: operands fetch)

d. Execution of the instruction in the data path (EX: execute)

e. Address and write the result to the data memory (WR: write)

The execution of a following instruction cannot be started until this step has been carried out. Thus, the time T_{INS} required for the execution of an instruction results from the sum of the execution times of the steps named above.

$$T_{INS} = T_{IF} + T_{ID} + T_{OF} + T_{EX} + T_{WR} \qquad (8.3.1)$$

To simplify the evaluation, it is assumed in this section that the execution time T_{EX} of the operations in the data path is the same for all operations. As shown in Figure 8.1.4, the effective throughput can be raised by pipelining. In the case of pipelining, registers are inserted between those hardware components that serve the execution of sub-steps. Due to the delay time of the registers for the reading and writing of their contents, a delay time T_{REG} must be additionally taken into account in the execution time. Thus, for the case of pipelining in accordance with Figure 8.1.4, the execution time for an instruction is given by

$$T_{INS,PIPE} = \max[T_{IF}, T_{ID}, T_{OF}, T_{EX}, T_{WR}] + T_{REG} \qquad (8.3.2)$$

The above equation holds for five pipeline stages. It is possible to reduce the number of pipeline stages by combining sequential sub-steps. The stages IF and ID or also EX and WR can be combined, for example. To increase the throughput, the maximal delay time can be further reduced by splitting. The time T_{EX} for the execution of an operation, for instance, can be divided into two clock phases. In the DSP 56000, five clock phases are used – one for IF and ID, one for OF, two for EX and one for WR.

If, for simplicity, an equal division of the execution time T_{INS} over N_{PIPE} pipeline stages is presumed, the time for a pipelined instruction is given by

$$T_{INS,PIPE} = \frac{T_{INS}}{N_{PIPE}} + T_{REG} \qquad (8.3.3)$$

The effective execution time of the pipelined instruction $T_{INS,PIPE}$ is simultaneously the timespan T_{CLK} of one clock period. The computation time $T_{INS,PIPE}$ can then only be achieved when the execution of an instruction is

independent of all instructions started fewer than N_{PIPE} clock cycles previously. Dependencies on previous instructions occur along with conditional
branches, conditional instructions and the use of previous results. The execution time of dependent instructions can be reduced through forwarding
[109]. To simplify the discussion, it is assumed here that, for dependent
instruction, all stages must be completed before the next one can be started.
Therefore, in this case, the execution time increases to

$$T_{INS,dep} = T_{INS} + N_{PIPE} \cdot T_{REG} \qquad (8.3.4)$$

Given that $P_{IND, INS}$ is the probability of an independent instruction in
a program, then the effective execution time in accordance with (8.3.3) and
(8.3.4) is

$$T_{INS,PIPE} = \left(\frac{T_{INS}}{N_{PIPE}} + T_{REG}\right)P_{IND, INS}$$
$$+ (T_{INS} + N_{PIPE} \cdot T_{REG})(1 - P_{IND, INS}) \qquad (8.3.5)$$

The throughput rate of a processor is the number of samples to be processed per unit of time. The effective time per instruction must thus still be
multiplied by the number of instructions per sample.

$$R_T = \frac{1}{T_{INS} \cdot n_{INS/OP} \cdot n_{OP/SAMPLE}} \qquad (8.3.6)$$

The number of operations per sample $n_{OP/SAMPLE}$ depends on the algorithm. For so-called low-level algorithms, this is a fixed number that can be
easily determined. Due to the fixed data dependencies, all the operations to
be carried out are known in advance for low-level algorithms. No data dependent branches or operations exist. Typical examples of low-level algorithms
are filters and transformations. For the two algorithms treated in section 8.2,

$$\begin{aligned} \text{FIR} - \text{Filter}: \quad & n_{OP/SAMPLE} = 2N \\ \text{FFT}: \quad & n_{OP/SAMPLE} = 5 \log N \end{aligned} \qquad (8.3.7)$$

The factor 5 in the FFT results from the 10 real-valued operations per butterfly element.

The number of instructions per operation $n_{INS/OP}$ is a static value that
depends on the data path of the processor and on the algorithm. The inner
loops of the FIR filter in the example programs yield the following values:

$$\text{MC68000}: \quad n_{INS/OP} = 4$$
$$\text{DSP56000}: \quad n_{INS/OP} = 0.5 \tag{8.3.8}$$

Based on butterfly elements they would yield:

$$\text{MC68000}: \quad n_{INS/OP} = 2.7$$
$$\text{DSP56000}: \quad n_{INS/OP} = 0.7 \tag{8.3.9}$$

These two examples clearly show the dependency on the processor type and the algorithm. The use of a characteristic value $n_{INS/OP}$ is not correct for an MC 68000 processor since the assumption of identical execution times for all instructions does not hold. The example program in Figure 8.2.6 shows considerable differences in the number of processor cycles for executing an instruction.

The actual number of instructions per operation exceeds the values of (8.3.8) and (8.3.9) since only the main routines were taken into consideration. The transport of the data blocks to the local memory and the particular initialization routines are missing.

If one presumes that the number of instructions per operation and the number of operations per sample are not influenced by the pipelining, the throughput with pipelining is given by

$$R_{T,PIPE} \sim \frac{1}{T_{INS,PIPE}} \tag{8.3.10}$$

This means that the effective time per instruction has the greatest influence on the throughput. Through examination of the zeros of the derivative of (8.3.5) with respect to N_{PIPE}, the optimal number of stages can be determined.

$$N_{PIPE,OPT} = \sqrt{\frac{1}{\delta_{REG}} \cdot \frac{P_{IND,INS}}{1 - P_{IND,INS}}} \tag{8.3.11}$$

δ_{REG} is the quotient T_{REG} / T_{INS}. Using typical values, a numeric, optimal value can be determined. For $P_{IND,INS} = 0.9$ and $\delta_{REG} = 1/30$, $N_{PIPE,OPT} \approx 16$. Thus one should strive for a large number of pipeline stages. It is to be noted that the number of maximally employable pipeline stages is normally limited by the semiconductor technology used. The inclusion of pipeline stages in memory, for instance, is only possible under very special conditions. In general, no pipeline stages are provided for the memory. Due to this limitation, no more than seven pipeline stages are employed in practice.

Through the inclusion of pipeline stages, the total costs are increased. Thus, one must again check if the efficiency, in accordance with (4.2.5), has been improved. This would also mean that the throughput per unit of active silicon area has been improved. From (4.2.13) it can be derived that the efficiency has improved as long as the relative gain in throughput is larger than the relative increase in area.

The increase in area for the pipelining is incurred through the additional register stages and is thus proportional to the number of stages.

$$A_{Si,PIPE} = A_{Si,1}[1 + (N_{PIPE} - 1)\alpha_{REG}] \qquad (8.3.12)$$

The gain in efficiency can be evaluated by combining (8.3.10) and (8.3.5). Using (4.2.13), the change in efficiency is given by

$$\frac{d\eta}{\eta} = \left(\frac{P_{IND,INS}}{N_{PIPE}^2} - \delta_{REG}(1 - P_{IND,INS}) - \alpha_{REG}\right)dN_{PIPE} \qquad (8.3.13)$$

For the optimal number of stages, the expression in parentheses must be zero. Equation (8.3.11) is modified to

$$N_{PIPE,OPT} = \sqrt{\frac{P_{IND,INS}}{(1 - P_{IND,INS})\delta_{REG} + \alpha_{REG}}} \qquad (8.3.14)$$

accordingly.

Using $\alpha_{REG} = 0.02$ and the previously used numbers for δ_{REG} and $P_{IND,INS}$, an optimal value of

$$N_{PIPE,OPT} \approx 6$$

results. This corresponds to the number of stages often used in practice.

8.3.2   Special arithmetic modules

In the previous section, it was assumed that each instruction has the same execution time T_{EX}. As particularly the example programs in Figures 8.2.1 and 8.2.6 show, this is not generally true. If the various clock cycle counts are to be regarded in the evaluation of the throughput, Equation (8.3.6) must include the clock cycles. Then

$$R_T = \frac{1}{T_{CLK} \cdot n_{CYC/OP} \cdot n_{OP/SAMPLE}} \qquad (8.3.15)$$

whereby the number of cycles per operation $n_{CYC(OP)}$ is, as before, a statistical variable (mean value) that depends on both the algorithm and the implementation of the data path. The number of cycles per operation can be evaluated as follows:

$$n_{CYC/OP} = \sum_{OP} n_{CYC}(OP)\, P(OP) \qquad (8.3.16)$$

Here, $n_{CYC}(OP)$ is the number of cycles for the particular operation OP and $P(OP)$ is the probability of occurrence of OP in an algorithm. The example programs have shown that, besides instructions that prompt operations on data, instructions exist that trigger no operation. Examples of this are pure data transfers, program branches and the initialization of program loops. These instructions are grouped as NOP (no operation) instructions. The number of cycles for the NOP instructions raises the mean number of cycles, i.e. (8.3.16) must be modified to

$$n_{CYC/OP} = \sum_{OP} n_{CYC}(OP)\, P(OP) + n_{CYC/NOP}\, h_{NOP/OP} \qquad (8.3.17)$$

where $h_{NOP/OP}$ is the ratio of NOP instructions to OP instructions.

To start, it is assumed that the data path contains only one ALU. Then the effective number of cycles is increased through all operations that require more than one cycle per operation in spite of the pipelining of the control circuitry, and the throughput is decreased accordingly. If the ALU is supplemented with further modules such as multipliers and barrel shifters, in particular the number of cycles for these operations is lessened. The change in the mean number of cycles through insertion of a multiplier and a barrel shifter in the data path amounts to

$$\begin{aligned} \Delta n_{CYC/OP} = {} & \Delta n_{CYC}(MUL) \cdot P(MUL) \\ & + \Delta n_{CYC}(SHIFT) \cdot P(SHIFT) \end{aligned} \qquad (8.3.18)$$

This equation demonstrates the obvious result that the gain in decreasing the mean number of cycles is associated with the probability of an operation occurring. If the multiplier is connected to an accumulator, two operations (MUL, ACC) can be completed in one cycle, and the mean number of cycles for coupled multiplication/accumulation operations is even further reduced.

Through insertion of additional modules in the data path, the minimal period length of a clock cycle is increased, upon which the maximal delay depends. The increase in the delay results from the additional multiplexers and the increased delay of the multiplier in comparison to the ALU.

Using numeric values, the achievable gains are to be demonstrated. An FIR filter is presumed. In this case, only multiplications and additions with an identical probability of 1/2 occur. It is assumed that the word width of the input data and coefficients is m and that the ALU requires m cycles for an $m \times m$-bit multiplication.

Assuming a ratio of $h_{NOP/OP} = 1$, from (8.3.17) it follows that

$$n_{CYC/OP} = \frac{1}{2} + m\frac{1}{2} + 1$$

Here it is assumed that each instruction that does not contain an operation requires one cycle.

Through use of a MAC unit with a half cycle per operation, the new cycle count per operation amounts to

$$n'_{CYC/OP} = \frac{1}{2} \cdot \frac{1}{2} + \frac{1}{2} \cdot \frac{1}{2} + 1$$

If, as the DSP 56000 example shows, the number of *NOP* instructions in the filter loop is brought down to zero through supplementary measures such as parallel data transfer, additional address generation units and hardware loop counters,

$$n'_{CYC/OP} = \frac{1}{2} \cdot \frac{1}{2} + \frac{1}{2} \cdot \frac{1}{2} = \frac{1}{2}$$

If it is further assumed that the data path with the MAC unit requires twice as long for one cycle in comparison to the ALU, then this example possesses the following ratio for the throughputs for $m = 16$ bits.

$$\frac{R'_T}{R_T} = \frac{1\left(\frac{1}{2} + 16 \cdot \frac{1}{2} + 1\right)}{2 \cdot \frac{1}{2}} = 9.5$$

This numeric example yields an increase in the throughput by a factor of approximately 10.

The expansion of the data path leads to an increase in area. Now it is questionable if the efficiency of the new configuration has risen, i.e. if the increase in throughput exceeds the increase in area. Designs carried out in a 1 μm CMOS process have yielded the following approximate areas:

16 bit ALU	0.6 mm^2
16 bit MUL	1.3 mm^2
40 bit ACC	0.85 mm^2
40 bit SHIFTER	0.2 mm^2

These figures show a significant increase in the area for the data path through multipliers, accumulators and shifters. For the evaluation of the efficiency, however, the area of the total configuration must be taken into account. The main portion of the chip area of a signal processor is required for control circuitry and local memory. Given a total chip size of 80 mm^2, an additional MAC unit and a shift increase the area only by roughly 3%. This moderate increase in area is met with a considerable gain in throughput.

It should be noted that the computations demonstrated are only easily comprehensible for simple models. Complex algorithms require complex computations and a more exact examination of the architecture parameters.

8.3.3 On-chip memory

The influence of local, on-chip memory for signal processing chips is to be briefly discussed. The access times of on-chip memory are designed such that they fit the temporal behaviour of the other modules. Access to storage cells of the local memory is just as fast as to the registers of internal register files. Since, in addition to this, memory calls and operations in pipelined DSP chips must occur simultaneously, the memory calls lead to an increase in the effective number of cycles per operation.

The influence of external memory access is to be explained using the FIR filter and FFT examples discussed above. For this, it is assumed that the access time for external memory is larger than for internal memory by an integer factor of $n_{MEM,ACC,EX/IN}$. For SRAMs, this factor is roughly 2 and can be about 5 for DRAMs. For the programming of DSPs, the increased access time for external memory is compensated via the insertion of wait cycles.

An FIR filter with an impulse response of length N is assumed. Each input sample is then required for the computation of N output values. Thus, $2N$ read operations are necessary for each input value since the filter coefficients must also be read. Counting one write operation for the result, the number of memory calls per sample amounts to

$$n_{ACC,FIR} = 2N + 1 \qquad (8.3.19)$$

For each MAC operation, two operands must be read. This means one memory call per operation. Thus, the cycle count per operation for external memory calls increases by

$$\Delta n_{CYC/OP} = n_{MEM, ACC, EX/IN} - 1 \qquad (8.3.20)$$

In particular for processors with an optimized data path of 1/2 cycle per operation, this leads to a noticeable decrease in throughput.

Tables 8.1.2 and 8.1.3 show that new DSP chips support local storage of input data and coefficients for FIR filters with up to 1000 coefficients. Thus, for practical sizes of filters, one can presume internal memory access. This means that N internal memory calls occur for each external memory call to an input sample. Since, furthermore, the coefficients are locally stored, each input data item has to be read once from the external memory and each output has to be written once into the external memory. Thus, there are two external memory calls per sample and $2N$ operations per sample. As a result of the local memory the number of cycles (8.3.20) is reduced to

$$\Delta n'_{CYC/OP} = \frac{1}{N}\left(n_{MEM, ACC, EX/IN} - 1\right) \qquad (8.3.21)$$

The memory calls to external memory in this case have practically no influence on the throughput.

For the FFT, it is assumed that N input values are to be converted into the frequency domain. According to Table 7.3.2, $(N/2)\log N$ butterfly PEs are necessary for N input values. For each butterfly PE, three complex values (A,B,C) must be read and two complex values (A', B') must be written. It follows that 10 memory calls must be carried out per butterfly PE. Thus, the number of memory calls per sample amounts to

$$n_{ACC,FFT-PE} = 5\log_2 N \qquad (8.3.22)$$

This means that the number of memory calls is identical to the number of operations. Thus, in the case of externnal memory calls, the same increase in the cycle count per operation holds for an FFT as holds for an FIR filter (Equation (8.3.20)). In the case of a sufficiently sized local memory the total number of accesses to the external memory is restricted to $4N$ data (real and imaginary part of x and y) per block. In total there are $5N\log N$ operations to be performed per block. Thus, the cycle count per operation reduces to

$$\Delta n'_{CYC/OP} = \frac{4}{5\log_2 N}\left(n_{MEM, ACC, EX/IN} - 1\right) \qquad (8.3.23)$$

The influence of local memory is not as dominant as with FIR filters. Yet even here, the throughput is noticeably improved through local memory.

It should be remembered that on-chip memory is advantageous for all algorithms in which a data item is accessed several times. Many signal processing algorithms have this characteristic. One strives to incorporate the largest possible memories onto the DSP chip. In current implementations, the on-chip memory requires approximately 50% of the area of the total processor. Due to the memory's large contribution to the area, on-chip memory leads to a smaller improvement in efficiency than is given by the extension of the data path.

8.3.4 Software measurement

The investigations in the previous sections require characteristics of signal processing algorithms such as statistics on operations, operation sequences, data access and data lifetime. Such statistics may be derived by analysis of high-level software programs describing the algorithm [126], [133].

Sophisticated signal processing algorithms are frequently developed and verified by non-real-time simulations with representative data on a computer. In such a case the algorithms are defined by a high-level software program. Measurement on such a software program could provide statistics on arithmetic operations applied to certain data types, data lifetime as a basis for memory capacity and memory access pattern. Keeping track of the control and data flow during program execution could ease the design of the control unit. The particular advantage of measurements on software programs during run-time with real data is that data dependent parts of the algorithms are incorporated. Only the statistics of low-level algorithms with predefined sequence of operations can be investigated without software measurements.

Software profiling is a method that gathers statistics on an executable program during run-time. The executable program is modified at compilation- or link-time to include profiling code. This additional profiling code periodically samples the program status and saves the sampled information on program exit. The profiling information is analysed by a postprocessor. At least a profiling tool gives the number of calls to a function and its duration. Often a function call graph is also generated that shows which function was the caller.

The information obtained by this simple profiling is not suited to describing the algorithm's computational or memory requirements. Refined profiling techniques try to analyse the software with respect to its computational and memory requirements. One approach is to perform a high-level analysis of the arithmetic and memory operations found in the source code. These operations are combined with a run-time execution profile, to get the numbers of operations performed in a function. In a low-level approach the machine instructions in an executable program are analysed and combined with the execution profile. This leads to a fine-grained but machine-dependent result. Both approaches depend highly on the algorithm implementation; low-level profiling may additionally incorporate inaccuracies introduced by inefficient code generated by a compiler.

As has already been shown in previous chapters, a signal processing task may be implemented with a different sequence of operations and different intermediate results. Alternatives for implementations of the DFT and FIR filters are examples. Thus, software instrumentation provides statistical data on a specific implementation of the algorithm and not a general result. There is a large variation between straightforward implementations in high-level programming languages and implementations of the same processing scheme that is optimized with respect to the underlying hardware. To get the most out of profiling, the software under investigation has to be structured in a way that fits a hardware partitioning. Furthermore, the software should be already optimized with respect to the targeted hardware to get meaningful execution times which serve as a starting point for possible hardware/software partitioning.

8.4 Data Path Synthesis

Adaptation of the DSP architecture and in particular the data path to the signal processing schemes is a key issue to achieve high performance. Because of the large variations of processing schemes a representative scheme has to be selected and investigated. Having a processing scheme as a composition of several tasks, a task with the highest demands on signal processing time may be chosen for analysis and subsequent data path design for performance increase. Mapping of a specified representative algorithm onto a data path structure may be the design procedure. Data path synthesis at the register transfer level will be presented in the following.

In section 8.1.2 suitable data path architectures were derived by a more or less *intuitive* method. These were developed through analysis of the signal processing schemes. A basis of adaptation of the hardware was the statistics of operations and data access. As opposed to this *intuitive* method, *systematic* approaches for deriving suitable structures for the data path exist in the literature. Such a systematic procedure is termed data path synthesis and is a part of the so-called high-level synthesis (HLS) [127], [128], [129]. This is used to designate the transformation of a behavioural description to a structural representation. Its aim is to determine a suitable hardware structure as automatically as possible from a behavioural description.

In the following, only data path synthesis at the register transfer level as an element of high-level synthesis will be dealt with. Data path synthesis allows systematic derivation of the structure of a data path to be synthesized.

The goal is to select, from a given set of logic units, those that represent the most favourable implementation for an algorithm while holding to limiting criteria such as computational performance, throughput and maximal area. At the same time, the operations of the algorithm must be assigned to time slots and in turn mapped to functional units such as adders, multipliers, ALUs, etc. Furthermore, the data and register transfers must be allotted to connection units such as buses and multiplexers. Finally, it must be determined in which storage units, i.e. registers, register files, RAMs, etc., both the variables and the intermediate values of the algorithm are to be stored.

In a preparatory step, the behavioural description of the algorithm is compiled into a graphical representation such as a control/data flow graph (CDFG), capturing all dependencies between the operations o_i. The CDFG is a directed acyclic graph (DAG), with each node corresponding to a single operation of the algorithm and each directed arc corresponding to a data/control dependence. An example of a behavioural description and the related CDFG is sketched in Figure 8.4.1.

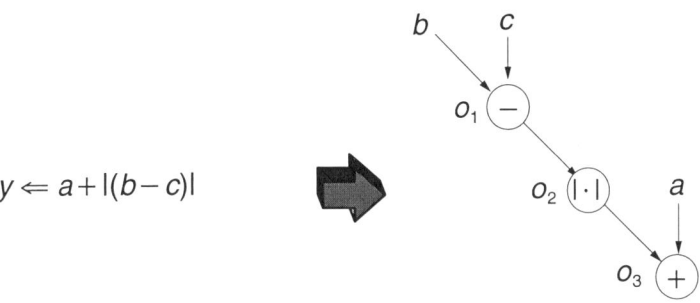

Figure 8.4.1: Behavioural description and related data flow graph (CDFG)

Note that there are some similarities of the CDFG when compared to the dependence graph (DG) and the signal-flow graph (SFG), derived in Chapter 5. The key difference is that each node of a CDFG corresponds to exactly one atomic operation, whereas a node of a DG or SFG, in general, comprises several operations. Thus, a CDFG node can be regarded as a special case of a DG/SFG node with one additional constraint, i.e. each node comprises exactly one operation.

Starting from a CDFG, the data path synthesis comprises three major subtasks: (1) derivation of a scheduling scheme, i.e. assigning operations to time or control steps, taking into consideration the dependences between operations, (2) selection of functional and storage units and (3) binding of the units. However, the three subtasks are closely related. An optimal solution to the data path synthesis problem has to solve the three subtasks concurrently, leading to a high computational effort. For this reason, the three subtasks are often solved sequentially. Especially, the scheduling task has a large impact on the quality of the synthesized data path in terms of area expense and achievable performance.

Concerning scheduling a multitude of algorithms have been developed in the past: The basic algorithms are scheduling the operations as soon as possible (ASAP) and as late as possible (ALAP). The number of control steps are specified by ASAP and ALAP as a lower bound and an upper bound respectively. More sophisticated algorithms with additional constraints have been developed in the high-level synthesis community. When an upper bound on the number of control steps is given, the scheduling scheme is denoted as time-constrained scheduling. When constraints are given in terms of either the number of functional units or the maximum silicon area, the scheduling scheme is denoted as resource-constrained scheduling. The related best-known scheduling algorithms are based on exact methods, such as ILP-based scheduling (ILP = integer linear programming), as well as heuristic approaches, such as force-directed scheduling, list-based scheduling, etc. A more detailed discussion concerning scheduling algorithms will not be given here. The interested reader is referred to [128].

In general, after the scheduling subtask has been solved, the unit selection and binding are performed. In order to limit the complexity of the data path allocation problem, a fixed target architecture with parallel processing units is often presumed whose input operands are read from register files and whose results are written back to one or more register files. The connections between logic and storage units are either bus or multiplexer based (Figure 8.4.2).

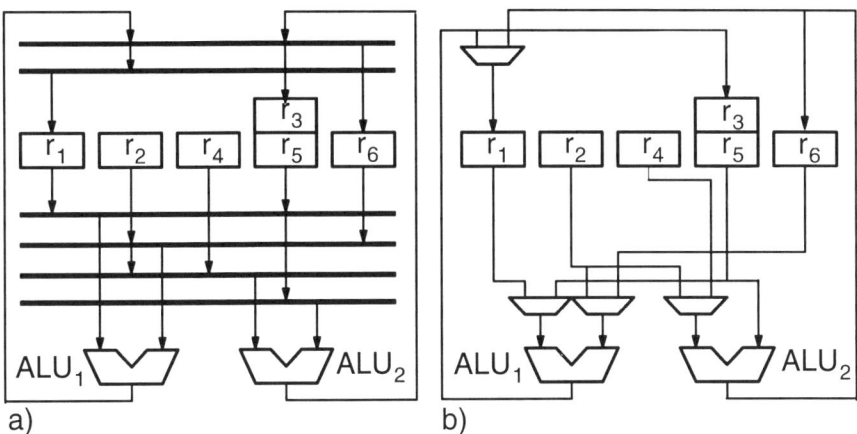

Figure 8.4.2:   Data path architecture: (a) bus structure; (b) multiplexer structure

Data path allocation is typically split into two major subtasks: (1) selection of logic and storage units and (2) binding of the units. Unit selection essentially comprises the decision of which and how many logic and storage units will be used. On the other hand, unit binding consists of assigning operations and data transfers to the logic, storage and connection units. During this process, several boundary conditions must be met in order to guarantee correct execution of the algorithm on the hardware architecture. Two operations, for instance, that are to be processed at the same point in time t cannot be assigned to the same logic unit. In addition, only those algorithm variables whose *lifetimes* do not overlap may be allotted to the same storage cell.

In addition to guaranteeing correct sequencing of the data flow graph, global boundary conditions for the design in terms of maximal delay, area or even power consumption must often be met. For solving these allocation problems, three approaches exist in the literature:

● constructive approaches [130]

● decomposition approaches [131], [132], [133]

● iterative approaches

Constructive approaches begin with an empty data path that is expanded step by step with logic, storage and connection units by sequentially working through the data flow graph. For each operation, the (hitherto constructed) data path is examined to determine if it contains a unit that allows execution of the particular operation. In this case, the operation is assigned to the unit with the lowest hardware costs; otherwise, the data path is supplemented with a new unit.

Constructive approaches are truly relatively simple, yet the attainable results, in terms of an optimal solution, are in general relatively poor. Decompositional approaches, on the other hand, yield significantly better results. They are based on dividing the allocation problem into a series of independent sub-problems. These sub-problems are then – each for itself – optimally solved using graph theory methods such as *clique partitioning* [131], *left-edge* [132] and weighted, bipartite matching [133].

Finally, iterative approaches represent an extension of the first two approaches in which one attempts to achieve improvement in the data path in terms of its area or delay characteristics through stepwise alteration of the allocation.

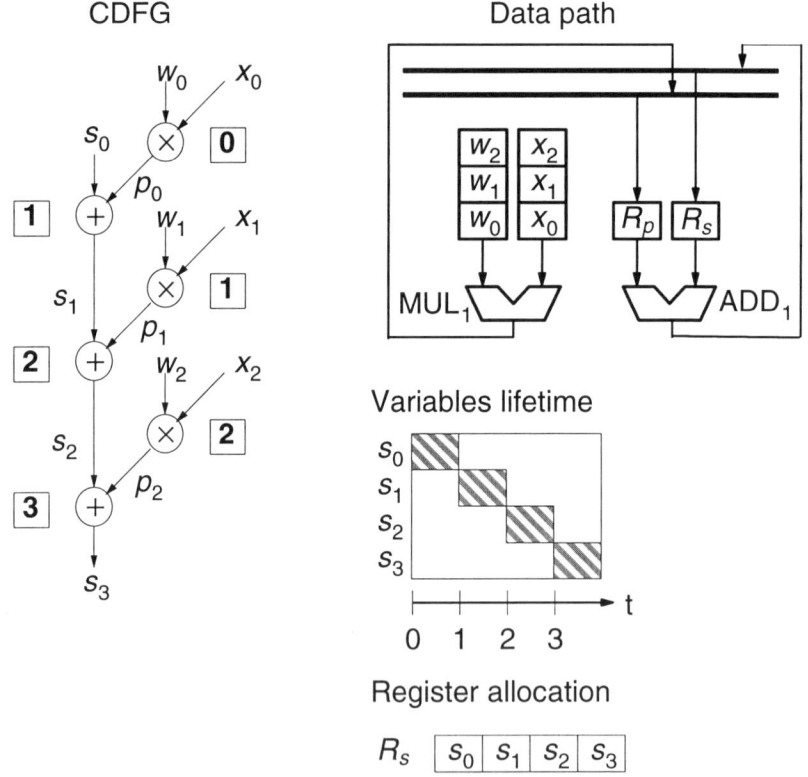

Figure 8.4.3:   Scheduling and allocation for a simple CDFG

In order to demonstrate the data path synthesis problem, including the three major subtasks, i.e. scheduling, allocation, and binding, a simple example is given in Figure 8.4.3 with a data flow graph comprising six nodes, three multiplications and three additions. The scheduling is indicated by the boxed numbers. Furthermore, two functional units are selected, performing a multiplication and an addition respectively. The input variables ($w_0...w_2$, $x_0...x_2$) as well as the intermediate sums ($s_0...s_3$) and products ($p_0...p_2$) have to be stored in register files. For example, based on the scheduling of Figure 8.4.3, all intermediate sums have non-overlapping lifetimes, therefore only a single register (R_s) has to be allocated. The same holds for the intermediate products ($p_0...p_2$) which are stored in a single register (R_p).

Again, the tasks of scheduling and allocation for a given target architecture have some similarities when compared to finding a schedule vector as well as a projection vector for mapping the nodes of a DG/SFG onto processor elements, as described in Chapter 5. The key differences are:

1. In data path synthesis every scheduling scheme for the nodes of the CDFG is admissible as long as the dependencies between the operations are taken into consideration. In contrast to this, finding a scheduling scheme for a DG/SFG is in most cases restricted to admissible linear schedules only. This is due to the fact that the mapping approach is based on algebraic linear time–space transformations.

2. In data path synthesis the allocation and binding of hardware resources is based on a target architecture with a predefined maximum number of function units. In contrast to this, the mapping approach for a DG/SFG is based on projecting identical nodes onto a single processor element. Thus, data path synthesis can be regarded as a more general approach to mapping a behavioural description onto hardware units, especially in the case of irregular algorithms, which cannot be described in terms of a linear shift-invariant dependence graph.

In recent years, researchers have developed a vast number of CAD programs for data path synthesis that differ in terms of their area of application, their underlying target architecture and the algorithm they use. A good overview can be found in [129]. Two CAD tools will be listed as examples:

*   CASTLE/SYDIS [134]
*   CATHEDRAL [135]

CASTLE/SYDIS
In the SYDIS/CASTLE project of GMD, Bonn, Germany, the design of a signal processing system begins with a behavioural description at the level of a sequential, high-level language (C/C++). From this description, the degree of exploitable instructional parallelism can be extracted. The CASTLE system's underlying processing unit is a parametrizable data path controlled by

a very long instruction word (VLIW) and an attached register file. The VLIW specifies several operations in one long word. So-called "multiple execution units" must be implemented that enable simultaneous execution of all operations in a VLIW. The VLIW also contains a field for loading and storing data.

CATHEDRAL

CATHEDRAL is a high-level synthesis system developed at IMEC, in Leuven, Belgium. The definition of the system behaviour is carried out in CATHEDRAL in the applicative language SILAGE. As opposed to an imperative programming language such as C or Pascal, for instance, the point in time for executing a function is not defined by the function call, but rather, its executability is made possible by the availability of its input data. Due to this form of description, it is possible not only to use the parallelisms at the operation level but to enable simultaneous execution of entire processing steps using different processing units. The basic architecture of the processing system is hierarchically structured, accordingly. It can consist of several processing units that are interconnected via bus systems and joined with various storage systems such as register files or on- and off-chip RAM. The processing units themselves consist of a custom-fitted data path and register files that are controlled by a microprogrammed controller.

8.5   Design Support for DSP Systems

Software tools as well as development systems and development boards are offered to support the design of DSP systems. In particular, the manufacturers of DSPs offer support for software design. Assemblers, linkers, ANSI C compilers and software simulators that run on PCs or workstations are generally provided.

An assembler translates the mnemonic code and macros into object code. An assembler links the object code of various program modules and creates executable machine code. ANSI C compilers serve to support the development of DSP programs in a high-level language. Optimized library modules for standard routines and special applications increase the efficiency of the runnable program. Software development on common platforms such as PCs and workstations is made possible through software simulation. Simulators support interactive design at the instruction level using a typical, window-oriented graphical interface. The operations and the memory contents can be followed cycle for cycle.

Using in-circuit emulators, DSP chips can be examined in the target system. In this case, both the DSP software and the interaction between the

DSP and the external hardware components such as memory, for example, are tested. Emulation systems make it possible to examine the course of a program in single-step mode. Breakpoints can be set, register and memory contents can be displayed and the program code can be directly modified.

Complete DSP boards are also offered commercially. In addition to the DSP chip, these boards contain A/D and D/A converters as well as interfaces to the signal sources, the signal sink and a host. This type of board is usually pre-configured for applications in speech or audio signal processing. The advantage is that no hardware development is needed and the manufacturer often also provides the software in addition to the hardware. This type of board enables quick evaluation of special signal processing applications.

8.6 Exercises

3. Two processors for the implementation of special algorithms are to be compared. To simplify the comparison, only the cycles necessary to carry out arithmetic operations are to be compared. The cycle counts of several important operations are listed below:

		Cycle count	
Operation		proc.1	proc.2
ADD	addition	1	1
SUB	subtraction	1	1
MUL	multiplication	16	1
MAC	MUL/ACC	16	1
ABS	absolute value	1	1
ASH k	shift k places	k	1
CMP	compare	1	1

a. The implementation of a matching function is to be examined.

$$\sum_{i=1}^{N}\sum_{j=1}^{N}[x\,(i,j) - y\,(i,j)]^2$$

Which ratio between the cycle counts results?

b. The algorithm of the matching function is altered as shown below.

$$\sum_{i=1}^{N}\sum_{j=1}^{N}|x\,(i,j) - y\,(i,j)|$$

How do the results of a change?

    c.   The implementation of a vector rotation according to (3.5.18) is to be examined. The trigonometric functions sin and cos are to be implemented through polynomial approximation as in (3.5.6). Which cycle counts result?

    d.   The vector rotation is to be alternatively implemented according to the CORDIC method. Integer shift sequences with 16 steps apiece are to be used. The scaling of the amplitude can also be implemented with a multiplication. The necessary values for $\tan^{-1}$ are stored in memory. How does the result from c change?

4.   The significant sections of a signal processing algorithm are given to be a matrix–vector multiplication and the matching function of exercise 1b (MAD method). For each block of $N \times N$ pieces of data, $2N$ matrix–vector multiplications must be carried out. The matrix has $N \times N$ elements. For a minimization of the distance, the matching function must be evaluated at $(1+2N)^2$ search points. Given is $N=8$.

    a.   Determine the frequency distribution of the operations and, from it, the cycle count per operation for the two processors given in exercise 1.

    b.   The number of operations is to be determined from the facts about the two algorithms. Which throughput results for a processing clock rate of 50 MHz?

    c.   The throughput is to be more exactly determined by multiplying and accumulating the number of operations with their corresponding cycle counts. What leads to the deviation from the results of 2b?

    d.   The search method is assumed to be modified such that only $(1 + 8 \log_2 N)$ search points must now be examined. How do the results of a and b change?

    e.   The execution of the matching method is to be accelerated by supplementing the arithmetic unit with a data path adapted to the matching function. The circuit structure of the data path is to be given. By how much does the throughput increase?

5.   The needed logic units for extensions of a data path for specialized instructions should be investigated. The data path consists of a standard ALU. By an additional multiplexer an accumulation operation is supported. The additional hardware for the following specialized instructions should be specified:

    a.   Average of two operands
$$r = (a + b)/2$$

    b.   Saturate a to $N-1$ if $a \geq N$ where N is a power of 2
$$r = sat(a, N)$$

c. Controlled ADD/SUB
   $r = a + sign(a) \times b$
   For positive a addition of b, for negative a subtraction of b.
d. Sum of the magnitude of differences
   $$r = \sum_i |a_i - b_i|$$

9 Multiprocessor Systems

The advances in semiconductor technology and processor architectures have led to ever increasing computational performance for programmable processors. In spite of this progress, the requirements of many signal processing applications exceed the computational performance provided by single processor systems. An obvious solution is the use of several processors. Programmable processors with several data paths are designated as multiprocessors here. Multiprocessors are architectures that offer large flexibility since various algorithms can run under program control.

Multiprocessor systems are, in regard to their implementation, more complicated than application specific array processors. They are always the proper solution when flexibility is a foremost aspect. This is given in systems in which a multiplicity of very different algorithms must be processed by the same hardware. Another field is the development of algorithms and the demonstration of signal processing under real-time conditions. Such a development environment is called rapid prototyping.

The field of multiprocessors will not be treated in depth within the bounds of this book. Interested readers are referred to the standard textbooks by Hwang and Briggs [136] and Hwang [137]. Multiprocessors especially for signal processing tasks are treated in [117].

In the following section, basic structures are presented and classified. In a further section, the performance of various multiprocessor structures is compared based on simple model-like computation.

9.1 Architectures of Multiprocessors

Multiprocessors serve to increase the number of simultaneous operations. As was shown in previous sections, this occurs through parallel processing and pipelining. Solutions for algorithms with fixed data dependencies were derived in Chapters 5, 6 and 7. In this case, the course of processing is predefined and is not dependent on the data. In Chapter 8 it was shown how parallel processing can be implemented in single processor systems. Several data paths are used simultaneously in multiprocessor systems.

In multiprocessor systems, the simple data and instruction streams are replaced by multiple streams. Special names for the basic structures of multiprocessor systems after Flynn are typical [116]. If several data streams are used, the corresponding configuration is designated SIMD (single instruction stream, multiple data streams). The use of multiple instruction streams is designated MISD (multiple instruction streams, single data stream), accordingly. Consequently, the designation MIMD is employed for multiple data and instruction streams. The principal structures of the respective multiprocessor systems are sketched in Figures 9.1.1, 9.1.2 and 9.1.3. The abbreviations correspond to those of Figure 8.1.15.

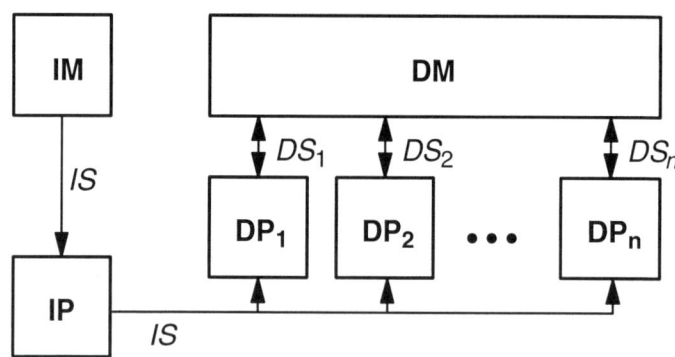

Figure 9.1.1: SIMD multiprocessor system

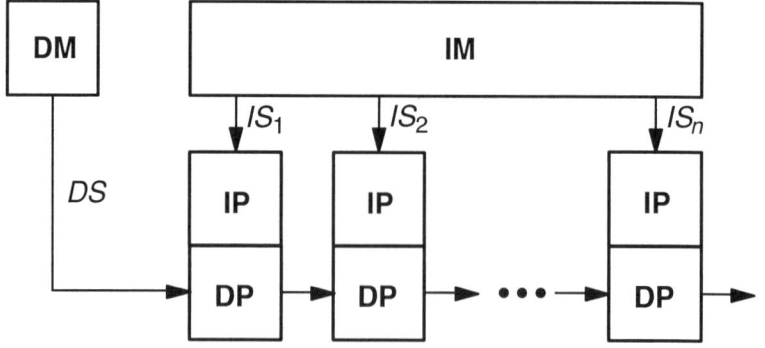

Figure 9.1.2: MISD multiprocessor system

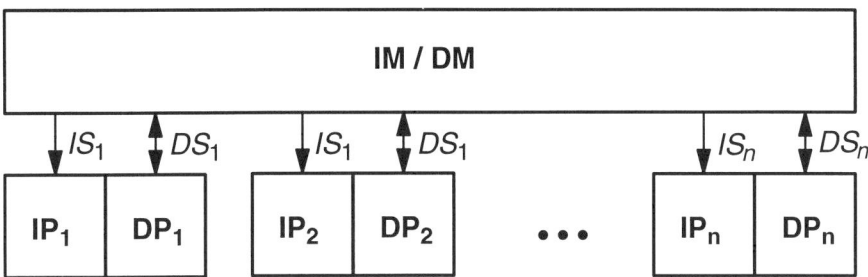

Figure 9.1.3: MIMD multiprocessor system

Algorithm models can be associated with the individual processor systems. Thus, SIMD configurations are favourable for algorithms in which the data are divided up in accordance with the number of data paths and all data are processed with the same sequence of instructions. An example of such an algorithm is matrix multiplication. In this case, the product of (5.2.1) is divided into several matrix vector products (5.2.3).

$$C = [c_0 c_1 \ldots c_{N-1}] = A[b_0 b_1 \ldots b_{N-1}]$$
$$= [Ab_0, Ab_1 \ldots Ab_{N-1}] \tag{9.1.1}$$

The matrix B is split into N vectors. Different vectors b_i are assigned to the data paths and the computation of the matrix vector product Ab_i is carried out in parallel in all data paths. It is obvious that the highest performance is achieved only when the number n of data paths is an integer divisor of the dimension N of the matrix B.

Sequential processing of partial processes can be associated with MISD configurations. They represent a pipeline. If not single units of data, but rather data blocks are processed sequentially, this is called macro-pipelining. Programmable MISD configurations are seldom used in practice since a full use of the capacities of the individual data paths is generally not given. In the case of pipelining, an identical number of predetermined operations to be processed sequentially must be given for a large number of algorithms. For macro-pipelining, the number of partial processes and the number of operations in the partial processes must be roughly identical in size and known in advance.

A MIMD configuration is the most general for a multiprocessor. It supports both parallel processing with identical and differing partial programs as well as macro-pipelining. Nonetheless, this is also the most complicated form since the control unit and the instruction memory must be multiply implemented.

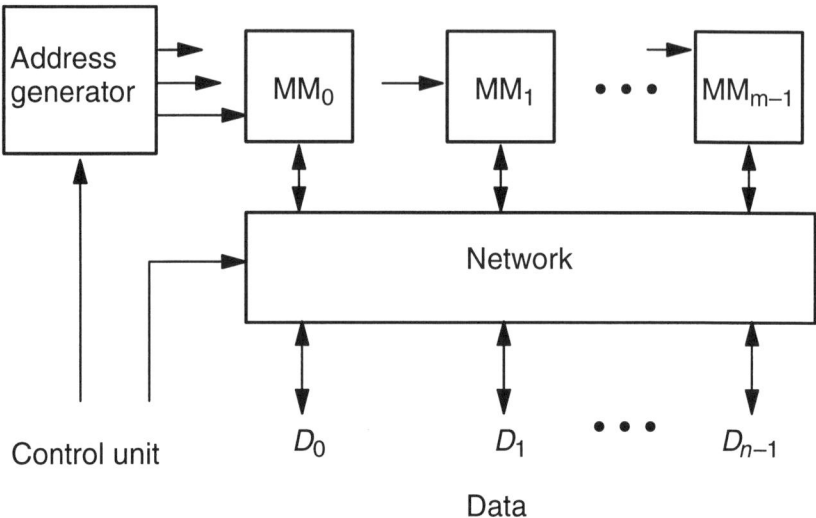

Figure 9.1.4:   Shared memory of a multiprocessor system with parallel access to several memory modules

The basic structures of multiprocessors show parallel access to common memory for data and instructions. These memories are called shared memory or parallel RAM (PRAM). A common parallel memory can be implemented by several memory modules (MM) and a connecting network. The number of mutually independent read and writeable memory modules must be equal to or larger than the number of ports to the data paths. In general, the memory modules have only one port, i.e. only one piece (word) of data can be read or written at the same time. Figure 9.1.4 shows the principal configuration of a parallel memory.

Several implementation forms exist for the connecting network. Complete networking yields a so-called crossbar network (Figure 9.1.5). Each of the m inputs can be connected to n outputs. The switches of the network cannot be switched entirely arbitrarily. An output may only be assigned the data of one input and not of several inputs. However, it is possible that one piece of input data is assigned to several outputs (broadcasting). If the connecting network is used in both directions (reading from and writing to memory), the dissymmetry of usage must be respected since the meaning of input and output changes. Parallel memory requires complex control since memory addresses must be provided in parallel for all modules in addition to the switching information for the network. In most cases, an address computation unit that supplies the same address to all memory modules is used for simplicity. Individual memory modules can be activated and deactivated via additional wiring. A crossbar network is very costly since a total of $n \times m$ bidirectional switches is required. Alternative networks exist that succeed with fewer switches.

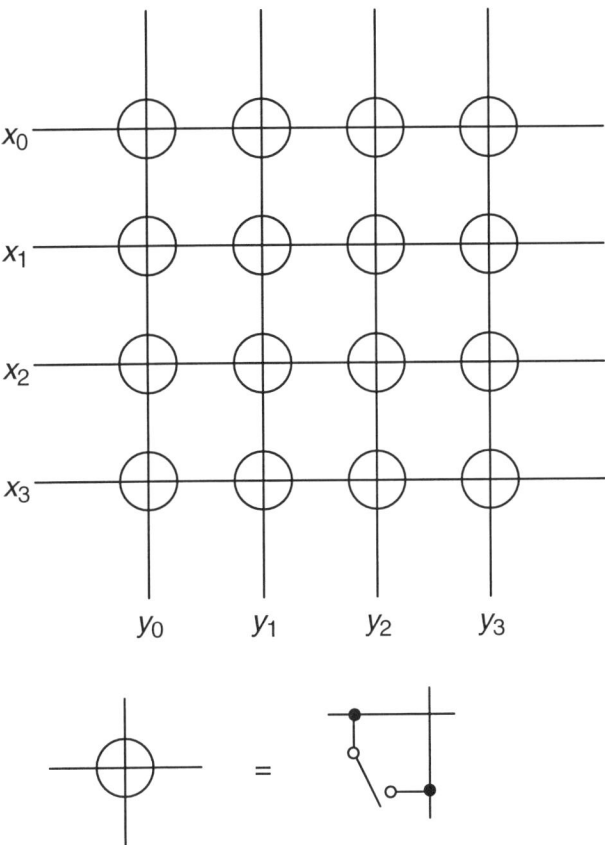

Figure 9.1.5: Crossbar network $(n = m = 4)$

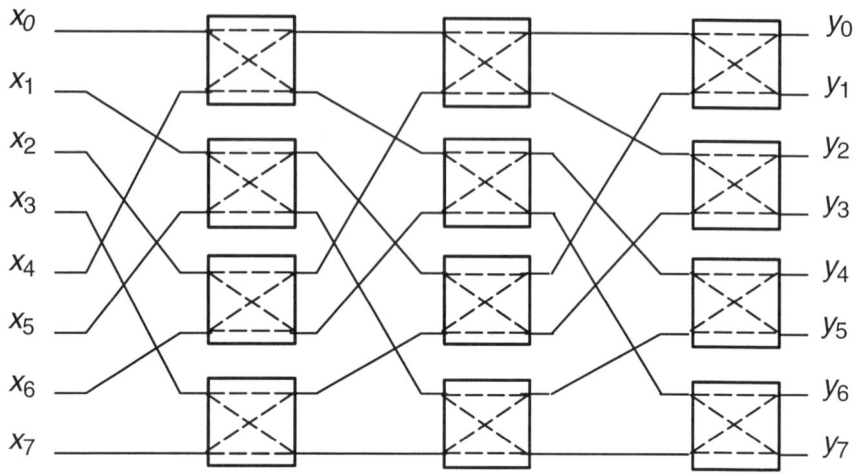

Figure 9.1.6: Exchange network $n = 8$ (omega network)

An example of this is simple, multistage shuffle networks. Such shuffle networks can also be implemented multistage using simple switch functions. Figure 9.1.6 shows an omega network [136], [138]. It has $\log n$ stages, and $n/2$ basic switches in each stage. The data are exchanged between each stage in accordance with a perfect shuffle. For a pure shuffle function, the basic switch must provide two functions (parallel transfer, cross shuffle). In principle, an omega network can also fulfil broadcast functions. In this case, each basic switch must provide four functions. The additional functions are the broadcasting of the upper or lower input value. An omega networks seems quite complex. Yet the complexity is of the order $O(n \log n)$ whereas a crossbar network is of the order $O(n^2)$.

The previously discussed, global, parallel shared memory allows data exchange between the individual processor nodes. Since the shared memory forms a unit together with the data paths and the control units, such configurations are termed close coupled multiprocessors. Loosely coupled multiprocessors consist of several complete processors in which data exchange is carried out blockwise (packet transfer) over networks in asynchronous mode. Each processor node has a data path, an instruction processor and local memory for data and instructions. Thus, loosely coupled multiprocessors are MIMD processors.

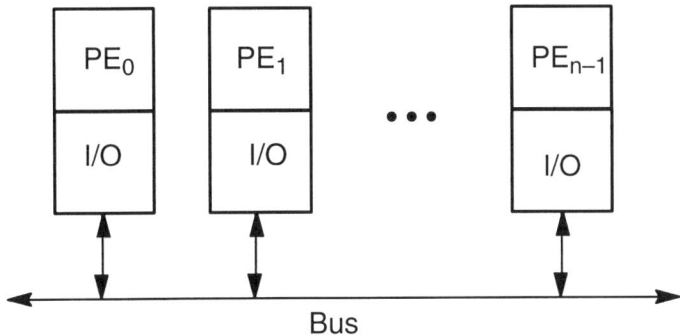

Figure 9.1.7: Coupling of processors via a bus

A bus system represents the simplest form of networking loosely coupled processors (Figure 9.1.7). Bus systems exist in which additional control and address lines ensure that no conflicts occur through simultaneous transfer from several processors and that the target processor for the data transfer is uniquely determined. To reduce the number of bus lines, bus systems without additional control lines are also implemented. In this case, each data packet is given a header that contains the most important information. Amongst other things, the target processor is a component of the header. Since no separate control lines regulate bus occupancy, particular measures are necessary to avoid occupancy conflicts. Each processor must contain a unit that monitors the bus and ensures that a transfer is initiated only when the bus is free. It can nonetheless happen that several units simultaneously begin header transfer. Due to the change in special information positions in the header, this can be detected by the bus monitoring units, and the transfer can be aborted. After a special time delay, the processor units can start the transfer anew. Ethernet is such a transfer system with serial transfer of the data and header bits. In Ethernet, coaxial cable is used for the transfer. To increase the transfer rate, bit-parallel bus systems are used for bridging short distances between the processors.

The bus system described above allows data transfer to only one processor at a time. To assist the simultaneous data transfer of several processors, networks are used. A 2-D grid net is shown in Figure 9.1.8. In this case, simultaneous data transfer to neighbouring processors is possible. A transfer to more distant processor nodes is carried out via the intermediate processors. To improve the networking, the grid can be extended to a 3-D, cube-formed net or high-dimensional hypercubes [136].

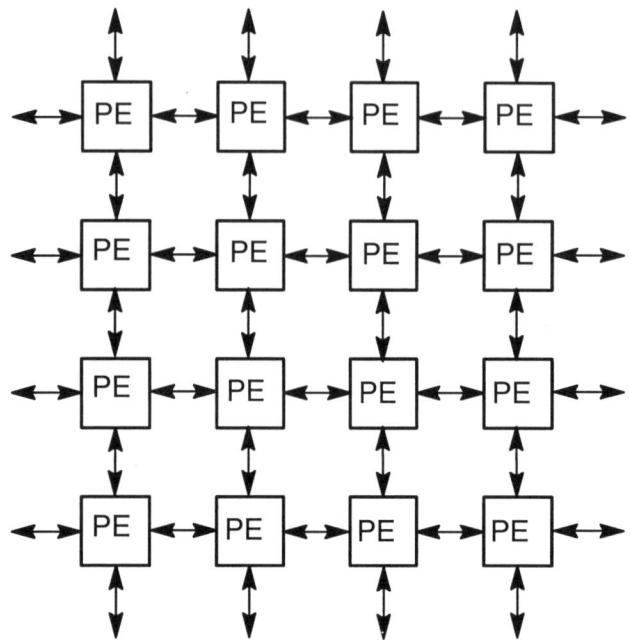

Figure 9.1.8: Processor connection network based on a 2-D mesh

Processors exist that contain units for supporting the connection networks. The INMOS transputer (T800, T9000) contains four units on the chip for bit-serial, asynchronous data transfer [139], [140]. The bit clock rate is 20 MHz during the transfer. Due to the four ports, networking of the transputer in a 2-D grid is possible. The data exchange is possible independently of the actual processor, i.e. the data transfer takes place simultaneously with the operations.

The TMS320C40 is a DSP processor with six data transfer units on one chip [141]. Thus, this processor also allows 3-D cubed-shaped nets. The data transfer is carried out on 8-bit parallel data lines. Four additional lines serve the controlling of the data transfers. At a maximal clock rate of 20 MHz, each communication port offers an asynchronous transfer rate of up to 20 Mbytes/s. The operations in the data path are executed independently of the DMA transfer via the connection network. This retains the high computational performance of the processor in spite of the data transfer. The data path of the C40 is implemented for 40-bit floating point operations at a clock rate of 25 MHz. To achieve high computational performance, the data path contains dedicated implementations of a multiplier and a barrel shifter besides an ALU.

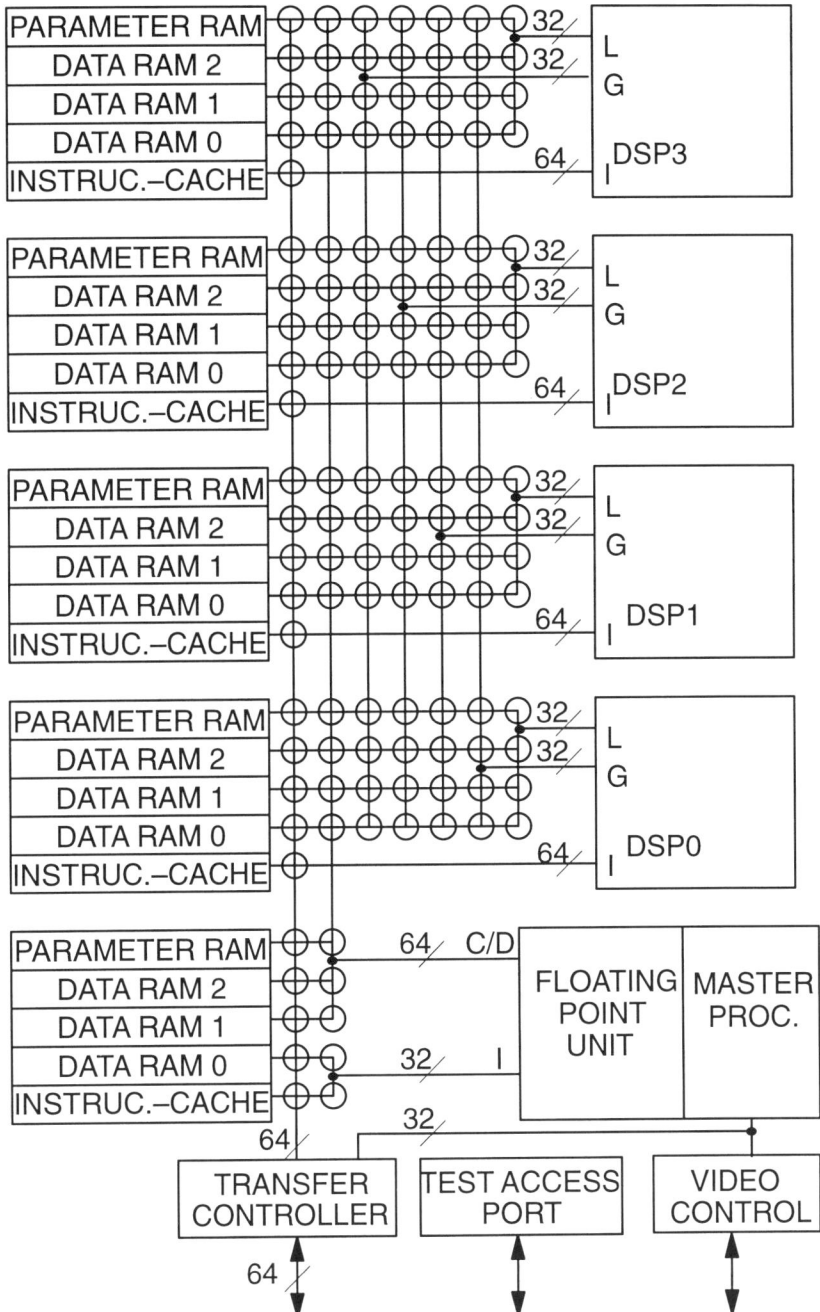

Figure 9.1.9: Block diagram of the TMS 320C80 multiprocessor

Loosely coupled processors are only suited for special parallel algorithms due to the asynchronous, packet-oriented data exchange. In the field of signal processing, many low-level algorithms are used that are better suited for a close coupled system with shared memory. The advances in semiconductor technology have now made it possible to monolithically implement several signal processors. An example of this is the TMS 320C80 [142]. Figure 9.1.9 shows a block diagram of this multiprocessor. It has an MIMD structure with four fixed point signal processors. A crossbar network allows fast data exchange between the memory units and the processors. The on-chip memory has a total capacity of 50 kbytes. The master processor is a general RISC processor with floating point computation unit. The master processor carries out the control tasks. It controls the task distribution among the parallel DSPs and monitors the communication with external processors. Furthermore, it carries out the partial tasks that require floating point computation. The four parallel DSPs have the typical structure of fixed point DSPs. They support the computation of 8-, 16- and 32-bit data.

Besides the multiprocessors that are a pure combination of programmable modules, combinations of dedicated modules and programmable processors are also used. These structures are also termed heterogeneous multiprocessors. They are always appropriate when the signal processing method contains sub-methods with both particularly high computational performance requirements and regular structures. A typical example of such a sub-method is a filter. In a filter, all data dependencies and operations to be executed are predetermined. Since a fixed number of operations is to be carried out for each sample, a relatively high computational performance results that is proportional to the sampling rate. As shown in Chapter 6, more cost-effective structures can be extracted due to the regularity of the process. Since the sub-process can be implemented cost-effectively, the hardware costs for the total process are also lessened. Since the dedicated modules in heterogeneous systems are fixed in terms of their architecture and the process, and only few parameters can be changed, such configurations display less flexibility for changes in the process. A heterogeneous system for implementing a video-codec is shown as an example in Figure 9.1.10 [143]. It consists of a dedicated module for motion estimation (BM module), a dedicated module for the discrete cosine transform (DCT module) and a programmable processor that carries out all remaining tasks. Since the inverse DCT (IDCT) and the DCT are structurally identical, the DCT module also serves the execution of the IDCT.

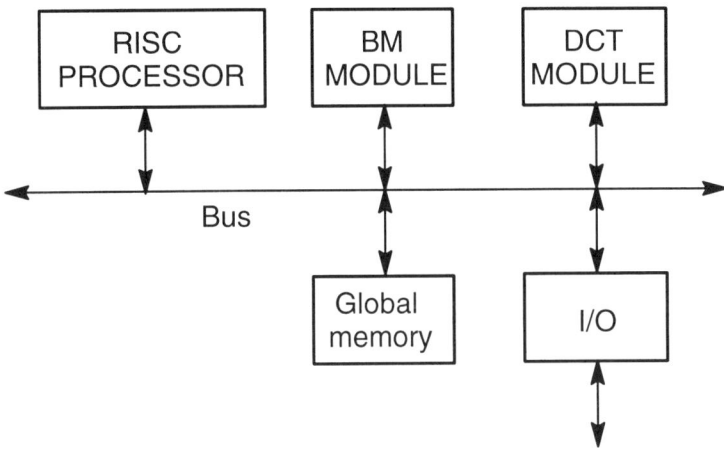

Figure 9.1.10: Heterogeneous video processor with a programmable processor (RISC) and dedicated modules

A special form of combining dedicated modules and programmable modules is coprocessor configurations. In this case, a special unit, the coprocessor, is assigned to each programmable processor. The coprocessor works largely self-sufficiently. Data exchange between the programmable processor and the coprocessor can be carried out via dual-port memory or via a bus. Few lines are generally needed to control the coprocessor. Figure 9.1.11 shows the principal configuration of such a multiprocessor system.

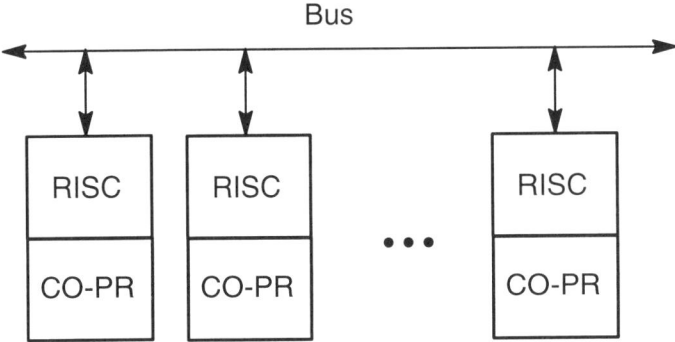

Figure 9.1.11: Multiprocessor with processor nodes comprising a programmable processor (RISC) and a dedicated coprocessor

Figure 9.1.12 shows a special coprocessor for convolution-like operation. This coprocessor supports algorithms with

$$\sum_{i=1}^{N} a_i b_i \tag{9.1.2}$$

as its kernel function. Algorithms for FIR filters, linear transformations and correlation functions have such a kernel. Whenever such algorithms represent the main portion, such a structure is useful.

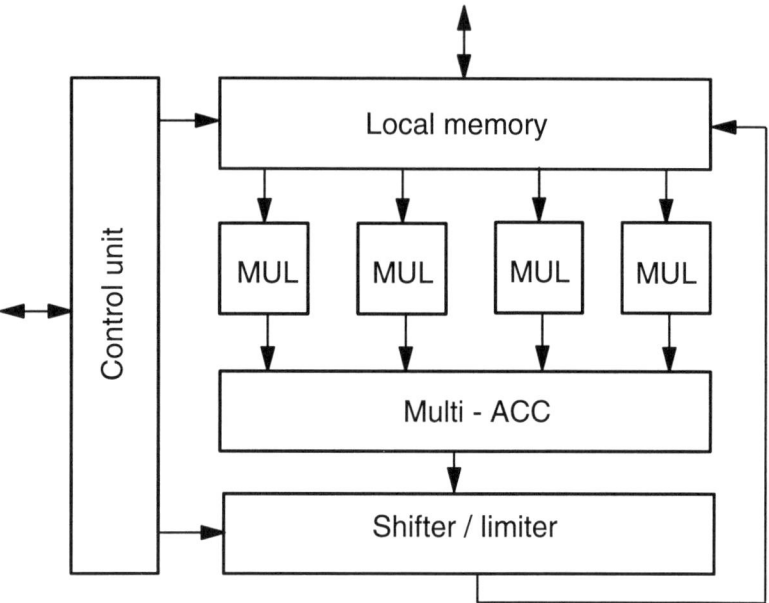

Figure 9.1.12: Coprocessor for convolution-like operations

A comparison with the C80 (Figure 9.1.9) makes the advantages clear. With four parallel data paths, this processor also supports such algorithms. However, the hardware costs are clearly higher. In the place of quite a simple controller there are four complex program sequencers and an instruction memory. The total costs for the data path are also less for the coprocessor.

The disadvantage of the coprocessor is easily seen. A change in the signal processing method with a clearly reduced proportion of sub-tasks with a kernel function as in (9.1.2) leads to a collapse in throughput since the processing proportions between master processor and coprocessor are no longer balanced. Such changes in method are supported by a configuration such as the C80 without problems.

From this, one can see that cost-efficient implementations can be achieved through tailoring to the signal processing method. Nonetheless, the flexibility is thus reduced, i.e. an adjustment to changes in method is only possible to a limited extent. Whether the hardware costs or the flexibility are to have priority must be decided in each case.

9.2 Performance Comparison of Multiprocessor Structures

As with the examinations of section 8.3, a quantitative comparison of SIMD and MIMD structures is carried out here based on simple models. The algorithms for this must also be modelled.

It is presumed that a sequence of samples must be continuously processed. The sequence of samples is assumed to be divided into blocks of fixed length, and each block is to be processed in the same manner. The block data are stored intermediately in data memory and are accessible in parallel for the processing unit. The processing time of a block determines the throughput. The possibility of overlapping processing of sequential blocks is not considered.

The total process is to be described by several characteristic intervals. In a first approach, the sub-tasks can be split into those without explicit parallelism (scalar processes) and those with explicit parallelism (parallel executable processes). Among the scalar sub-processes, there are those without data interdependencies that are thus mutually independently executable. These are to be indexed *SI*. Besides these, there are scalar sub-processes that display dependency on subsequent processes, i.e. subsequent ones are executable after completion of this sub-process. These are to be indexed *SS*. The parallel processable processes are to offer parallelism via division of the data. Parallel sub-processes that demonstrate identical instructions independent of the data are characterized by the index *P*. Furthermore, sub-processes with data dependent branches are to exist. A parallel splitting up of the data is to be possible; however, different instructions depending on the program branch are to be executed. These sub-processes are to be indexed with *D*.

The share of operations within the individual, characteristic intervals to the total operations is described by the relative frequency of occurrence *P*. For modelled computations, several simplifications are assumed. In each process interval, an identically large number of cycles per operation ($n_{CYC/OP}$) should occur for the given data path. Then only the number of operations per block weighs in the determination of the throughput. Furthermore, arbitrary parallelability is assumed for the parallel process. The number of operations per block to be sequentially processed is n_{SS}. The number of independent scalar processes is k_{SI}, and each of these processes has an equally large number of operations per block, given by n_P. The number of data independent branches is k_D and the number of operations per block in each branch is equally large and has the value n_D.

The number of operations to be executed per block can be split in accordance with the process model into four intervals:

$$n_{OP/BLOCK} = n_{SS} + k_{SI}\, n_{SI} + n_P + n_D$$
$$= n_{OP/BLOCK}\, (P_{SS} + k_{SI}\, P_{SI} + P_P + P_D) \qquad (9.2.1)$$

This accounts for the fact that the sum of the relative frequencies of the four shares is:

$$P_{SS} + k_{SI}\, P_{SI} + P_P + P_D = 1 \qquad (9.2.2)$$

The throughput of a single processor machine results from the inverse of the processing time for a block.

$$R_{T,1} = \frac{1}{T_{CLK} \cdot n_{CYC/OP} \cdot n_{OP/BLOCK}} \qquad (9.2.3)$$

In a multiprocessor with parallelism n_{PAR}, the operations n_P can run in parallel via data distribution. An MIMD structure allows various instructions in each processor. Thus, the operations n_D can also be executed in parallel. In the case of an SIMD structure, all processors receive the same instructions. Parallel processing of branches with differing instructions is not possible. The operations of each branch must be carried out sequentially. During processing of the process branches, some of the processors must be deactivated so that the operations are carried out on the correct data. Activation / deactivation is given by the knowledge of the data distribution and the criterion for program branching. The scalar portions n_{SS} and n_{SI} can only run sequentially in an SIMD structure. Due to the possibility of differing instructions in each processor of an MIMD structure, such a configuration also supports parallel processing for the scalar portions n_{SI}.

From these concepts, it follows that the throughput of an MIMD system is

$$R_{T,MIMD} = \frac{R_{T,1}}{P_{SS} + \frac{k_{SI} P_{SI}}{n_{PAR,SI}} + \frac{P_P + P_D}{n_{PAR}}} \tag{9.2.4}$$

$$n_{PAR,SI} = \min (n_{PAR}, k_{SI})$$

The throughput rises with increasing parallelism. Yet due to the non-parallelizable portions P_{SS}, a saturation characteristic appears. The portions P_{SI} have a maximal parallelization of k_{SI} .

In the case of an SIMD system, it is to be observed that the scalar operations cannot be carried out in parallel. Also, in the case of data dependent branches, the individual branches may need to be executed sequentially. The effective parallelism of the data dependent branches thus reduces with the number of branches k_D.

$$R_{T,SIMD} = \frac{R_{T,1}}{P_{SS} + k_{SI} P_{SI} + \frac{P_P + k_D P_D}{n_{PAR}}} \tag{9.2.5}$$

The dependency on the parallelism is similar to that of the MIMD system. However, due to the sequential processing of the independent, scalar sub-processes and the data dependent branches, the throughput is lower. A comparison of Equations (9.2.4) and (9.2.5) shows that an MIMD structure is to be preferred from the aspect of throughput. Yet since an MIMD system requires higher hardware investments, the question of efficiency must be evaluated.

In accordance with the structure of Figure 8.1.15, the chip size of a single processor system comprises four shares.

$$A_{Si,1} = A_{IM} + A_{IP} + A_{DP} + A_{DM} \tag{9.2.6}$$

For simplicity, it is now assumed that the shared memory in a multiprocessor increases proportionally to the number of parallel processing units. The size of the multiprocessors is then given by:

$$\begin{aligned} A_{Si,MIMD} &= n_{PAR} (A_{IM} + A_{IP} + A_{DP} + A_{DM}) \\ &= n_{PAR} A_{Si,1} \end{aligned} \tag{9.2.7}$$

$$\begin{aligned} A_{Si,SIMD} &= A_{IM} + A_{IP} + n_{PAR} (A_{DP} + A_{DM}) \\ &= (\alpha + n_{PAR} \beta) A_{Si,1} \end{aligned} \tag{9.2.8}$$

If the dependency of the efficiency on the number of parallel units is examined with the aid of (4.2.13), this yields for $n_{PAR} > k_{SI}$:

MIMD:

$$\frac{d\eta}{\eta} = \left[\frac{P_P + P_D}{n_{PAR} (P_{SS} + P_{SI}) + P_P + P_D} - 1 \right] \frac{dn_{PAR}}{n_{PAR}} \qquad (9.2.9)$$

SIMD:

$$\frac{d\eta}{\eta} = \left[\frac{P_P + k_D P_D}{n_{PAR}(P_{SS} + k_{SI} P_{SI}) + P_P + k_D P_D} - \frac{n_{PAR}\beta}{\alpha + n_{PAR}\beta} \right] \frac{dn_{PAR}}{n_{PAR}} \qquad (9.2.10)$$

Equation (9.2.9) shows that in MIMD systems, the efficiency continuously decreases with increasing n_{PAR}. In SIMD systems, there is first a rise in efficiency and then a decrease for large values of n_{PAR}. In SIMD systems, the optimal value for n_{PAR} is

$$n_{PAR,OPT} = \sqrt{\frac{\alpha}{\beta} \cdot \frac{P_P + k_D P_D}{P_{SS} + k_{SI} P_{SI}}} \qquad (9.2.11)$$

If the following parameter values are assumed,

$$\alpha = \beta = 0.5$$

$$P_{SS} = 0.1 \qquad P_{SI} = 0.05 \quad k_{SI} = 4$$

$$P_P = 0.6 \qquad P_D = 0.1 \quad k_D = 2$$

then

$$n_{PAR,OPT} \approx 2 \qquad (9.2.12)$$

For the assumed model parameters, it turns out that only SIMD structures with few processors are efficient. For this reason, SIMD clusters combining SIMD and MIMD have been suggested. Configurations with several SIMD systems that process in parallel are termed SIMD clusters, whereby each SIMD system contains only a few nodes.

Diagrams in which the throughput is presented with respect to the chip size are helpful in judging individual processor systems. Such a dependency can be derived by combining Equations (9.2.4) and (9.2.7) or rather (9.2.5) and (9.2.8). The interdependency of throughput and chip size for the above-specified data is illustrated in Figure 9.2.1. It is seen that the MIMD structures offer higher throughput for the same chip size. The reason for this more favourable behaviour is the processing shares *SI* and *D* that must be processed sequentially in the case of SIMD.

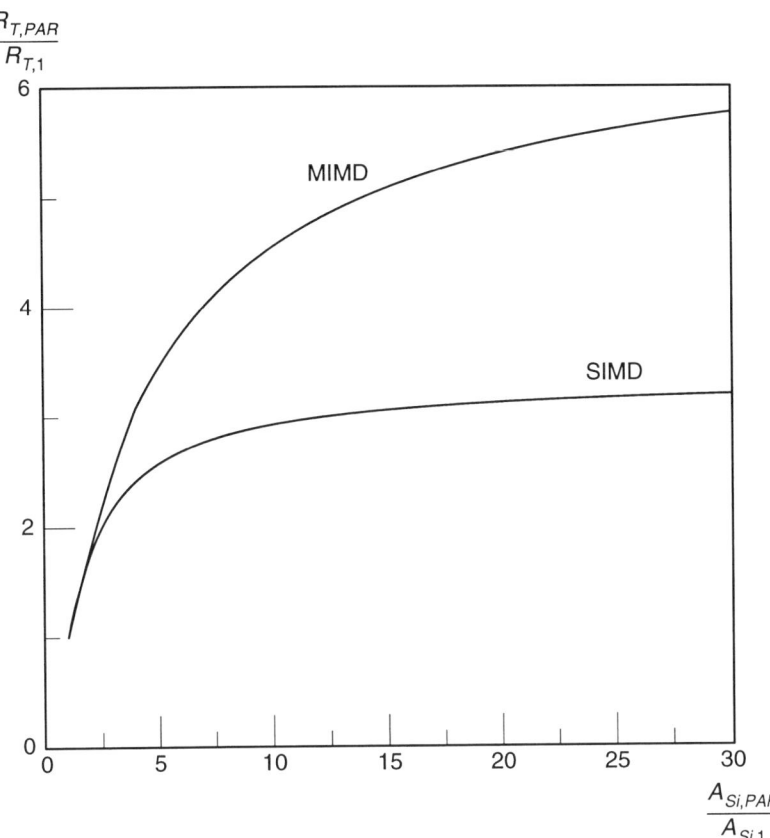

Figure 9.2.1: Throughput in respect to chip size for MIMD and SIMD structures. Model parameters: $\alpha = \beta = 0.5$; $P_{SS} = 0.1$; $P_{SI} = 0.05$; $k_{SI} = 4$; $P_P = 0.6$; $P_D = 0.1$; $k_D = 2$

A modification of the model parameters can lead to other results. The set of parameters

$$\alpha = \beta = 0.5 \qquad P_{SS} = 0.2$$
$$P_P = 0.6 \qquad P_D = 0.2 \qquad k_D = 2$$

yields as its result that an SIMD structure offers higher throughput for the same chip size. Figure 9.2.2 shows the corresponding relationship of throughput to chip size.

From these observations, it follows that the throughput as a function of the chip size can be substantially influenced by the model parameters of the algorithm. If the sub-procedures that are not parallelizable using SIMD structures comprise the dominant share, then MIMD structures are to be preferred. The decision must always be made individually for the particular procedure at hand.

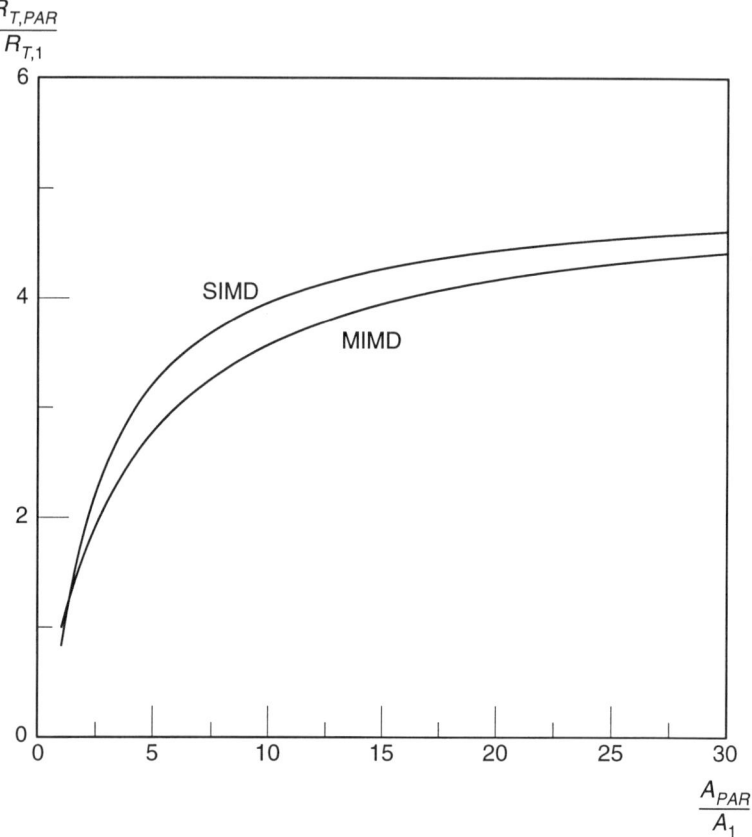

Figure 9.2.2:   Throughput as a function of the chip size for MIMD and SIMD structures. Model parameters: $\alpha = \beta = 0.5$; $nP_{SS} = 0.2$; $P_P = 0.6$; $P_D = 0.2$; $k_D = 2$

To illustrate the modelling procedure and optimization, a concrete example will be observed. The procedure consists of matrix multiplications followed by threshold operations. The input data are divided into square blocks measuring $N \times N$ samples. Each input block X is multiplied from the left and from the right by a matrix A.

$$Y = A^T X A \qquad\qquad (9.2.13)$$

The elements $y(i,j)$ of the matrix Y are to be compared with a threshold and, if the elements are larger than the threshold, multiplied by a factor b.

$$y'(i,j) = \begin{cases} y(i,j)b & \text{if } y(i,j) > q \\ 0 & \text{otherwise} \end{cases} \qquad (9.2.14)$$

Yet the threshold operations are to be carried out in a particular sequence and are to be aborted the first time the threshold is not reached. Only those elements $y'(i,j)$ for which $y(i,j)$ exceeded the threshold are transferred to a successive unit. Figure 9.2.3 shows a possible sequence, the so-called zig-zag scan. The characteristics of the matrix are such that, on a statistical average, it takes $4N$ values of $y(i,j)$ until the threshold is not reached.

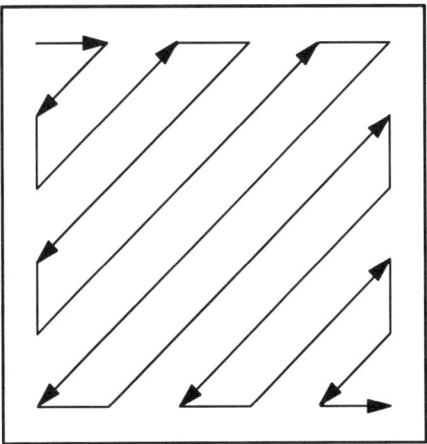

Figure 9.2.3: Zig-zag scan in a 2-D field

This procedure is modelled by two sub-procedures. The matrix multiplications can be parallelized to a large degree by splitting the data. For each block of input data, this part has

$$n_p = 2N^3 \qquad (9.2.15)$$

MAC operations. The second sub-procedure is deemed to be non-parallelizable due to the particular, data-dependent operations. Two operations (COMPARE, MUL) must be carried out for each matrix element that exceeds the threshold. Thus, this part can be modelled by

$$n_{SS} = 8N \qquad (9.2.16)$$

operations. Even for relatively small values of N, the parallelizable share is dominant. For $N = 8$, for example,

$$P_P = 0.94; \quad P_{SS} = 0.06$$

It is easy to see that, for this specific example, an MIMD structure offers no advantages since no sub-procedures exist for which MIMD is more favourable in terms of throughput.

Due to the special matrix products, it is practical to store the matrices X, A, XA and Y locally. Since the storage capacity does not change with the increase in parallel data paths, it is assumed, for simplicity, that only the data paths contribute to the increase in chip size. As long as a data path has 20% of the total area of a scalar processor, the parameters for the area are

$$\alpha = 0.8 \quad \beta = 0.2$$

Thus, in this case where $N = 8$, the optimal parallelism in accordance with (9.2.11) is

$$n_{PAR,OPT} = 8$$

Simple consideration proves that the developed SIMD processor does not represent the best solution. Only one processor is active during the processing of the scalar portion; the others are deactivated. This does not represent optimal usage of the resources. In the case of a multiprocessor, the scalar portions no longer contribute a negligible share. Improvement can be achieved through a cascade implementation of an SIMD processor. The SIMD multiprocessor is then only used for the matrix multiplications. In a cascade implementation, the slowest member determines the throughput. Thus, it does not make any sense to make the SIMD multiprocessor faster than the scalar processor. The throughput of both processors is identical when the SIMD processor has parallelism corresponding to the ratio of operations of both sub-procedures.

$$n_{PAR} = \frac{n_P}{n_{SS}} = \frac{N^2}{4} \tag{9.2.17}$$

For $N = 8$, the new configuration has a throughput increased by a factor of 16. Under the previous assumptions, the chip size increases in comparison to a scalar processor by a factor of 5 (4 times the size of the SIMD multiprocessor plus 1 scalar processor). The corresponding figures for a pure SIMD multiprocessor with 8-fold parallelism are a factor of 5.6 for the throughput and a factor of 2.4 for the size. Thus, in comparison to a scalar processor, a cascade implementation yields a 3.2-fold improvement in efficiency, where-as a pure SIMD configuration has only 2.3.

The cascade implementation discussed already indicates the yields available through heterogeneous structures. Due to their regularity, the ma-trix operations are suitable for dedicated implementation using systolic structures as in Chapter 5. Dedicated structures can be implemented with minimal possible expenditure. Thus the efficiency can be further increased. It follows that adaptations of the architecture to the problem lead to mini-mization of the costs.

9.3 Exercises

1. DFT implementations using an SIMD multiprocessor system are to be ana-lysed. A DFT is to be carried out sequentially for vectors x with N values. The number of data paths n and the vector length N are powers of 2, with $n < N$. The following questions are to be answered for the particular values $N = 16$ and $n = 4$. The shared data storage is to consist of n memory modules (MM) and a network for parallel access.

   a. Given that n vectors x are to be processed simultaneously, the vectors x and the matrix W (7.2.1) are to be meaningfully distributed among the memory blocks. Determine the sequencing of the memory calls and the operations. Which connections must the network provide at what time?

   b. Now only one vector x is to be processed at a time, i.e. the elements of a vector are to be distributed among the memory blocks. How does the sequencing of the network connections and the operations change?

   c. The processing is to be accelerated through use of the FFT. How does the sequencing of the network connections and the operations change for the case where n vectors x are processed simultaneously?

   d. The simultaneous processing of only one vector x is now to be analysed as in part b. Which problems occur during the computation of the FFT?

2. Two implementations of 2-D digital filters using SIMD multiprocessor sys-tems are to be analysed. The 2-D impulse response is given to be non-separa-ble and has $n_h \cdot n_v$ coefficients. The image data to be filtered are character-ized by $n_{P/L}$ pixels per line and $n_{L/F}$ lines per frame. The image repetition frequency is f_F. The following questions are to be worked out using the par-ticular values $n_h = n_v = 5, n_{P/L} = n_{L/F} = 512$ and $f_F = 25$ Hz. It is presumed that the image data are supplied line sequentially and that no spe-cial treatment of the image edges is carried out. A hierarchical storage config-uration with large external storage and small local storage, to which the data paths are assigned, is assumed.

a. An implementation is to be developed using task distribution, i.e. the data paths are to take over a portion of the operations for each result. Name a feasible distribution of the tasks. How much parallelism can be implemented? Describe the memory configuration and its corresponding data. Which data transfer rate is required between the multiprocessor and the external memory? How large is the minimal storage capacity of the local memory?

b. An implementation using data distribution is to be implemented, i.e. the data paths are to carry out all operations for a result, each however for separate input data. Name a sensible distribution of the data. What degree of parallelism can be achieved? Explain the accompanying concept for the memory and compute the resulting minimal storage capacity.

c. The input data are now to be stored intermediately in block line memory and then supplied to the processor using a meander scan. Block line memory is RAM storage for several rows. Meander scanning is characterized by the following sketch.

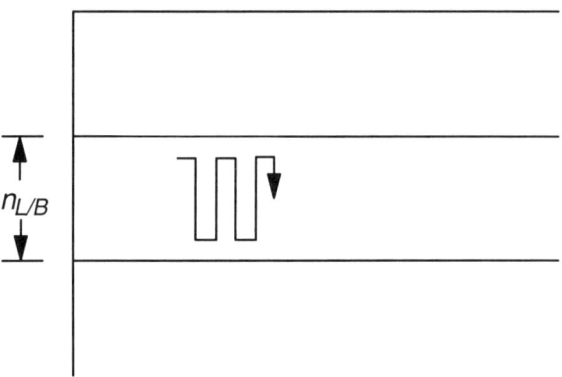

The number of rows per block $n_{L/B}$ must be larger than n_v. The particular value is presumed to be $n_{L/B} = 8$. The SIMD processor is to have shared memory. The minimal storage capacity of the shared memory is to be determined for the following number of processors:

$$n_{PE,a} = n_{L/B} - n_v + 1$$
$$n_{PE,b} = (n_{L/B} - n_v + 1)^2$$

What capacity must the block line memory have and which data transfer rate between the block line memory and the multiprocessor results?

3. A method to be implemented is block-oriented, i.e. the processing is periodically repeated for a block of data. The operations per block are characterized into two classes in accordance with section 9.2, namely into independent sca-

lar operations n_{SI} where $k_{SI} = 1$ and arbitrarily parallelizable operations n_P.

a. Implementations with pure SIMD and MIMD systems are to be compared. Equations (9.2.4), (9.2.5), (9.2.7) and (9.2.8) can be used for this purpose, whereby $\alpha = \beta = 0.5$ is assumed for (9.2.8). Show that, for a given chip size, SIMD systems offer the highest throughput.

b. An MIMD structure allows for different instructions in each processor node. One possible alternative implementation is that one processor node executes the scalar operations n_{SI} and the remaining processor nodes simultaneously execute the parallelizable operations n_P. Derive the equivalent equations to (9.2.4) and (9.2.7). Which processor node count n_{PE} achieves maximal throughput? For which count n_{PE} is this solution more efficient (higher throughput for a given chip size)? Assume $n_P/n_{SI} = 16$. Hint: One of the two portions of the operations dominates the processing time and thus the throughput.

c. The SIMD structure is to be replaced by a parallel implementation of a scalar processor and a SIMD processor. Derive the equivalent equations to (9.2.5) and (9.2.8). Which processor node count n_{PE} achieves maximal throughput? Which count n_{PE} yields the most efficient solution?

4. A block-oriented method to be implemented is given to have independent scalar operations n_{SI} ($k_{SI} = 1$) and parallelizable operations n_P. The parallelability is limited to $n_{PAR,max}$. For example, $n_{PAR,max} = 16$. Due to the special method, only powers of two are sensible for parallelizations under $n_{PAR,max}$, i.e. $n_{PAR} \in \{1, 2, 4, 8, 16\}$. The portions of the operations have the ratio $P_P/P_{SI} = 25$.

a. An implementation with pure SIMD systems is to be examined. How large is the maximally achievable throughput? Which of the allowable degrees of parallelism provides the highest efficiency? How large are the resulting throughput and the chip size?

b. A parallel implementation using a scalar processor and an SIMD processor is to be pursued. How large is the maximally achievable throughput? Which processor count yields the highest efficiency? How large are the resulting throughput and chip size?

10 Implementation Strategies

The development of DSP systems is carried out in a special sequence of design phases. Different design levels exist in each of the individual development phases. The system level is the highest level. It is where system partitioning and the embedding of the particular system into its environment are treated. An algorithm level follows in which the functional relationships between the interface data and the internal data are described. At the architectural level one level below, the interwiring of blocks for the execution of data operations and for storage is specified. The subsequent circuit level contains the interconnection of gates and transistors. The optimization of logic and the transistor sizes is included here. The layout level follows in which the physical design is described. At the lowest level, the technology level, the semiconductor technologies and the housing and mounting techniques for the implementation are considered.

In general, the designing is carried out in the order: system, algorithm, architecture, circuit. Yet, as many examples of the previous chapters have shown, interaction between these various levels takes place. Due to the mutual influence of the description levels, modifications that lead to a reduction in the hardware costs should be determined at an early stage, i.e. at the algorithm level. The respect to which modifications to the algorithm have an influence on the implementation and/or the processing time depends on the type of implementation. An implementation with full custom designed integrated circuits allows tweaking down to the bit level. Implementations using commercially available DSPs allow program optimization only at the instruction level.

In the following section, architecture modifications that reduce the hardware costs and/or the processing time will be presented in several examples. A second section will show implementation alternatives and their corresponding design flow.

10.1 Algorithm-Dependent Architecture Modifications

Sequential processing of signals under particular time constraints leads to the throughput being the critical measure. In addition to the throughput, further characteristics that have relevance to the implementation can be specified. Examples of these are:

- computational rate
- storage requirements
- regularity and modularity
- parallelism

For a given algorithm, the number of operations can be counted and, through correlation with the time constraints, the computational rate can be determined. In computing the performance, all operations are weighted equally, i.e. differences in the implementation costs and the processing time are not considered. In spite of this weakness, the evaluation of the computational rate yields useful estimates. One can derive from the algorithm how long a piece of data must be available for processing after its first occurrence (lifetime). The minimal storage requirements can be deduced from the lifetime of all the pieces of data. The regularity can be deduced from the repeated use of functional modules in algorithms. The design costs are also simplified due to modularity. One can derive from the algorithm which pieces of data can be processed independently of one another and which functional units are not mutually dependent, whence possible strategies for and degrees of parallelization can be derived. This discussion should demonstrate how key values for the implementation can be determined from algorithms given in the form of either a program or a graph.

Generally, a large variety of architectural alternatives exists for individual implementations of specific subtasks. The derivation of a particular architecture can be based on the following principles:

- Regularity. The regularity of algorithms can be exploited in the mapping process onto array processors.

- Algebraic axioms, linearity. The application of algebraic axioms on the target algorithm can lead to architectural alternatives: distributivity enables factorization of operands; associativity enables one to change the sequence of operations. Linearity allows the factorization of subfunctions resulting in a different computation structure.

- Hierarchy. Computations can be performed with different granularity, e.g. on bit level, word level, or block level.

- Table lookup. Instead of repetitively calculating results at runtime, pre-calculated results may be stored in lookup tables for all possible combinations of operands, thus replacing calculations by memory accesses.

- Numerical characteristics. Special numerical characteristics of algorithms may be exploited to reduce the number of operations or the word length for results.

- Sequential functions. A realization based on iterative networks can reduce the hardware effort by sequentializing the computations.

Dependence graph   Signal flow graph

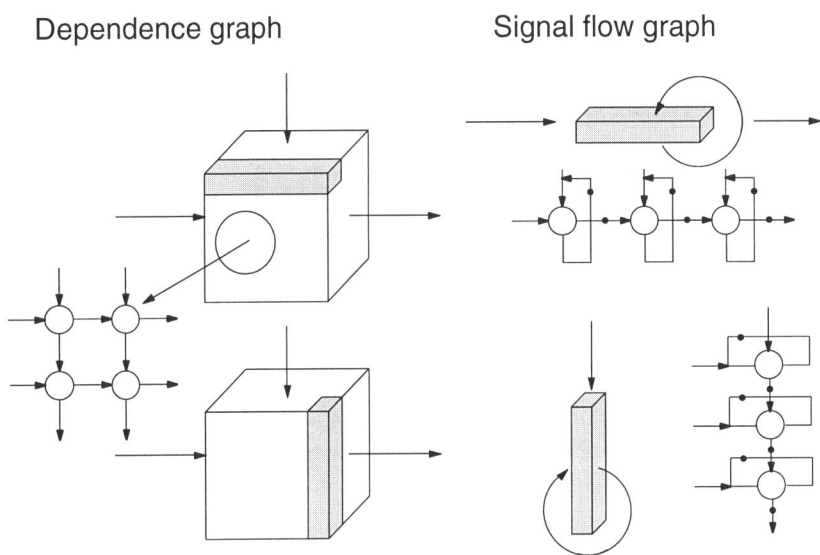

Figure 10.1.1: Mapping the DG onto a SFG. Projecting the maximally expanded graph of operations and data dependencies onto a graph of lower dimensionality involves the problem of allocation of PEs and scheduling of operations.

First the derivation of architectures for regular algorithms will be discussed. The regular structure of the dependence graph (DG) allows several alternative mappings between the DG and the signal flow graph (SFG). As indicated in Figure 10.1.1 a partition of the DG nodes may be formed. A repetition of the DG partition in the DG index space provides all needed operations. A mapping of the DG onto a SFG with reduced number of processing elements (PEs) requires time sequential operations of DG nodes. The PEs of the SFG may be specified by the nodes of the partition of the DG (allocation process). Thus the repetition in the DG index space is replaced by a SFG where the partition operates in a specific sequence in time (scheduling). In addition a retiming within the SFG may be applied to get systolic features.

Because of the regularity there is a large variety in selecting a partition of nodes and specifying a temporal order in which the DG nodes are processed. The allocation and scheduling process influences the internal data exchange and the interface between the array processor and the input/output data stream. Matching between the data streams and the array minimizes the memory expense for the array and the interface.

Due to algebraic axioms and linearity, algorithms can be structurally modified without changing the result. These modifications lead to a change in the signal processing characteristics and, in a desired sense, to a reduction of the costs and/or the processing time.

Firstly, transformations based on distributivity and associativity will be observed. The distributive law states that

$$a \cdot b + a \cdot c = a \cdot (b + c) \tag{10.1.1}$$

The left side requires two multiplications and one addition, whereas the right side requires but one multiplication. The transformation from the left to the right spares one operation. It should be noted that this profit also occurs for other algebras. For example, algorithms with ADD-Compare-Select-Logic exist. One operation can also be spared here in the sense of distributivity. It is given by:

$$\max\{a + b, a + c\} = a + \max\{b, c\} \tag{10.1.2}$$

In this case, the addition adopts the role of the $\cdot$ operation and the selection of the maximum adopts the role of the + operation.

The associative law states that the order in which the corresponding operations are executed does not play a role. For addition, this law is given by:

$$(a + b) + c = a + (b + c) \tag{10.1.3}$$

By changing the order, no operation can be spared. Nonetheless, in special cases it can be exploited to remove an operation from the time-critical path. Figure 10.1.2 shows an example of this. The implied, time-critical path is reduced by one operation. In the case where a and b are constants and not variables, one operation can even be spared.

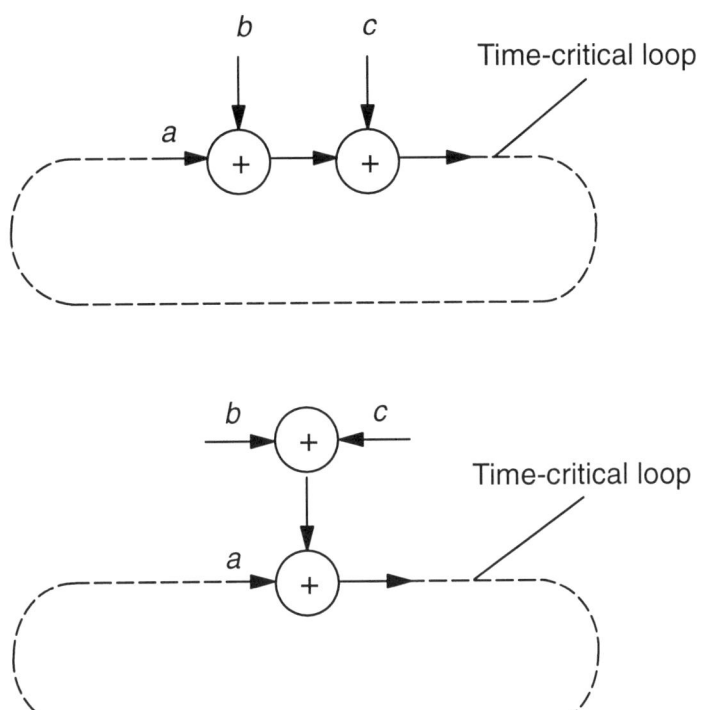

Figure 10.1.2: Modification of a time-critical loop by employing the associativity of operations

The algebraic axioms are also a basis for alternative representation of numbers. Let a set of coefficients $\{a_i\}$ be needed for an algorithm like filtering or transformation. Each coefficient can be represented by a sum of products.

$$a_i = \sum_k \sum_j b_k c_j \qquad (10.1.4)$$

The sum of product form may be chosen to reduce the implementation expense. A typical strategy is to determine common factors b_k for several coefficients a_i.

Example 10.1.1 Each data sample x shall be multiplied by a 10-bit coefficient $a = 0110110011$. A direct implementation of the multiplication according to Figure 6.2.9 requires five adders. The number of adders is determined by the non-zero bits minus 1. The CSD representation of the specific coefficient is $a = 100\bar{1}0\bar{1}010\bar{1}$. The number of non-zeros is reduced by 1 and the number of adders /subtractors accordingly. A factorization of the coefficient a into $b \cdot c$ provides $a = (0010010001) \cdot (11)$. Implementation with hardwired shifts and adders requires just three adders. The implementations are displayed in Figure 10.1.3.

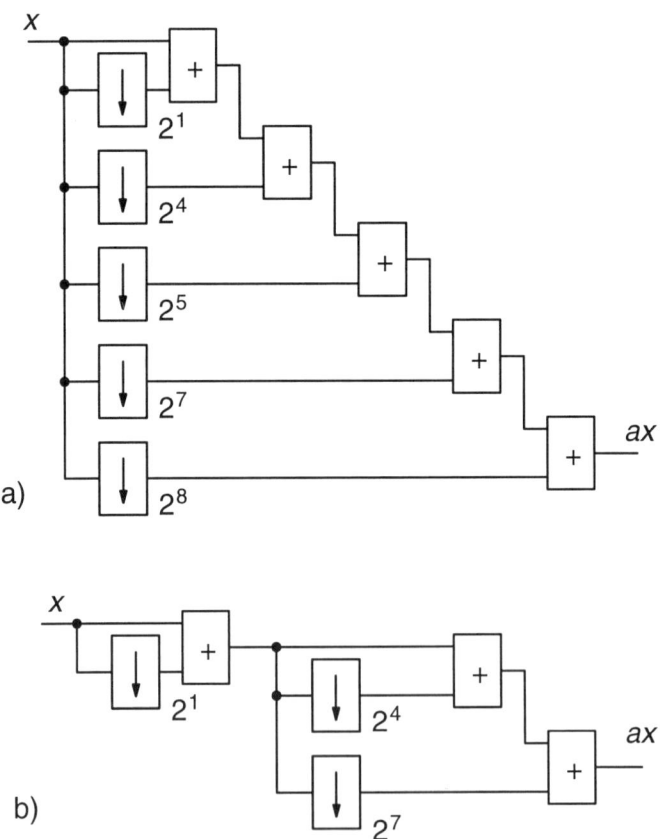

Figure 10.1.3: Dedicated implementation for Example 10.1.1 with hardwired shift/add technique: (a) direct implementation of the binary number; (b) factorization

The associativity of additions allows as an alternative the implementation of Figure 10.1.3 a in a tree structure which reduces the delay from five adder delays to three adder delays.

The algebraic axioms and their derived theories can also be applied to conglomerate functions. The complex multiplication of Figure 7.2.3 is an example of this. It shows an alternative that has one more adder and one less multiplier. In dedicated implementations, this reduces the costs of the data path. It should be noted that in the original structure two values (Re, Im) and in the modified structure three values (Re + Im, Im, Re – Im) must be stored for each coefficient. This implies an enlargement of the storage requirements. The reduced costs in the data path are thus balanced by increased costs in storage. For implementations using DSPs, the modified structure has no advantages since both multiplication and addition require the same number of cycles. It should be noted that the structure of Figure 7.2.3 also holds for rotations in a plane. The modifications shown at the operation level can also be extended to the function level. Figure 10.1.4 shows an example analogous to the distributive law.

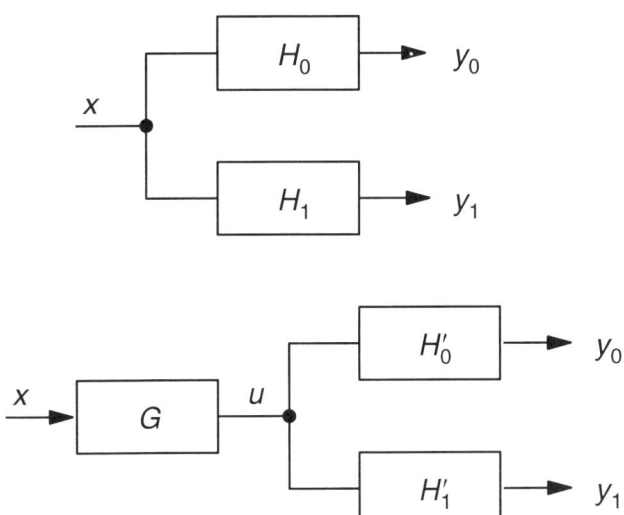

Figure 10.1.4: Modification of parallel processing units through the extraction of common sub-functions

Equation (10.1.1) can also be formulated such that common factors are factored out. In Figure 10.1.4, common sub-functions are factored out. The reductions in terms of operations are to be derived for implementations of FIR filters. Given are H_0 and H_1 FIR filters with N coefficients.

$$y_0(i) = \boldsymbol{h}_0^T \boldsymbol{x}\,(i)$$
$$y_1(i) = \boldsymbol{h}_1^T \boldsymbol{x}\,(i) \tag{10.1.5}$$

The factoring out of the partial filter G corresponds to a multiplication of the system functions in the frequency domain.

$$H_0(e^{j\omega}) = G(e^{j\omega})\, H_0{'}(e^{j\omega})$$
$$H_1(e^{j\omega}) = G(e^{j\omega})\, H_1{'}(e^{j\omega}) \tag{10.1.6}$$

The product in the frequency domain corresponds to cascading in the time domain.

$$u(i) = \boldsymbol{g}^T \boldsymbol{x}(i)$$
$$y_0(i) = \boldsymbol{h}_0{'}^T \boldsymbol{u}(i) \tag{10.1.7}$$
$$y_1(i) = \boldsymbol{h}_1{'}^T \boldsymbol{u}(i)$$

It is assumed that $\boldsymbol{h}_0^{'T}$ and $\boldsymbol{h}_1^{'T}$ each have $N - M$ coefficients. To achieve the original degree of filtering, $\boldsymbol{g}^T$ requires $M + 1$ coefficients, whereby one of the coefficients can always be set to 1. If the original structure according to (10.1.5) were to require $2N$ MAC operations, then the structure according to (10.1.7) requires only $2N - M$ MAC operations. Noticeable savings are thus achieved.

Further functions exist that allow a cascade implementation with simplified sub-functions. Many of the known linear transforms such as the discrete Fourier, the Hadamard and the discrete cosine transform [3] belong to this group. The DFT can be formulated in matrix notation according to (7.2.1).

The matrix $\boldsymbol{W}$ of the DFT can then be split into sub-matrices $\boldsymbol{W}_0, \boldsymbol{W}_1, ... \boldsymbol{W}_{\log N}$.

$$y = \boldsymbol{W}\boldsymbol{x}$$
$$= \boldsymbol{W}_{\log N} ... \boldsymbol{W}_1\, \boldsymbol{W}_0\, \boldsymbol{x} \tag{10.1.8}$$

Several alternatives exist for the sub-matrices. One possible solution can be derived via the FFT. In the case of the FFT according to Figure 7.3.1 ($N = 8$), for example, the matrix $\boldsymbol{W}_0$ is given by

$$
\mathbf{W}_0 =
\begin{bmatrix}
1 & 0 & 0 & 0 & 1 & 0 & 0 & 0 \\
0 & 1 & 0 & 0 & 0 & 1 & 0 & 0 \\
0 & 0 & 1 & 0 & 0 & 0 & 1 & 0 \\
0 & 0 & 0 & 1 & 0 & 0 & 0 & 1 \\
w^0 & 0 & 0 & 0 & -w^0 & 0 & 0 & 0 \\
0 & w^1 & 0 & 0 & 0 & -w^1 & 0 & 0 \\
0 & 0 & w^2 & 0 & 0 & 0 & -w^2 & 0 \\
0 & 0 & 0 & w^3 & 0 & 0 & 0 & -w^3
\end{bmatrix}
\qquad (10.1.9)
$$

The matrix $\mathbf{W}_0$ is sparse, i.e. it has few non-zero coefficients. It follows from the structure of $\mathbf{W}_0$ that a total of N complex additions/subtractions and $N/2 - 2$ complex multiplications are necessary for the implementation. This takes into consideration that a transformation in accordance with (10.1.1) is possible for the factors w^i and that $w^0 = 1$. Similar relationships hold for the other matrices $\mathbf{W}_k$. For all other matrices, the values shown in Table 7.3.1 are achieved as the total number of operations.

The modifications presented above were carried out at the word level. The special structures at the bit level within pieces of data can also be used for alternatives with reduced costs. This will be discussed for an FIR filter.

An FIR filter with N coefficients is presumed, whereby each coefficient has m bits. The costs for the operative section (excluding storage) can be specified in terms of the number of adders k_{ADD}. A direct multiplier implementation requires $m - 1$ adders. The total filter then has equivalent adder costs of

$$
k_{ADD} = N \cdot m \qquad (10.1.10)
$$

The particular characteristics of the filter can be used for a reduction in costs. The typical impulse response of an FIR filter is shown in Figure 10.1.5. The amplitude of the coefficients decreases, starting from the middle of the coefficients. Due to the decreasing amplitude, the maximal number of non-zero bits for representing the amplitudes also decreases. Characteristic filter functions have shown that the reduction in amplitude leads to a halving of the costs according to (10.1.10). A CSD representation of binary numbers forces at least one zero between non-zero digits. In the worse case for a signed digit, half as many non-zeros occur as the range of the amplitude requires, i.e. a CSD representation can lead to an additional halving. A further reduction is achievable by exploiting the symmetry characteristics of the impulse response (Figure 6.2.13). In this case, the number of multipliers is halved, yet at the same time additional adders are required for combining the symmetric samples.

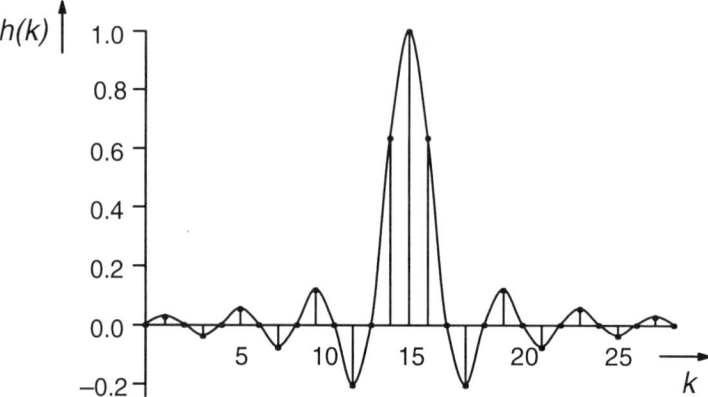

Figure 10.1.5: Typical impulse response of a low-pass filter

This means, in sum, that the amplitude reduction in combination with CSD coding and symmetry of the coefficients leads to total costs of

$$k_{ADD} < \frac{1}{8}Nm + \frac{N}{2} \tag{10.1.11}$$

For video signals with $m = 9$ bits, this yields a reduction by a factor of approximately 5 in comparison to (10.1.10). The necessary hardware costs in terms of the number of adders are presented in Table 10.1.1 for filter functions for video signals as published in the literature [144], [145]. Equation (10.1.11) is confirmed by these results. From these observations of filters, one can say that implementations employing coefficient resolution down to the bit level offer significant potential for savings.

Table 10.1.1: Examples of minimally required adder counts k_{ADD} for implementing FIR filters. The coefficients are coded with the lowest number of non-zero digits

Number of coefficients N	Word width m	Adder count k_{ADD}	Estimation of equation (10.1.11)	Reference
10	9	16	17	[144]
14	9	22	23	[144]
11	9	12	18	[145]
37	9	51	61	[145]

The examples discussed above show that the knowledge of special algorithmic characteristics can be used to reduce the costs. Examples of such characteristics are algebraic axioms, linearity, symmetry, separability, cascadability, and further, a priori knowledge. An exact analysis of the algorithms with respect to their implementation can reveal the possible cost reductions. The interaction between an algorithm and its architecture costs can also be used inversely, whereby the developer of the algorithm is instructed from the implementation side as to which characteristics the algorithm should preferably comprise.

10.2 Implementation Alternatives

The alternatives for the implementation of DSP systems can, in a first approach, be divided into two groups, i.e. dedicated hardware and programmable signal processors. The design limitations dictate the choice of alternative. Several design limitations are listed below:

- Prototype development
- Product development
- Production costs
- Development time
- Size
- Power consumption
- Flexibility

A technique is to be verified under real-time conditions in a target system via prototype development. Since the goal is the verification of the technique, the size and the power consumption do not play essential roles. Such prototypes are often implemented using programmable DSPs or programmable logic modules such as FPGAs, for instance. FPGAs are complex, programmable modules [146]. They are programmed by capping or creating connections. If MOS transistors controlled from memory are used as connection elements, multiple programming is possible. If the connection elements are so-called fuses or anti-fuscs, thcy can be programmed only once. The programmability of the elements allows for flexibility in changing the method, as is required in such a development phase.

Product development for large scale production with compact size and low power consumption, on the other hand, will have an implementation based on full custom designed ICs as a goal. An example would be a processor for a mobile telephone. Since many sub-tasks are carried out in such a processor (speech decoding, transmission protocols, modulation techniques, etc.), a programmable DSP would also be suitable. Due to the special design limitations (size, power consumption), a programmable processor specially adapted to the application would have to be implemented.

A further example of a production design for large scale production is a digital video decoder for television distribution systems. Digital video signal transmission with new service features is planned for cable networks and via satellite. Due to its mass usage and the competitive situation, low production costs are dominant. A dedicated implementation with full custom components is thus the obvious choice.

Product development cases also exist wherein the "time to market" is particularly important. This can be, for instance, the introduction of a new product to the market. In order to be one of the first on the market, a short development time is important. This can lead to product development using commercially available components that avoids the custom development of components. Indeed, the production costs cannot be brought to their lowest level in this way, but the production scale is normally not extremely large in the introductory phase.

This discussion of the design limitations should demonstrate that the decision regarding the type of system implementation is a complex task. System implementations can also be heterogeneously grafted together using programmable and dedicated components. Furthermore, custom developed components can be combined with commercially available ones. A decisive factor here is the balance between the various costs and the commercial availability of the components.

The examples of architectures for signal processors in Chapter 8 showed that many processing steps are executed in parallel. An algorithm written in a high-level language such as C cannot take such parallelism completely into consideration. This can be achieved by programming in assembler. However, the investments in programming then increase. An optimizing compiler can be effectively employed here [147]. Such a compiler improves the performance of a program without affecting the actual processing.

The methods for optimizing compilers are categorized into machine-dependent and machine-independent methods. The focus of machine independent methods is to reduce complicated instructions and/or functions.

This can usually be achieved through the use of temporary variables. In the following example, the number of function calls is halved.

Before:

$$x = \sin(a) + \cos(b)$$
$$y = \sin(a) - \cos(b)$$

After:

$$u = \sin(a)$$
$$v = \cos(b)$$
$$x = u + v$$
$$y = u - v$$

A further example shows the exchange of a division for a multiplication.

Before:

```
for ( i = 0; i < 20, i ++)
a [ i ] = b [ i ] / c;
```

After:

```
cinv = 1 / c;
for ( i = 0; i < 20, i ++)
a [ i ] = b [ i ] * cinv ;
```

The focus of machine-dependent code optimization is to substantially exploit the independence of instructions in order to avoid pipeline hazards. An example for reducing the pipeline hazards is shown below.

Before:

```
ADD        R1, R2 → R3
MUL        R3, R4 → R5
SUB        R6, R7 → R8
XOR        R9, R10 → R11
```

After:

```
ADD        R1, R2 → R3
SUB        R6, R7 → R8
XOR        R9, R10 → R11
MUL        R3, R4 → R5
```

In the original version, the result of one instruction is used immediately in the subsequent instruction. In the case of instruction pipelining, NOP cycles must be inserted since data can only be read after being written. In the second version of the program, NOP cycles are avoided through insertion of independent instructions.

Particular signal processing hardware can be implemented through a combination of standard components and programmable components such as FPGAs and ASICs (application specific ICs). Gate arrays, standard cell ICs and full custom designed ICs are possibilities for ASICs [148]. In addition to the typical standard cells, full custom designed modules are generally employed in standard cell ICs for regular elements such as memory, PLAs and multipliers. One characteristic of FPGAs and ASICs is that a cell library is normally provided. This cell library contains simple basic gates and registers and often complex cells such as adders and multipliers, too. The task thus exists of describing the hardware in terms of the elements of a given cell library. The mapping of an algorithm to cells of a given library is not directly possible. It takes place in several sub-steps.

Signal processing algorithms are normally developed on workstations using actual data. High-level languages such as C and, increasingly, special program packages such as MATLAB [149] are employed for this. Algorithm descriptions in this form do not sufficiently take hardware aspects into account. In order to evaluate special hardware design impacts such as word width limitation of languages and program packages are being increasingly employed that allow a description that is closer to the actual hardware. Examples of such are SILAGE [76] and COSSAP [150]. In addition to data types with predetermined word widths, timing information can be set via the data dependencies. Also hierarchical system descriptions are supported. The user can define functional units in the form of a high-level language. Predefined basic functions such as addition, multiplication, etc. are provided by the system.

The two program systems SILAGE and COSSAP are used in particular for system simulation. Compilers for conversion into hardware description languages such as VHDL also exist, however [151]. Silicon compilers also exist for SILAGE for bit-serial and bit-parallel processor structures and code generators for standard DSPs.

After the definition and simulation of the algorithm, the specification of the hardware follows as a next step. This specification is no longer carried out in the form of text, but rather in the formal form of a hardware description language. The best-known languages for this purpose are VHDL [151] and Verilog [152]. These languages support both behaviour-oriented descriptions and structural descriptions. Hierarchical structural descriptions from the architectural level to the gate level are possible. The two forms of description can also be mixed. The advantage of formal specifications is that, in addition to the documentation of the hardware, simulation is also supported.

```
library IEEE;
      use IEEE.std_logic_1164.all;
      use IEEE.std_logic_unsigned.all;

      entity ADDER is
      Port(A :        In STD_LOGIC_VECTOR(7 downto 0) ;
           B :        In STD_LOGIC_VECTOR(7 downto 0) ;
           Y :        Out STD_LOGIC_VECTOR(8 downto 0)  ) ;
      end ADDER;

      architecture BEHAVIORAL of ADDER is
      begin
           main: process( A, B )
           begin
             Y <= ( '0' & A ) + ( '0' & B ) ;
           end process main;
      end BEHAVIORAL;
```

Figure 10.2.1: Functional description of an 8-bit adder in VHDL. The corresponding interface declaration and the calls to the libraries used for the arithmetic functions are also illustrated.

```
architecture STRUCTURAL of ADDER is
component FA
      Port ( A    :    In STD_LOGIC) ;
             B    :    In STD_LOGIC;
             C_IN    : In     STD_LOGIC;
             S    :    Out    STD_LOGIC;
             C_OUT:   Out    STD_LOGIC ) ;
end component;
signal WIRE : STD_LOGIC_Vector (8 downto 0) ;
      begin
           WIRE (0) <= '0' ;
           G_FA: for I in 7 downto 0 generate
           I_FA: FA port map ( A(I), B(I), WIRE(I), Y(I), WIRE(I+1) );
           end generate
           Y(8) <= WIRE(8) ;
end STRUCTUR
```

Figure 10.2.2: Structural description of an 8-bit adder in VHDL

Figure 10.2.1 shows an adder at the behavioural level as an example. A large section is taken up by the interface definitions of the variables. The actual body corresponds more or less to a description in a high-level language. The structural description of this adder in an architecture as in Figure 3.2.1 is usually carried out hierarchically. The adder is comprised of full adders and each full adder is comprised of gate elements. Figure 10.2.2 shows a structural definition of the full adder.

The full adders used here can be specified via logical relationships as listed in Figure 10.2.3. Delays can also be assigned to the resulting values. This is not taken into consideration in the examples shown.

```
entity FA is
        Port ( A    :        In STD_LOGIC) ;
               B    :        In STD_LOGIC;
               C_IN     : In      STD_LOGIC;
               S    :        Out      STD_LOGIC;
               C_OUT:    Out      STD_LOGIC ) ;
end FA;


architecture BOOLEAN of FA is
begin
        main: process( A, B, C_IN )
        begin
                S    <=  A xor B xor C_IN;
                C_OUT <= (A and B) or (A and C_IN) or (B and C_IN) ;
        end process main;
end BOOLEAN;
```

Figure 10.2.3: Description of a full adder cell via Boolean equations in VHDL

The advances in CAD tools support automatic logic synthesis using the cell libraries of several manufacturers [153]. The starting point for such synthesis is a system description in VHDL or Verilog at the register transfer level (RTL). The combinatoric logic between the registers that is to be synthesized can be designated by the designer in terms of arithmetic operators, logical functions or a table. The synthesis tool creates a netlist of simple gates from the given library. It is furthermore possible to select interactively the best one from several predesigned operation modules. The criteria for the selection are the costs in terms of reference gates or transistors and the delay time as a substantial measure of the throughput. In addition to CAD tools that support synthesis starting from an RTL description, new tools exist that also support mappings to cell libraries based on a behavioural description. To optimize the throughput, the delay characteristics of the gate elements of the combinatoric logic must be known at the time of synthesis. The individual logic elements are assigned fixed delay times resulting from the function of the cell and the number of connections. In this phase of design, exact evaluation of the capacitive loads is not possible. Using the manufacturer's models of the technology, the delay times are estimated via statistical models.

The result of the presented synthesis is a netlist. The next step is the physical design, for which the entire netlist is partitioned into interconnected larger units. The placement of these units on the chip is set using a floor-planning tool. A decision as to the exact position of all the cells follows with the aid of placement and routing tools. The goal is short connection lines, in particular for the critical path. After completing placement, the exact data of the connection lines is known, and a timing simulation for evaluating the behaviour of the entire circuit can be carried out. If not all requirements in terms of the delay characteristics are fulfilled, a synthesis must be carried out anew based on specific changes at the RT level.

This course of design holds for implementations using FPGAs, gate arrays and standard cells. Such synthesis is based on an architectural description at the RT level. Detailing of the structure down to the transistor level, as partially shown in Chapter 3, is not necessary.

Particular requirements in terms of the throughput and the production costs can, in some cases, only be fulfilled by full custom circuits. In this case, the layout structures of the integrated circuits are resolved in hierarchical, modular configurations down to the transistor level. Several such circuit configurations were shown in Chapters 3 and 5. Full custom design has also been eased by available CAD tools.

In summary, due to the frequency of error and the high design costs, design by hand is increasingly being superseded by the employment of CAD tools. Tools exist that, starting from an algorithmic description, produce executable code for standard DSPs. Designs for pre-manufactured integrated circuits such as FPGAs and for pre-designed integrated circuits such as standard cell ICs can be carried out with the aid of synthesis tools and the manufacturers' "plug-ins". The design process is continuously supported from the algorithm to the layout. This process is not automatic, but rather is carried out interactively using several tools. Tools for easing manual design are even available for the full custom design of integrated circuits.

10.3 Exercises

1. Alternative circuit structures are to be extracted by applying algebraic axioms. This is to be carried out for the exemplary logic function

    $$y = ab \lor ac \lor ad$$

    a. A two-stage NAND–NAND implementation is to be derived from the given logic function. How many transistors does the corresponding CMOS implementation require? Hint: Complement the y function twice and apply De Morgan's theorem.

    b. Extract all terms with the common variable a by applying the distributive law. Apply De Morgan's theorem to yield a NOR–NOR structure. How does the transistor count change in comparison to a?

    c. The given function is to be implemented as complex gates. What effect does the factoring out of the variable a in the pull-up and pull-down paths have? How many transistors are needed with and without factoring?

2. An architecture for determining the range of a set of data is to be worked out. The maximum and the minimum are to be determined. A pseudoprogram for this task is:

    ```
    amin  =  a(1),
    amax  =  a(1);
    for i  =  2 to n do
    begin
    amin  =  min (amin, a(i) );
    amax  =  max (amax, a(i) );
    end;
    ```

    a. Construct a dependency graph for the sequential extraction of the maximum and the minimum in accordance with the program (sketch for $n = 8$). Which circuit elements are to be used for implementing the nodes?

    b. Minima and maxima are associative operations. The linear configuration of a is to be converted to a tree structure (one tree per maximum and minimum) based on this associativity. The costs and the delay characteristics of the linear structure and the tree structure are to be compared. Hint: Addition is also associative. Analogously to addition, the structures correspond to multi-operand addition.

    c. Instead of separate nodes for determining maxima and minima, each node is now to determine the maximum and the minimum of two operands simultaneously (see Figure 5.5.2). The two trees of b can now be combined with one another. What structure results for the given task? Determine the number of nodes of the new structure and compare the implementation costs with the solution from b.

3. A dedicated implementation of an 8-tap FIR filter is to be investigated. The specified coefficients are

$$h = \frac{1}{128}[-12, -14, 51, 103, 103, 51, -14, -12]$$

   a. The number of adders/subtractors needed for the complete FIR filter is to be determined if each coefficient is individually implemented by hardwired shifts and add/subtract.

   b. The coefficients are to be represented by a sum of products and appropriate common factors are to be determined. The number of adders/subtractors needed for a new structure utilizing common factors is to be specified. Hint: The incoming samples multiplied by the common factors are introduced into the filter section.

References

[1]   A. V. Oppenheim, R. W. Schafer: *Digital Signal Processing,* Prentice Hall, 1975

[2]   L. R. Rabiner, B. Gold: *Theory and Application of Digital Signal Processing,* Prentice Hall, 1975

[3]   W. K. Pratt: *Digital Image Processing,* John Wiley, 1978

[4]   A. V. Oppenheim, A. S. Willsky, I. T. Young: *Signal and Systems*, Prentice Hall, 1983

[5]   A. Rosenfeld, A.K. Kak, *Digital Picture Processing*, Academic Press, 1982

[6]   R. N. Bracewell: *The Fourier Transform and its Applications*, McGraw–Hill, 1978

[7]   N. S. Jayant, P. Noll: *Digital Coding of Waveforms,* Prentice Hall, 1984

[8]   D. H. Ballard, C. M. Brown: *Computer Vision*, Prentice Hall, 1982

[9]   A. Papoulis: *Probability, Random Variables, and Stochastic Processes,* McGraw–Hill, 1991

[10]   P. R. Adby, M. A. H. Dempster: *Introduction to Optimization Methods,* Chapman and Hall, 1974

[11]   A. Ralston, H. S. Wilf: *Mathematical Methods for Digital Computers,* Vol. I and Vol. II, John Wiley, 1960 and 1967

[12]   A. Mukherjee: *Introduction to NMOS and CMOS VLSI Systems Design*, Prentice Hall, 1986

[13]   D. A. Hodges, H. G. Jackson: *Analysis and Design of Digital Integrated Circuits*, McGraw–Hill, 1987

[14]   C. Mead, L. Conway: *Introduction to VLSI Systems*, Addison–Wesley, 1980

[15]   L. A. Glasser, D. W. Dobberpuhl: *The Design and Analysis of VLSI Circuits,* Addison–Wesley, 1985

[16]   N. Weste, K. Eshraghain: *Principles of CMOS VLSI Design,* Addison–Wesley, 1985

[17]   H. Klar: *Integrierte digitale Schaltungen MOS/BICMOS,* Springer, 1993

[18]   J. Wakerly: *Digital Design:Principles and Practices,* Prentice–Hall, 1990

[19]   S. Muroga: *Logic Design and Switching Theory,* John Wiley, 1979

[20]   H. T. Nagle, B.D. Carroll, J. D. Irwin: *An Introduction to Computer Logic,* Prentice–Hall, 1975

[21]   A. Vladimirescu, et al.: *SPICE2 Version 2G Users Guide,* Department of Electrical Engineering and Computer Science, University of California, Berkeley, 1981

[22]   S. Gollnisch: *Simulation und Modellierung des Schaltverhaltens von Basiszellen und arithmetischen Modulen,* Diploma Thesis, University of Hannover, 1994

[23]   M. Shoji: *CMOS Digital Circuit Technology,* Prentice–Hall, 1988

[24]   T. Lin, C. Mead: Signal delay in general RC networks, *IEEE Trans. on Computer Aided Design,* Vol. CAD–3, No. 4, pp. 331–349, 1984

[25]   E. Elmore: The transient response of damped linear networks with particular regard to wideband amplifiers, *Journal of Applied Physics,* pp. 55–63, January 1948

[26]   A. P. Chandrakasan, R. W. Brodersen: *Low Power Digital CMOS Design,* Kluwer, 1995

[27]   A. Bellaaouar, M. I. Elmasry: *Low Power Digital VLSI Design, Circuits and Systems,* Kluwer, 1995

[28]   J. M. Rabaey, M. Pedram: *Low Power Design Methodologies,* Kluwer, 1996

[29]   H. L. Garner: Number systems and arithmetic, in *Advances in Computers,* Vol. 6, Academic Press, pp. 131–194, 1965

[30]   I. Flores: *The Logic of Computer Arithmetic,* Prentice Hall, 1963

[31]   A. Avizienis: Signed digit number representations for fast parallel arithmetic, *IRE Trans. on Electronic Computers,* Vol. EC–10, pp. 389–400, 1961

[32] G. W. Reitwiesner: Binary arithmetic, in *Advances in Computers*, Vol. 1, pp. 261–265, Academic Press, 1960

[33] A. D. Booth: A signed binary multiplication technique, *Quart. J. Mech. Appl. Math.*, Vol. 4, Part 2, pp. 236–240, 1951

[34] L. P. Rubinfeld: A proof of the modified Booth's algorithm for multiplication, *IEEE Trans. on Computers*, Vol. C–24, pp. 1014–1015, 1975

[35] A. Weinberger, J. L. Smith: *A Logic for High Speed Addition*, National Bureau of Standards Circular 591, pp. 3–12, 1958

[36] M. Lehmann, N. Burla: Skip techniques for high speed carry–propagation in binary arithmetic units, *IRE Trans. Electron. Comput.*, Vol. EC–10, No. 4, pp. 691–698, 1961

[37] O. J. Bedrij: Carry select adders, *IRE Trans. Electron. Comput.*, Vol. EC–11, pp. 340–346, 1962

[38] J. Slansky: Conditional sum addition logic, *IRE Trans. Electron. Comput.*, Vol. EC–9, pp. 226–231, 1960

[39] T. Kilburn, D.B. Edwards, D. Aspinall: Parallel addition in digital computers: A new fast carry circuit, *IEE Proc.*, Vol. 106, Part B, pp. 464–466, 1959

[40] R. P. Brent, H. T. Kung: A regular layout for parallel adders, *IEEE Trans. on Computers*, Vol. C–31, No. 3, pp. 260–264, 1982

[41] E. E. Swartzlander: *Computer Arithmetic,* Vol. I and Vol. II, IEEE Press, 1990

[42] K. Hwang: *Computer Arithmetic*, John Wiley, 1979

[43] J. J. F. Cavanagh: *Digital Computer Arithmetic*, McGraw–Hill, 1985

[44] J. M. Rabey: *Digital Integrated Circuits, A Design Perspective*, Prentice Hall, 1996

[45] S. Waser: High–speed monolithic multipliers for real–time digital signal processing, *IEEE Computer Magazine*, Vol. 11, No. 10, pp. 19–29, 1978

[46] E. L. Braun: *Digital Computer Design*, Academic Press, 1963

[47] O. L. MacSorley: High speed arithmetic in binary computers, *Proc. IRE*, Vol. 49, pp. 91–103, 1961

[48] C. S. Wallace: A suggestion for a fast multiplier, *IEEE Trans. on Electron. Comput.*, Vol. EC–13, pp. 14–17, 1964

[49]   L. Dadda: Some schemes for parallel multiplication, *Alta Frequenca,* Vol. 34, pp. 349–356, 1965

[50]   L. Dadda: On parallel digital multipliers, *Alta Frequenca*, Vol. 45, pp. 574–580, 1976

[51]   S. D. Pezaris: A 40 ns 17–bit–by–17–bit array multiplier, *IEEE Trans. on Computers*, Vol. C–20, No. 4, pp. 442–447, 1971

[52]   C. R. Baugh, B. A. Wooley: A two's complement parallel array multiplication algorithm, *IEEE Trans. on Computers*, Vol. C–22, No. 1–2, pp. 1045–1047, 1973

[53]   K. J. Dean: Binary division using data dependent iterative arrays, *Electronics Letters*, Vol. 4, pp. 283–284, 1968

[54]   H. H. Guild: Some cellular logic arrays for nonrestoring binary division, *Radio and Elec. Engr.*, Vol. 39, pp. 345–348, 1970

[55]   J. E. Robertson: A new class of digital division methods, *IEEE Trans. on Computers*, Vol. C–7, pp. 218–222, 1958

[56]   J. F. Hart, et al.: *Computer Approximations*, John Wiley, 1968

[57]   R. Z. Goldschmidt: *Applications of Division by Convergence*, MSc Thesis, MIT, Cambridge, MA, June 1964

[58]   T. C. Chen: Automatic computation of exponentials, logarithms, ratios and square roots, *IBM Journal Res. and Dev.*, pp. 380 – 388, July 1972

[59]   J. E. Volder: The CORDIC trigonometric computing technique, *IRE Trans. Electron. Comput.*, Vol. EC–8, pp. 330–334, 1959

[60]   J. S. Walther: A unified algorithm for elementary functions, *Proc. Spring Joint Computer Conf.*, pp. 379–385, 1971

[61]   E. F. Deprettere, P. Dewilde, R. Udo: Pipelined CORDIC architectures for fast VLSI filtering and array processing, *Proc. ICASSP 1984*, pp. 41A6.1–41A6.5, 1984

[62]   C. E. Leiserson, F. M. Rose, J. B. Saxe: Optimizing synchronous circuitry by retiming, *Proc. Caltech VLSI Conference*, Pasadena, CA, 1983

[63]   S. Y. Kung: *VLSI Array Processors*, Prentice Hall, 1988

[64]   C. L. Seitz: Concurrent VLSI architectures, *IEEE Trans. on. Computers*, Vol. C–33, No. 12, pp. 1247–1265, 1984

[65]   J. R. Jump, S. R. A. Ahuja: Effective pipelining of digital systems, *IEEE Trans. on Computers*, Vol. C–27, No. 9, pp. 855–865, 1978

[66]   T. G. Hallin, M. J. Flynn: Pipelining of arithmetic functions, *IEEE Trans. on Computers*, Vol. C–21, pp. 880–886, 1972

[67]   A. Schlegel, T.G. Noll: The effect of glitches on the power dissipation of CMOS circuits, Internal Report, EECS Department, RWTH Aachen, February 1997

[68]   A. Schlegel, T.G. Noll: Entwurfsmethoden zur Verringerung der Schaltaktivität bei verlustoptimierten digitalen CMOS–Schaltungen; *DSP Deutschland'95*, pp. 61–74, Munich, September 1995

[69]   C. L. Hwang, A.S.M. Masud: *Multiple Objective Decision Making – Methods and Applications*, Springer, 1979

[70]   R. R. Yager: Fuzzy decision making including unequal constraints, in *Fuzzy Sets and Systems 1*, North Holland, pp. 87–95, 1978

[71]   D. I. Moldevan: *Parallel Processing: From Applications to Systems*, Morgan Kaufmann, 1993

[72]   H. Kung, C. Leiserson: Systolic arrays for VLSI, *SIAM Sparse Matrix Proceedings*, pp. 245–282, Philadelphia, 1978

[73]   H. T. Kung: Why systolic architectures?, *IEEE Computer Magazine*, pp. 37–45, January 1982

[74]   K. M. Chandy, J. Misra: *Parallel Program Design*, Addison-Wesley, 1988

[75]   L. Thiele: Mapping algorithms onto VLSI architectures, in P. Pirsch (Editor), *VLSI Implementations for Image Communications*, pp. 69–116, Elsevier, 1993

[76]   P. Hilfinger, J. Rabaey, D. Genin, C. Scheers, H. DeMan: DSP specification using SILAGE language, *Proc. IEEE Int. Conference on Acoustics, Speech, and Signal Processing*, pp. 1057–1060, 1990

[77]   J. Teich, L. Thiele: Partioning of processor arrays: a piecewise regular approach, *Integration,* Vol. 14, pp. 297–332, 1993

[78]   S. K. Rao, T. Kailath: Regular iterative algorithms and their implementation on processor arrays, *Proc. IEEE*, Vol. 76, No. 3, pp. 259–269, 1988

[79]   R. Sedgewick: *Algorithms*, Addison–Wesley, 1988

[80]   U. Vehlies: DECOMP – A program for mapping DSP algorithms onto systolic arrays, in *Transformational Approaches to Systolic Design*, G. M. Megson, (Editor), Chapman and Hall, 1993

[81] D. I. Moldevan: ADVIS: a software package for the design of systolic arrays, *IEEE Trans. on Computer–Aided Design*, Vol. CAD–6 (1), pp. 33–40, 1987

[82] V. van Dongen, M. Petit: *PRESAGE: a tool for the parallelization of nested loop programs*, in L. J. M. Claesen (Editor), *Formal VLSI Specification and Synthesis*, Vol. 1, pp. 341–360, North Holland, 1990

[83] L. Thiele: Compiler techniques for massive parallel architectures, in P. Dewilde, S. Vandevalle (Editors), *Computer Systems and Software Engineering*, pp. 101–150, Kluwer, 1992

[84] P. Frison, P. Gachet, P. Quinton: Designing systolic arrays with diastol, in S. Y. Kung, R. E. Owen, J. G. Nash (Editors), *VLSI Signal Processing II*, pp. 93–105, IEEE Press, 1986

[85] D. I. Moldevan, R. A. B. Fortes: Partitioning and mapping of algorithms into fixed size systolic arrays, *IEEE Trans. Computers*, Vol. C–35, pp. 1–12, 1986

[86] J. J. Navarro, J. M. Llaberia, V. Mateo: Partitioning: an essential step in mapping algorithms into systolic array processors, *Computer Magazine*, Vol. 20, pp. 77–89, 1987

[87] J. V. McCanny, K. W. Wood, J. G. McWhirter: The relationship between word and bit level systolic arrays as applied to matrix and matrix multiplication, *Proc. SPIE, Techn. Symp. Real Time Signal Processing VI*, 1983

[88] H. Kwakernaak, R. Sivan: *Modern Signal Systems*, Prentice Hall, 1991

[89] T. W. Parks, J. H. McClellan: Chebyshev approximation for nonrecursive digital filters with linear phase, *IEEE Trans. CT*, Vol. 19, pp. 189–194, 1972

[90] *Programs for Digital Filter Design*, IEEE Press, 1979

[91] T. G. Noll: High throughput digital filters, in P. Pirsch (Editor), *VLSI Implementations for Image Communications*, pp. 171–215, Elsevier, 1993

[92] P. R. Cappello, K. Steiglitz: A note on free accumulation in VLSI filter architectures, *IEEE Trans. on Circuits and Systems*, Vol. CAS–32, pp. 291–296, 1985

[93] P. B. Denyer, D. J. Myers: Carry–save arrays for VLSI processing, *1st Int. Conf. on VLSI*, Edinburgh, pp. 151–160, 1981

[94] C. S. Burrus: Digital filter structures described by distributed arithmetic, *IEEE Trans. on Circuits and Systems*, pp. 674–680, 1977

[95] N. Fliege: *Multirate Digital–Signal processing,* John Wiley and Sons, 1994

[96] R. E. Crochiere, L. R. Rabiner: Interpolation and decimation of digital signals–a tutorial, *Proc. IEEE*, Vol. 69, pp. 300–331, 1981

[97] D. Esteban, C. Galand: Applications of quadrature mirror filters to split band voice coding schemes, *ICASSP'77*, pp. 191–195, 1977

[98] M. J. Smith, T. P. Barnwell: A procedure for designing exact reconstruction filter banks for tree structured sub–band coders, *ICASSP'84*, pp. 27.1.1–27.1.4, 1984

[99] T. Claasen, et.al.: Effects of quantization and overflow in recursive digital filters, *IEEE Trans. on Acoustics, Speech, and Signal Processing*, Vol. 24, pp. 517–529, 1976

[100] A. Fettweis: Wave digital filters: theory and practice, *Proc. IEEE*, Vol. 74, No. 2, pp. 270–327, 1986

[101] S. C. Knowles, J. G. McWhirter, R. F. Woods, J. V. McCanny: Bit–level systolic architectures for high performance IIR filtering, *Journal of VLSI Signal Processing*, Vol. 1, pp. 9–24, 1989

[102] S. E. McQuillan, J. V. McCanny: A systematic methodology for the design of high performance recursive digital filters, *IEEE Trans. on Computers*, Vol. 44, No. 8, pp. 971–982, 1995

[103] K. K. Parhi, D. G. Messerschmitt: Pipeline interleaving and parallelism in recursive digital filters – Part I & II, *IEEE Trans. on Acoustics, Speech, and Signal Processing*, Vol. 37, No. 7, pp. 1099–1134, 1989

[104] D. E. Pearson: *Transmission and Display of Pictorial Information*, Pentech Press, 1975

[105] R. E. Blahut: *Fast Algorithms for Digital Signal Processing*, Addison–Wesley, 1985

[106] A. Wenzler, E. Lüder: New structures for complex multipliers and their noise analysis, *ISCAS'95*, pp. 1432–1435, 1995

[107] K. Grüger: *Kaskadierte Registerschaltungen für die Datenformatkonvertierung bei der digitalen Videosignalverarbeitung*, VDI Verlag, Reihe 9: Elektronik, Nr. 192, 1994

[108]   H. S. Stone (ed.): *Introduction to Computer Architecture*, Science Research Associates, 1980

[109]   J. L. Hennessy, D. A. Patterson: *Computer Architecture: A Quantitative Approach*, Morgan Kaufmann, 1990

[110]   G. D. Kraft, W. N. Toy: *Mini/Microcomputer Hardware Design*, Prentice Hall, 1979

[111]   J. A. Fisher: Very long instruction word architectures and the ELI–512, *Proc. 10th Symp. Computer Architecture*, pp. 140–150, ACM Press, New York, 1983

[112]   S. Dutta, A. Wolfe, W. Wolf, K.J. O'Connor: Design issues for very–long–instruction–word VLSI video signal processors, *VLSI Signal Processing IX*, IEEE Press, pp. 95–104, October 1996

[113]   W. W. Hwu, R. E. Hank, D. M. Gallagher, S. A. Mahlke, D. M. Lavery, G. E. Haab, J. C. Gyllenhaal, D. I. August: Compiler technology for future microprocessors, *Proc. IEEE*, Vol. 83, No. 12, pp. 1625–1676, December 1995

[114]   R. B. Lee: Subword–parallelism with Max–2, *IEEE Micro*, Vol. 16, No. 4, pp. 51–59, August 1996

[115]   U. Weiser: Trade–off considerations and performance of Intel's MMX Technology, *Proc. HOT CHIPS VIII Conference*, pp. 147–161, August 1996

[116]   M. J. Flynn: Very high speed computing systems, *Proc. IEEE*, Vol. 54, No. 12, pp. 1901–1909, 1966

[117]   V. K. Madisetti: *VLSI Digital Signal Processors*, IEEE Press, 1995

[118]   *TMS32010 User's Guide*, Texas Instruments Inc., 1983

[119]   *TMS320C25 User's Guide*, Texas Instruments Inc., 1986

[120]   *DSP56000 Digital Signal Processor User's Manual*, Motorola Inc., 1986

[121]   *Motorola's 16–, 24–, and 32–Bit Digital Signal Processing Families*, Motorola Inc., 1995

[122]   *DSP/MSP Products Reference Manual*, Analog Devices, 1995

[123]   *Single Chip Microcomputer Data*, Motorola Inc., 1984

[124]   I. S. Mackenzie: *The 68000 Microprocessor*, Prentice Hall, 1995

[125]   D. Alpert, D. Avnon: Architecture of the Pentium microprocessor, *IEEE Micro, Hot Chips IV*, pp. 11–21, June 1993

[126] C. Müller–Schloer: D. Qian, ArcSim – an execution–driven simulator for multiprocessor systems, *Proc. ESM 92,* June 1992, York, pp. 8

[127] R. Camposano, W. Wolf: *High–Level VLSI Synthesis*, Kluwer, 1991

[128] D. Gajski, N. Dutt, A. Wu, S. Lin: *High–Level Synthesis – Introduction to Chip and System Design*, Kluwer, 1992

[129] R.A. Walker, R. Camposano: *A Survey of High–Level Synthesis Systems*, Kluwer, 1991

[130] K. Kucukcakar, A.C. Parker: Datapath tradeoffs using MABAL, *Proceedings of the 27th Design Automation Conference*, pp. 511–516, 1990

[131] C.J. Tseng, D.P. Siewiorek: Automated synthesis of data paths on digital systems, *IEEE Transactions on Computer–Aided Design of Integrated Circuits and Systems*, Vol. CAD–5, No. 3, pp. 379–395, 1986

[132] A. Hashimoto, J. Stevens: Wire routing by optimizing channel assignment within large apertures, *8th Design Automation Conference Workshop*, pp. 155–169, 1971

[133] C.Y. Huang, Y.S. Chen, Y.L. Lin, Y.C. Hsu: Data path allocation based on bipartite weighted matching, *Proceedings of the 27th Design Automation Conference*, pp. 499–504, 1990

[134] J. Wilberg, R. Camposano, M. Langevin, P. Plöger, T. Vierhaus: Cosynthesis in CASTLE, in G. Saucier, A. Mignotte (Editors), *Logic and Architecture Synthesis –State–of–the–Art and Novel Approaches*, Chapman & Hall, pp. 355–366, 1995

[135] J. Rabbaey, H. De Man, J. Vanhoof, G. Goossens, F. Catthoor: CATHE-DRAL–II : a synthesis system for DSP multiprocessor systems, in D.D. Gajski (Editor), *Silicon Compilation*, Addison–Wesley, pp. 311–360, 1988

[136] K. Hwang, F. A. Briggs: *Computer Architecture and Parallel Processing*, McGraw-Hill, 1985

[137] K. Hwang: *Advanced Computer Architecture*, McGraw-Hill, 1993

[138] D. H. Lawrie: Access and alignment of data in an array processor, *IEEE Trans. on Computers*, Vol. C–31, pp. 435–442, 1982

[139] *The Transputer Databook*, Inmos Ltd, 1989

[140] M. E. C. Hull, D. Crookes, P. J. Sweeney: *Parallel Processing, the Transputer and its Applications*, Addison–Wesley, 1994

[141]   *TMS320C4x User's Guide*, Texas Instruments Inc., 1993

[142]   *TMS320C80 Multimedia Video Processor (MVP), Technical Brief,* Texas Instruments Inc., 1994

[143]   P. Pirsch, N. Demassieux, W. Gehrke: VLSI architectures for video compression – a survey, *Proc. IEEE*, Vol. 83, No. 2, pp. 220–246, 1995

[144]   M. Winzker, K. Grüger, W. Gehrke, P. Pirsch: VLSI chip set for HDTV subband filtering with on–chip line memories, *IEEE Journal of Solid–State Circuits*, Vol. 28, No. 12, pp. 1354–1361, 1993

[145]   D. Chiappano, D. Raveglia: Anti–aliasing VLSI digital filters for video signal coders, *ISCAS'88*, pp. 709–713, Espoo, Finland, May 1988

[146]   S. M. Trimberger: *Field Programmable Gate Array Technology*, Kluwer, 1994

[147]   A. V. Aho, R. Sethi, J. D. Ullmann: *Compilers: Principles, Techniques, and Tools*, Addison–Wesley, 1986

[148]   S. Goto: Design methodologies, *Advances in CAD for VLSI*, Vol. 6, North–Holland, 1986

[149]   *MATLAB User's Guide*, The Mathworks Inc., 1992

[150]   *COSSAP Reference Manual*, CADIS GmbH, 1994

[151]   *IEEE Standard VHDL Language Reference Manual*, IEEE Std. 1076–1987, IEEE, 1988

[152]   D. Thomas, P. Moorby: *The Verilog Hardware Description Language*, Kluwer, 1991

[153]   Synosys: *Design Compiler Reference Manual*, Version 1997.08, July 1997

Index

A

Accumulator, 92, 261, 317, 341

Adder, 64
  adder tree, 91, 102, 104, 105, 117
  binary lookahead carry, 78, 87
  bit serial, 68, 171
  carry lookahead, 76
  carry propagate, 91
  carry ripple, 69, 95, 99, 162
  carry save, 91, 259
  carry select, 81
  conditional sum, 83, 88
  delay, 69, 74, 76, 81, 82, 85
  full, 65, 67, 98
  half, 98
  multi-operand, 91
  number of transistors, 66, 68, 76, 81, 82, 85
  transmission gate, 73, 87
  vector merging, 91, 99

Address generation unit, 321

Algebraic axiom, 382, 391
  associativity, 233, 250, 257, 258, 384
  commutativity, 257
  distributivity, 159, 384

Algorithm, 1, 191, 382
  high-level, 2
  localization, 193
  low-level, 2, 345
  medium-level, 2
  recursive, 192
  regular, 191
  signal processing, 1, 329

Allocation, 348, 350

ALU (arithmetic logic unit), 6, 307, 315, 341
  operations, 309

AND, 23

Approximation of functions
  Chebyshev approximation, 131, 249
  polynomial approximation, 131
  Taylor series, 131

Architectural assessment, 164
  efficiency measure, 165
  multicriteria, 179

Array divider, 123
  delay, 126, 128
  non-restoring, 127
  number of transistors, 125, 128
  restoring, 125

Array multiplier, 95, 162, 173, 315

Array processor, 187
  cellular, 188
  design process, 238